AF693063

Advances in the Synthesis and Applications of Transition/Noble Metal Oxide Photocatalysts

Advances in the Synthesis and Applications of Transition/Noble Metal Oxide Photocatalysts

Guest Editor

Nina Kaneva

MDPI

Basel • Beijing • Wuhan • Barcelona • Belgrade • Novi Sad • Cluj • Manchester

Guest Editor
Nina Kaneva
Faculty of Chemistry and Pharmacy
University of Sofia
Sofia
Bulgaria

Editorial Office
MDPI AG
Grosspeteranlage 5
4052 Basel, Switzerland

This is a reprint of the Special Issue, published open access by the journal *Catalysts* (ISSN 2073-4344), freely accessible at: https://www.mdpi.com/journal/catalysts/special_issues/ZX90NTFW85.

For citation purposes, cite each article independently as indicated on the article page online and as indicated below:

Lastname, A.A.; Lastname, B.B. Article Title. *Journal Name* **Year**, *Volume Number*, Page Range.

ISBN 978-3-7258-8005-8 (Hbk)
ISBN 978-3-7258-8006-5 (PDF)
https://doi.org/10.3390/books978-3-7258-8006-5

Contents

About the Editor

Nina Kaneva

Nina Kaneva is an Associate Professor and PhD at the Faculty of Chemistry and Pharmacy, Sofia University St. Kliment Ohridski, Bulgaria. Her research interests are focused on heterogeneous catalysis, photocatalysis, nanostructured metal oxides, semiconductor materials, and advanced oxidation processes for environmental applications, particularly wastewater treatment and the removal of organic pollutants. Her scientific work includes the synthesis, characterization, and application of transition and noble metal oxide photocatalysts with enhanced activity under UV and visible light irradiation. She has participated in numerous national and international research projects related to sustainable environmental technologies and advanced functional materials. Dr. Kaneva is the author and co-author of numerous scientific publications in peer-reviewed international journals and has contributed significantly to the development of photocatalytic materials for environmental protection and green chemistry applications.

Preface

This reprint presents recent advances in transition and noble metal oxide photocatalysts for environmental applications, especially wastewater treatment. It focuses on improving charge separation, reducing recombination, and enhancing photocatalytic efficiency, offering valuable insights for researchers in photocatalysis and environmental science.

Collectively, these studies aim to advance photocatalytic environmental remediation through the design of novel or modified semiconductor materials that overcome the limitations of traditional photocatalysts. The ultimate goal is to develop efficient, solar-driven solutions for the purification of water and air contaminated by recalcitrant organic pollutants.

Nina Kaneva
Guest Editor

catalysts

MDPI

Editorial

Editorial Catalysts: Special Issue on "Advances in the Synthesis and Applications of Transition/Noble Metal Oxide Photocatalysts"

Nina Kaneva

Laboratory of Nanoparticle Science and Technology, Department of General and Inorganic Chemistry, Faculty of Chemistry and Pharmacy, University of Sofia, 1 James Bourchier Blvd., 1164 Sofia, Bulgaria; nina_k@abv.bg

Heterogeneous photocatalysis, due to its high efficiency, safety and profitability, has become an effective technology for solving environmental problems, for example, in wastewater treatment, for the removal of organic pollutants [1,2]. Despite the many advantages of semiconductor photocatalysts (e.g., TiO_2, ZnO, Fe_2O_3, CdS and ZnS), there are also disadvantages, for example, the fast recombination rate of photogenerated electron–hole pairs, low quantum yield and high band gap value. The rapid recombination of the photogenerated pairs ($e-$/h+), following energy activation of the catalyst, inhibits the redox process, and thus, this results in reduced photocatalytic activity [3–5]. It is necessary to study a method for optimizing the semiconductor structure in order to improve separation efficiency and to inhibit recombination efficiency toward enhancing the photocatalytic properties of the semiconductor. The recombination process decreases when the surface of the semiconductor is modified by transition/noble metals. Hence, photocatalysts modified with transition/noble metal oxides can enhance the photocatalytic decomposition of organic pollutants in wastewater. Therefore, it is important to continue the search for new and more efficient photocatalysts that offer an improved performance. Key accomplishments to date are compiled in this Special Issue, as summarized below.

A nanomaterials composite with graphitic carbon nitride was studied for the photodegradation of organic compounds because they help to optimize the degradation process of organic pollutants. In this way, Lai et al. [6] prepared a g-C_3N_4/porphyrin nanocomposite by the self-assembly of monomeric Tetrakis (4-carboxyphenyl) porphyrin (TCPP) molecules with g-C_3N_4 nanomaterials. The band gap energies of the hybrid materials are estimated to be 2.38 and 2.7 eV, which could indicate that the hybrid material is an efficient photocatalyst. The authors concluded that the hybrid C_3N_4/TCPP material exhibited enhanced photodegradation activity toward Rhodamine B at a degradation speed of up to 3.3×10^{-2} min^{-1}, which was superior to other porphyrin photocatalysts. Sert et al. [7] also synthesized a graphitic carbon nitride composite, but they implemented a facile calcination method utilizing urea and zinc nitrate hexahydrate as initiators. The authors investigated several effects—molar ratio of ZnO, amount of catalyst, concentration of organic pollutant and concentration of H_2O_2—in the photocatalytic degradation of crystal violet dye. Their experimental results show that the molar ratio of ZnO in g-C_3N_4 (0.05 mmol ZnO and 0.10 g/L catalyst) has an important role, because in 2 h, the photocatalytic degradation reaches 95.9% in the presence of UV irradiation. Photocatalysis compared to adsorption (32.3%) has a dominant role in crystal violet decolorization. Finally, the results showed no significant decrease in the degradation efficiency of crystal violet in the presence of the g-C_3N_4/ZnO photocatalyst after five consecutive cycles.

Musielak et al. [8] report a new, different method, which is a promising alternative to classical AOP methods—photodynamic therapy (PDT), using curcumin. Curcumin has low solubility in water and high photosensitivity. Therefore, the authors chose a suitable carrier—a suitably processed commercial zeolite of the FAU type. Studies related to the release of curcumin from zeolites were carried out in simulated gastrointestinal conditions using two methods: the first is the classical method of releasing the active substance in the open circulation, and the second is photodynamic release. Musielak et al. found that the photodynamic release of curcumin from zeolites proved to be a more cost-effective and efficient method. The reason for this is that curcumin is very sensitive to light.

The exceptional properties of modified semiconductor nanomaterials (i.e., TiO_2 and ZnO) have attracted attention in the field of environmental remediation and water treatment. Gervasi et al. [9] investigated the optical, structural and photocatalytic properties of pure and iron-doped (0.05, 1.0 and 2.5 wt%) anatase/brookite TiO_2 using sol–gel synthesis. The specific surface area increases with increasing amount of iron—from 236 m^2/g (undoped TiO_2) to 263 m^2/g (2.5 wt% Fe)—while the band gap energy decreases from 3.10 eV (undoped TiO_2) to 2.85 eV (2.5 wt% Fe). All of these results, established by Gervasi et al., have an impact on the photocatalytic properties—the sample with 0.05 wt% iron has the highest efficiency and fastest degradation of simazine (77%). The sol–gel method was also used for the preparation of titanium dioxide by another research group—Bachvarova-Nedelcheva et al. [10]—but they doped the semiconductor with TeO_2. The nanopowders TiO_2/TeO_2 were obtained and annealed in the temperature range of 200–700 °C. The photocatalytic properties of the as-prepared samples were tested for the degradation of Malachite Green in the presence of UV light—pure titanium dioxide degrades the dye about 35%, while the powder doped with TeO_2 about 60%. The authors also investigated antimicrobial properties of sol–gel powders. The doped powders exhibited good antimicrobial activity against *E. coli* K12, as the samples heated at higher temperature (400 °C) showed better antibacterial activity (68%) compared to those heated at 200 °C (50%).

The results obtained by Sisay et al. [11] using photocatalytic membranes—multiple nanoparticles (TiO_2, carbon nanotubes, $BiVO_4$) into polyvinylidene fluoride—demonstrates real dairy wastewater treatment. These membranes exhibited lower filtration resistance, better flux and higher FRR than the pristine membrane. The authors found that the factors that affect the filtration resistance are salinity, pH and concentration of lactose. The presence of lactose and higher pH values increased the irreversible resistance and severely reduced chemical oxygen demand rejection. The TiO_2-carbon nanotubes-$BiVO_4$-polyvinylidene fluoride membrane containing all constituents showed the best regeneration performance, exceeding that of the pristine membrane by 30%.

Salomatina et al. [12] studied photocatalytic efficiencies of poly(titanium oxide) dispersed in optically transparent polymeric matrices of different natures under UV and visible light on aqueous solutions of azo dyes and phenols. During the photocatalytic process in poly(titanium oxide) material, a one-electron transition to $Ti^{4+} + e^- \rightarrow Ti^{3+}$ was established, as well as the formation of electron–hole pairs and active oxygen species. Doping with gold and silver nanoparticles reduces the poly(titanium oxide) band gap from 3.11–3.35 to 2.11 eV. Therefore, doping with noble metals leads to the improvement of the photocatalytic properties of the nanocomposites under UV irradiation, and this is the reason for their high activity under the influence of visible light. In this case, azo dyes and phenol were found to be degraded by 90%.

Jaison et al. [13] report the current developments in the degradation of pollutants in the two types of photocatalytic oxidation, using TiO_2—namely, the influence of environmental conditions and catalyst deactivation—and possible solutions. The authors investigated the potential effects of the nature of the reactant, catalyst support, light intensity and relative humidity. The mechanisms of deactivation of photocatalysts and the possibility of reducing catalyst deactivation were also discussed.

Heterogeneous photocatalysts to generate sulfate radicals ($SO_4^{\bullet-}$) from peroxydisulfate ion ($S_2O_8^{2-}$) were successfully prepared by a solvothermal method. Alapi et al. [14] reported the preparation of BiOX (X = Cl, Br, and I) to increase the activity of the peroxydisulfate ion and faster degradation of organic substances in the presence of ultraviolet and visible light illumination. The BiOI efficiency highly exceeds that of BiOBr and BiOCl for PDS activation, even under visible irradiation.

In conclusion, this Special Issue on "Advances in the Synthesis and Applications of Transition/Noble Metal Oxide Photocatalysts" presents research related to the latest advances in the preparation and characterization of photocatalytic materials, with the aim of improving their performance in the removal of organic pollutants from water. We believe that our Special Issue will serve to inspire researchers in this field.

We thank all of the authors for their valuable contributions, and we also thank the editorial team of *Catalysts* for their kind support in making this Special Issue possible.

Conflicts of Interest: The authors declare no conflict of interest.

References

1. Daneshvar, N.; Rabbani, M.; Modirshahla, N.; Behnajady, M. Kinetic modeling of photocatalytic degradation of Acid Red 27 in UV/TiO_2 process. *J. Photochem. Photobiol. A* **2004**, *168*, 39–45. [CrossRef]
2. Behnajady, M.; Modirshahla, N.; Hamzavi, R. Kinetic study on photocatalytic degradation of C.I. Acid Yellow 23 by ZnO photocatalyst. *J. Hazard. Mater. B* **2006**, *133*, 226–232. [CrossRef] [PubMed]
3. Sakthivel, S.; Shankar, M.; Palanichamy, M.; Arabindoo, B.; Bahnemann, D.; Murugesan, V. Enhancement of photocatalytic activity by metal deposition: Characterisation and photonic efficiency of Pt, Au and Pd deposited on TiO_2 catalyst. *Water Res.* **2004**, *38*, 3001–3008. [CrossRef] [PubMed]
4. Utami, F.; Rahman, D.; Sustini, E.; Abdullah, M. Immobilization of TiO_2 on transparent plastic and its application in photocatalytic wastewater treatment. *J. Phys. Conf. Ser.* **2019**, *1171*, 6–13. [CrossRef]
5. Yang, X.; Qiu, L.; Luo, X. ZIF-8 derived Ag-doped ZnO photocatalyst with enhanced photocatalytic activity. *RSC Adv.* **2018**, *8*, 4890–4894. [CrossRef] [PubMed]
6. Lai, H.; Nguyen, G.; Tran, N.; Nguyen, T.; Tran, C.; Nguyen, D.; Chang, S.; Chung, W.; Nguyen, D.; Thi, H.; et al. Assembled porphyrin nanofiber on the surface of g-C_3N_4 nanomaterials for enhanced photocatalytic degradation of organic dyes. *Catalysts* **2022**, *12*, 1630. [CrossRef]
7. Sert, B.; Bilici, Z.; Ocakoglu, K.; Dizge, N.; Rad, T.; Khataee, A. Preparation of S-scheme g-C_3N_4/ZnO heterojunction composite for highly efficient photocatalytic destruction of refractory organic pollutant. *Catalysts* **2023**, *13*, 485. [CrossRef]
8. Musielak, E.; Feliczak-Guzik, A.; Jaroniec, M.; Nowak, I. Photodynamic light-triggered release of curcumin from hierarchical FAU zeolite. *Catalysts* **2023**, *13*, 394. [CrossRef]
9. Gervasi, S.; Blangetti, N.; Freyria, F.; Guastella, S.; Bonelli, B. Undoped and Fe-doped anatase/brookite TiO_2 mixed phases, 2 obtained by a simple template-free synthesis method: Physico-3 chemical characterization and photocatalytic activity towards 4 simazine degradation. *Catalysts* **2023**, *13*, 667. [CrossRef]
10. Bachvarova-Nedelcheva, A.; Iordanova, R.; Naydenov, A.; Stoyanova, A.; Georgieva, N.; Nemska, V.; Foteva, T. Sol-gel obtaining of TiO_2/TeO_2 nanopowders with biocidal and 2 environmental applications. *Catalysts* **2023**, *13*, 257. [CrossRef]
11. Sisay, E.; Kertész, S.; Fazekas, Á.; Jákói, Z.; Kedves, Z.; Gyulavári, T.; Ágoston, Á.; Veréb, G.; László, Z. Application of $BiVO_4$/TiO_2/CNT composite photocatalysts for membrane fouling control and photocatalytic membrane regeneration during dairy wastewater treatment. *Catalysts* **2023**, *13*, 315. [CrossRef]
12. Salomatina, E.; Shelud'ko, P.; Smirnova, L. Photocatalytic decomposition of azo dyes and phenols using polymer composites containing nanostructured poly(titanium oxide) doped with gold or silver nanoparticles under light irradiation in a wide wavelength range. *Catalysts* **2023**, *13*, 423. [CrossRef]
13. Jaison, A.; Mohan, A.; Lee, Y. Recent Developments in Photocatalytic Nanotechnology for Purifying Air Polluted with Volatile Organic Compounds: Effect of Operating Parameters and Catalyst Deactivation. *Catalysts* **2023**, *13*, 407. [CrossRef]
14. Alapi, T.; Veres, B.; Náfrádi, M.; Farkas, L.; Pap, Z.; Covic, A. Application of BiOX photocatalyst to activate peroxydisulfate ion-investigation of a combined process for the removal of organic pollutants from water. *Catalysts* **2023**, *13*, 513. [CrossRef]

catalysts

MDPI

Review

Recent Developments in Photocatalytic Nanotechnology for Purifying Air Polluted with Volatile Organic Compounds: Effect of Operating Parameters and Catalyst Deactivation

Augustine Jaison [1], Anandhu Mohan [2,*] and Young-Chul Lee [1,*]

[1] Department of BioNano Technology, Gachon University, 1342 Seongnam-Daero, Sujeong-Gu, Gyeonggi-Do, Seongnam-Si 13120, Republic of Korea
[2] Department of Industrial and Environmental Engineering, Graduate School of Environment, Gachon University, 1342 Seongnam-Daero, Sujeong-Gu, Gyeonggi-Do, Seongnam-Si 13120, Republic of Korea
* Correspondence: anandhumohangcm@gmail.com (A.M.); dreamdbs@gachon.ac.kr (Y.-C.L.)

Abstract: Photocatalytic oxidation (PCO) is a successful method for indoor air purification, especially for removing low-concentration pollutants. Volatile organic compounds (VOCs) form a class of organic pollutants that are released into the atmosphere by consumer goods or via human activities. Once they enter the atmosphere, some might combine with other gases to create new air pollutants, which can have a detrimental effect on the health of living beings. This review focuses on current developments in the degradation of indoor pollutants, with an emphasis on two aspects of PCO: (i) influence of environmental (external) conditions; and (ii) catalyst deactivation and possible solutions. TiO_2 is widely used as a photocatalyst in PCO because of its unique properties. Here, the potential effects of the operating parameters, such as the nature of the reactant, catalyst support, light intensity, and relative humidity, are extensively investigated. Then the developments and limitations of the PCO technique are highlighted, especially photocatalyst deactivation. Furthermore, the nature and deactivation mechanisms of photocatalysts are discussed, with possible solutions for reducing catalyst deactivation. Finally, the challenges and future directions of PCO technology for the elimination of indoor pollutants are compared and summarized.

Keywords: indoor air pollutants; VOC; photocatalysts; environmental conditions; catalyst deactivation

1. Introduction

Volatile organic compounds (VOCs) are hazardous air pollutants that can be poisonous and mutagenic, and the continued release of VOCs elevates the ozone level of the atmosphere and leads to photochemical haze. VOCs are released by various sources, including electronic devices, construction materials, smoke, consumer goods, occupant-related activities, indoor fuel gas, combustion wastes, and so on [1–3]. Halogenated hydrocarbons, ketones, aldehydes, alcohols, and aromatic compounds are some of the several kinds of VOCs that are used in the industry and normally found in emissions. Consumer goods often result in continuous VOC release and pungent gases like ammonia are released by food wastes, which can markedly lower the well-being of the people in indoor spaces [4,5]. VOCs that are easily diffused and volatile when released into the atmosphere are considered serious contributors to air pollution. These VOCs often generate undesired by-products that can be more hazardous than the parent compound and are associated with serious health risks [6–9].

Adsorption, membrane separation, biological degradation, thermal and non-thermal plasma treatment, and photocatalytic oxidation (PCO) have all been explored so far as means for reducing harmful pollutants in indoor air. PCO has garnered the most interest among all these techniques because of its high success rate in the removal of typical low-concentration indoor pollutants, with maximum efficiency and minimum energy consumption. Photocatalysis is a promising technology because it can directly convert gaseous atmospheric pollutants (especially VOCs) to less harmful CO_2 and H_2O molecules under optimum conditions. The attractive nature of air cleaning by photocatalysts under UV-visible light has prompted the synthesis of novel nano-catalytic materials for efficient VOC removal [1]. However, photocatalytic oxidation continues to have limitations, such as rapid catalyst deactivation, less efficiency under normal operating parameters, and production of undesirable by-products [3,6].

Conventional semiconductor nanomaterials often have poor pollutant adsorption capabilities at low concentration levels (ppb-ppm) [10]. Numerous studies have shown the potential of TiO_2 as a photocatalyst among various other catalysts (ZrO_2, SnO_2, Fe_2O_3, WO_3, CeO_2, ZnS, etc.) owing to its excellent photocatalytic efficiency under UV light irradiation and ideal energy gap between the valence and conduction bands (3.2 eV) [11–14]. TiO_2 catalysts have been synthesized in various morphologies, including nanoparticles, nanotubes, hollow fibers, thin films, and microporous and mesoporous forms. However, TiO_2 normally absorbs wavelengths less than 400 nm, and it is ineffective in enclosed spaces, owing to the lack of visible light absorption capability. Therefore, effective visible light active photocatalysts must be developed for air cleansing applications, especially in the indoor environment [15,16]. Similarly, the deactivation-resistant catalyst also plays an important role in air purification. The synthesis and immobilization of these catalysts on a stable solid support is crucial for superior photo-activity. Reactive intermediates from the breakdown of gaseous reactants may build up on the surfaces of catalysts over time, obstructing the active sites and eventually leading to catalyst deactivation [17,18]. In addition, the photocatalytic effectiveness and by-product generation depends on the operating parameters employed during VOC oxidation. A thorough evaluation of the catalytic activity with a wide range of operating conditions, such as relative humidity (RH), flow rate, light intensity, reactant concentration, and catalyst support, is required to achieve the maximum photocatalytic efficiency for air purification [19–21].

In this review, we focus on two main areas of photocatalytic VOC oxidation: (i) the impact of various operating parameters on catalytic activity; and (ii) photocatalyst deactivation. First, we summarize most of the published articles discussing the effects of various reaction parameters on VOC oxidation. Furthermore, strategies for synthesizing deactivation-resistant photocatalysts, the nature of deactivated photocatalysts, and different techniques for analyzing the extent of photocatalyst deactivation are reviewed and discussed. Finally, the current status and major drawbacks of photocatalytic air purification are discussed, and a few recommendations are suggested for future studies.

2. Air Pollutants and VOCs

Epidemiological studies have shown that exposure to air pollution is the main cause of various unfavorable health effects, including headaches, allergies, and eye irritation. Gaseous particles released into the air by various manmade and natural sources, as well as photochemical transformation processes, make up the complex phenomena known as air pollution [22]. CO_2, SO_2, NO_x, particulate matter (1°), NH_3, CH_4, volatile organic compounds, and hydrocarbons are classified as primary pollutants, because these compounds have a direct impact on air quality. Examples of secondary pollutants include O_3, secondary particulate matter, and volatile organic by-products, which are formed by the redox interaction between primary pollutants. Several water contaminants, including heavy metals such as nickel, mercury, and arsenic, can also be present in the air, particularly in industrial areas, where they will lead to fatal illnesses [23].

The most concerning indoor air pollutants (IAPs) are VOCs, which are carbon-containing chemicals that are readily volatile at very low temperatures. VOCs have a vapor pressure higher than 1 mmHg at room temperature, and common examples of VOCs are benzene, toluene, formaldehyde, ethanol, trichloroethylene, etc. These target VOCs in the atmosphere are further categorized into aromatic, aliphatic, chlorinated, and oxygenated hydrocarbons [24,25]. Exposure to VOCs such as benzene and other cyclic aromatic hydrocarbons can cause cancer in some individuals; however, a direct link connecting these two has not yet been conclusively proven at typical indoor air concentrations. However, some studies have shown that the risks of cancer are very high under exposure to very low concentrations of these pollutants, with symptoms like headache, drowsiness, weariness, and disorientation [26]. Due to their high bioaccumulation potential and toxic nature, halogenated VOCs are one of the critical classes of indoor air pollutants. Polychloromethanes (PCMs) are the most prevalent and common halogenated VOCs. The remains of water filtration and evaporation systems are the main sources of PCMs, and examples of PCMs include chloroform, dichloromethane, trichloromethane, and carbon tetrachloride. Chlordane ($C_{10}H_6Cl_8$) is another organochlorine VOC that is known to be carcinogenic to humans. It is a white solid, categorized as an organic pollutant and detrimental to human health and environment, just like other chlorinated cyclodiene compounds [27]. Aldehydes make up a sizable portion of all the active VOCs present in the environment, with formaldehyde and acetaldehyde being the important ones [28,29].

Petroleum products, such as gasoline and other fuels made from crude oil, including benzene, toluene, and ethyl benzene, are considered as aromatic VOCs. A considerable quantity of aromatic compounds is released into the environment because of the incomplete combustion of liquid fuels in automobiles. These compounds are also found in a wide range of consumer goods, such as gasoline, paint, drugs, detergents, etc. In addition to being poisonous and cancer-causing, aromatic chemicals can cause serious harm to the ozone layer [30]. Alcohols and ketones constitute another group of VOCs that are frequently found in cosmetics and personal care items, such as hair spray, nail polish remover, cologne, varnishes, paint thinners, and adhesives. Ethyl alcohol, isopropyl alcohol, and benzyl alcohol are examples of common alcohol-based VOCs and ketone-based VOCs include acetone, methacrylate, ethyl acetate, and methyl ethyl ketone [29]. Ethylene is an important organic VOC and when fruits and vegetables are exposed to ethylene during food storage, they undergo physical and chemical transformations, causing food decay. Because of its easy availability and strong solubility, ethylene is widely employed as a solvent in the production of varnishes, resins, adhesives, printing ink, and pharmaceutical products [31]. Table 1 lists the important VOCs and their major indoor sources [32,33].

Table 1. Indoor sources of VOCs.

VOCs	Major Indoor Sources
Toluene	Paints, gasoline, varnishes, solvents, polishes, anti-freezing materials, pesticides
Benzene	Glues, adhesives, detergents, tobacco smoke, gypsum board
Naphthalene	Mothballs, dyes, smoke, insulating materials
Ethylbenzene	Paints, gasoline, inks, glues, candles, insecticides
Chloroform	Solvents, glues, chlorinated water
Ethylene	Food products, gasoline
Formaldehyde	Wood products, wallpaper and paints, flooring items, foam, wallpaper paste, hair conditioners
Tetrachloroethylene	Water repellents, wood cleaners, adhesives, dry-cleaned clothes
Acetaldehyde	Wood products, deodorants, building materials and furnishings, carpets, gypsum
Trichloroethylene	Metal cleaners, adhesives, flooring materials, paint removers
Styrene	Paints, detergents, plasters, adhesives, modelling clay
Acetone	Solvents, sprays, cosmetics, floor coverings, resins, disinfectants, air fresheners

3. Types of Nano Photocatalysts

Semiconducting photocatalysts made of TiO_2-based nanomaterials continue to form the core foundation of PCO technology, producing a significant number of excited electrons and holes when irradiated with light of suitable wavelengths [34]. Because of the large bandgap (3–3.2 eV), TiO_2 does not work well for all photo-catalytic processes, because it uses only 5% of the total solar energy that impinges on the surface of the earth. Over the past few decades numerous efforts have been made to alter the TiO_2 surface for energy band reduction by doping with elements such as N, C, and S. Researchers have attempted to develop various hetero-junction catalysts by fusing titanium with metals such as Pt/Pd or with semiconductors like NiO, WO_3, CdS, and quantum dots for improved light utilization [35]. To increase the efficiency of acetone photocatalytic degradation and light absorption, Hu et al. developed a single-atom Au-supported TiO_2 photocatalyst. Similarly, He and colleagues described a method to significantly increase the photocatalytic removal of hydrophilic VOCs (formaldehyde and acetone) from the air by coating hygroscopic periodic acid on WO_3 [36,37]. Sydorenko et al. synthesized transparent TiO_2 thin films and successfully coated them on borosilicate glass by spray pyrolysis. The removal efficiency of the synthesized thin film was tested using acetaldehyde, acetone, heptane, and toluene in 9 ppm VOC mixtures under visible light, and the results showed 100, 100, 78, and 31% conversion, respectively [38].

Materials like SiO_2 have a mesoporous framework (2–50 nm), and these compounds can be used as supports in the synthesis and functionalization of photocatalytic materials. Some advantages of this system are listed below: (1) Pore diameters and structures can be altered at the atomic levels; (2) the pore size makes it easier for reactants to diffuse into the channels and access the active sites; (3) exceptional transparency in the broad UV-visible region provides a photochemically inactive region for the catalyst; (4) with numerous silanol groups in the structure, functionalization with various modifiers is easily achievable [34,39]. Single-site catalysts with tetrahedral geometry are another group of compounds used for photocatalysis. This group includes isolated metal (Ti, V, Cr, and Mo-oxide) moieties with strong solid support, such as mesoporous silica or microporous zeolite. Similarly, replacing Si atoms in the SiO_2 matrix with Ti gives single-site titanium oxide photocatalysts, which can be achieved by several anchoring techniques, including sol–gel method, hydrothermal synthesis, and chemical vapor deposition [39,40].

Another type of metal-free semiconductor used in PCO is graphitic carbon nitride (g-C_3N_4 or CN), which has a bandgap between 2.6 and 2.8 eV. Although carbon nitride is a well-known photocatalyst, its low VB potential is insufficient to produce hydroxyl radicals, resulting in poor activity in terms of VOC degradation. To overcome this, Sun et al. synthesized a g-C_3N_4-based catalyst that was combined with $BiVO_4$. Under visible light, the toluene conversion rate of $BiVO_4$/CN was 4.5 times higher than that of the parent carbon nitride [41–43]. Another common carbon-based material used in VOC oxidation is activated carbon fiber (ACF), synthesized from organic precursors through pyrolysis and activation in the presence of carbon dioxide and water vapor. The increased ability of ACFs to absorb VOCs like chloroform and isobutane is attributed to their high specific surface area and superior micro-sized pore volume. Metal-free conducting polymer composites are a cost-effective and environmentally friendly alternative to rGO-based metal oxides/chalcogenides composites for air purification [44–46]. Yan et al. synthesized a novel composite using rGO aerogel and polypyrrole via a hydrothermal process. The result showed that the composite exhibited superior phenol adsorption, owing to the organic nature of the composite material [47].

The formation of electron–hole pairs plays a critical role in semiconductor PCO and requires suitable light energy absorption with the simultaneous promotion of electrons from the valence band (VB) to the conduction band (CB) (as shown in Figure 1). In the following steps, the photogenerated charge carriers combine with oxygen and water molecules to form extremely reactive intermediate species such as hydroxyl radicals. In photocatalytic oxidation reactions, the mass transfer of reactants from the gas phase to the solid phase is crucial. The overall effectiveness and the rate of reaction are significantly impacted for pollutants in the gaseous phase by the net mass transfer from the bulk to the exterior surface of the photocatalyst. Following diffusive mass transfer, the reactive species encounter pollutant molecules at the surface, causing them to break down into compounds with lower molecular weights, which ultimately break down into CO_2 and water [48–51].

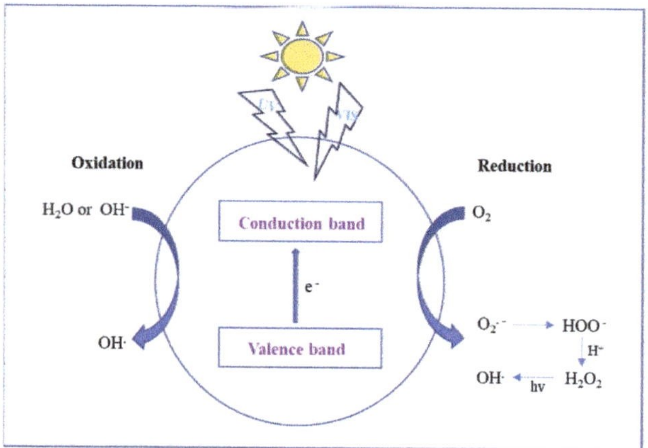

Figure 1. Schematic diagram of electron transfer from CB to VB.

4. Influence of Environmental Conditions

4.1. *Effect of Relative Humidity*

In recent years, a considerable focus has been given to studying how humidity influences the removal of volatile organic compounds (VOCs) with different photocatalysts. The surface-bound water molecules and OH groups produced by the dissociative chemisorption of moisture on the catalyst surface can be quite important in the photo-oxidation of VOCs. In the total absence of water vapor, the complete mineralization to CO_2 does not occur and the photocatalytic breakdown of organic compounds like toluene or formaldehyde is significantly slowed down. However, as water molecules start occupying more and more active sites, the reaction rate decreases, because of the considerable build-up of water vapor on the catalyst surface. Therefore, finding an optimal level of humidity is a crucial step in photocatalytic reactions [52].

The photo-oxidation in a dry atmosphere as well as under higher humidity conditions was usually less effective. Quici et al. found that the ideal humidity levels for the breakdown of toluene both with and without a photocatalyst are around 10% RH. Thevenet et al. observed a continual decline in the reaction rate with a rise in humidity for the photocatalytic oxidation of a relatively high concentration of acetylene (1000 ppm) [38]. This trend, however, cannot be generalized for all reactants or with different experimental setups. For example, Durme et al. showed a higher conversion rate for toluene (500 ppb) at 26% RH by non-thermal plasma treatment, which is a complementary method for gas-phase PCO [53,54]. Korologos et al. investigated the effects of water vapor concentrations on the photo oxidation of benzene, toluene, ethylbenzene, and *m*-xylene (BTEX). The effect of humidity on the PCO showed strong dependence on the type of the BTEX and the water vapor content. The increase in residence time resulted in a significant increase in the conversion for every compound [55].

Humidity has both beneficial and detrimental effects on the ability of TiO_2 photocatalysts to catalyze reactions. For example, the total oxidation of toluene to CO_2 is increased at higher humidity levels because of the production of hydroxyl radicals, after which the catalytic activity becomes stable over time. However, then the VOCs would face competition from the high water content in the atmosphere, which can decrease the selective adsorption, and thereby reduce the rate of reaction [56]. Like toluene, the amount of moisture had a similar impact on the rate at which formaldehyde is oxidized. The oxidation rate increased to a maximum when the humidity level was gradually raised, and then the rate was decreased with a further rise in the water vapor concentration [57]. Assadi et al. performed VOC oxidation of trimethylamine and isovaleraldehyde and observed the impact of humidity on the degradation rate. Water vapor can enhance the degradation through the formation of OH• radicals when the humidity level is less. But as the water vapor content increases, competition between water vapor and VOCs for the active sites at the surface increases, and as a result, the rate of degradation slows down. From the experiments, the ideal RH levels for isovaleraldehyde and trimethylamine were found to be 40 and 25%, respectively [58]. Vildozo et al. performed photo-oxidation of 2 propanol-toluene mixtures at indoor air concentrations at two different relative humidity values (0 and 60%) at 25 °C using a TiO_2 catalyst. The study showed that (1) the competition of 2-propanol-toluene binary mixture with water vapor for active sites is insignificant under dry conditions, (2) the lesser the solubility of the organic pollutant, the more the inverse impact of humidity on the adsorption of the pollutant on the TiO_2 surface, and (3) the degradation of VOC increases with a decrease in water vapor content, with the maximum conversion achieved at 0% RH [59].

Along with the impact on degradation efficiency, RH can also play a major role in the selectivity of products and the types of by-products formed during PCO. According to Vincent et al., a high relative humidity (30%) has a favorable impact on the conversion of 1-propanol while also stifling the production of unwanted by-products (such as propionaldehyde) owing to the higher OH radical concentration at higher humidity levels [60]. Different results were obtained when Debono et al. studied the degradation of decane and observed that water vapor slows down the removal of unwanted impurities and drastically increases their quantity around the catalyst surface [61]. The above findings were further supported by the oxidation of binary mixtures of 2-propanol and toluene. Higher humidity levels led to significant loss of overall conversion (40 to 70%), with a varying initial concentration ratio of 2-propanol to toluene [59]. Kibanova et al. showed that the photo-oxidation of formaldehyde was affected by the competition between water vapor and reactants for active sites. They compared the performance of clay-TiO_2 nanocomposites to the reference catalyst P25 under controlled conditions of RH [62] and the results are shown in Figure 2.

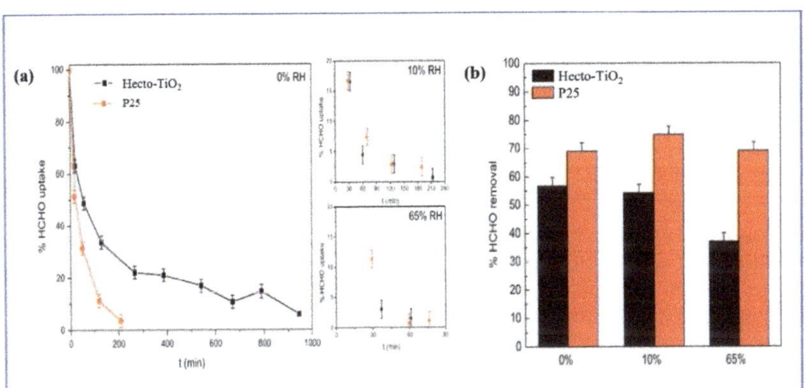

Figure 2. (**a**) Formaldehyde uptake under 0, 10, and 65% RH. (**b**) Percentage of formaldehyde removal with RH. Reproduced with permission from [62]. Copyright 2012 Elsevier Ltd.

The nature of water to compete with reactants can considerably diminish the ability of catalysts to adsorb VOCs. However, selective sorption of VOCs to water vapor can be accomplished by hydrophobically altering the catalyst surfaces. According to Yu et al., increasing the RH (30–70%) enhances the catalytic oxidation of toluene more than that of formaldehyde, presumably because toluene is more hydrophilic than formaldehyde [62,63]. Figure 3a shows the impact of relative humidity on toluene degradation efficiency. Even though the photocatalyst and pollutant remain the same, no specific pattern was observed in these experiments. Instead, the conversion rate was found to mostly depend on the operating parameters employed in each study [51].

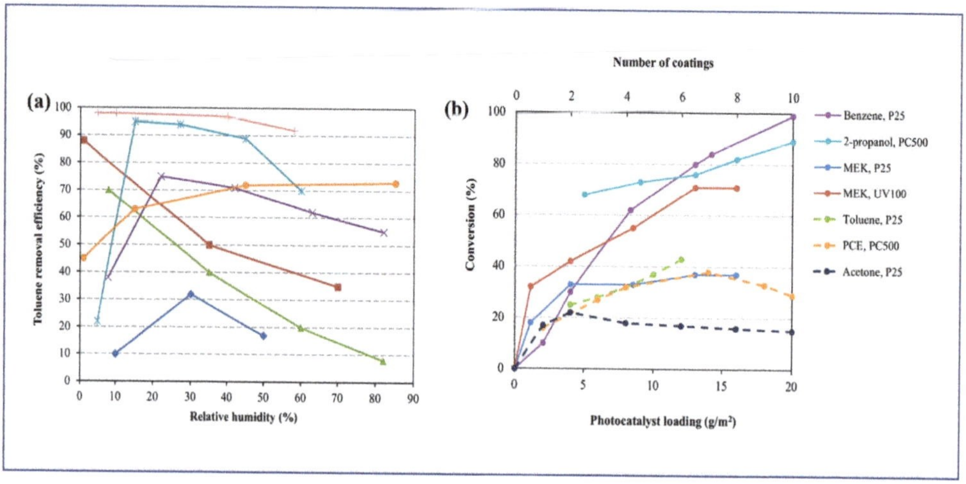

Figure 3. (**a**) Effect of RH on the toluene removal rate using the P25 catalyst. (**b**) Impact of catalyst loading on VOC conversion rate. Reproduced with permission from [51]. Copyright 2017 Elsevier Ltd.

4.2. Effect of VOC Type and Concentration

The type of reactants (VOCs) used and their inlet concentration considerably affect the PCO of air pollutants. There is a widespread acceptance that increasing VOC concentrations would lead to lower removal efficiency for different classes of VOCs [51]. Mamaghani et al. performed VOC oxidation using commercial TiO_2 catalysts to study the impact of various operational parameters. It was found that the target pollutant's removal rate is always greater at lower concentrations (100 ppb) than at higher concentrations (1 ppm). This is because, at higher VOC concentrations, the ratio of reactive species to VOC molecules decreases, because of which a large portion of VOCs remains unreacted. They compared the removal efficiency of toluene and methyl ethyl ketone (MEK) and observed that on almost all titania catalysts, the MEK removal rate was lower than that of toluene. One possible reason given for the high reaction rate of toluene is the better interaction of toluene with the intermediate OH radicals, which leads to a higher removal efficiency [19]. The same group of researchers studied the effect of MEK inlet concentration over various titanium catalysts, and like previous results, the VOC removal rate was decreased with an increase in the inlet concentration (Figure 4a). This was expected, given that, when the inlet concentration increases, the proportion of active sites to the incoming MEK molecules drops. Thus, MEK molecules are prevented from participating in the degradation mechanism. Figure 4b illustrates the concentration of by-products produced during the VOC oxidation with various MEK inlet concentrations. A potential reaction route is shown in Figure 4c, based on the products/by-products formed in the PCO of MEK [64].

Jafarikojour et al. performed PCO of toluene and the data showed that the conversion fell from 37 to 27% when the initial concentration of toluene increased from 20 to 100 ppm. Similarly, Mo et al. found that raising the inlet concentration of toluene from 1 to 4 ppm caused a 30% reduction in the breakdown efficiency of toluene [51]. According to Yu et al., rate constants of toluene, *p*-xylene, and *m*-xylene are linearly related to their hydroxyl radical reaction rate constants. In a different work, Vildozo et al. demonstrated that the mineralization rate of 2-propanol was reduced from around 90 to 63% by raising the initial concentration of 2-propanol from 100 to 700 ppb [65]. Through their study, Boulamanti and Philippopoulos established the following order for the VOC oxidation rate of alkanes: *n*-pentane < *i*-pentane < *n*-hexane < *i*-hexane < *n*-heptane. Based on this progression, they hypothesized that branched alkanes had a higher reaction rate because they possess a tertiary carbon atom [66,67].

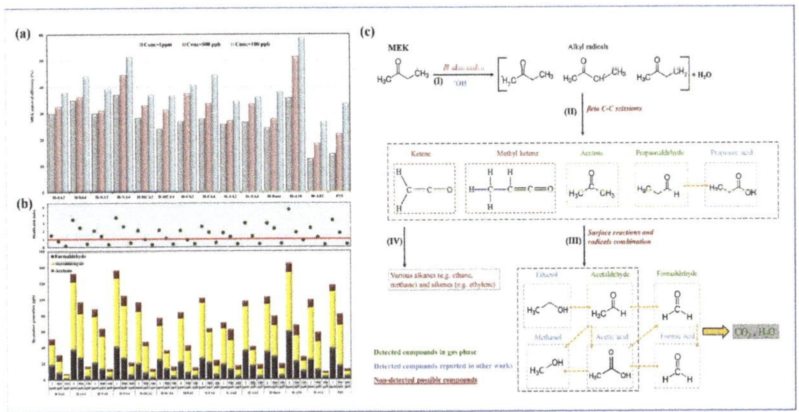

Figure 4. (**a**) Influence of inlet concentration on MEK oxidation. (**b**) By-products formation over various MEK inlet concentrations. (**c**) Predicted reaction route for the PCO of MEK. Reproduced with permission from [64]. Copyright 2019 Elsevier Ltd.

For aromatic compounds, the removal efficiency increases in the following order: benzene < toluene < *o*-xylene < ethylbenzene over bare P25, and benzene < *o*-xylene < ethylbenzene < toluene over the P25/AC catalyst. This result suggests that the presence of activated carbon on TiO_2 affects the conversion of toluene more than that of any other aromatic hydrocarbons [68]. According to Li et al., the catalyst activity dropped from 93 to 24% when the initial toluene concentration increased from 200 to 1000 ppb. Similarly, the adsorption amount of toluene on the g-C_3N_4/TiO_2/ACF catalyst decreased with an increase in the initial concentration of toluene. Thus, the reaction rate is maximum when the number of reactant molecules is less than or equal to the number of active sites on the catalyst. After the photocatalyst reaches its maximum capacity, the initial concentration does not affect the rate of reaction anymore [2,69]. So far, researchers have found that a higher initial VOC concentration can increase the number of adsorbed pollutants on the catalyst surface. However, once the initial concentration reaches a saturation point where there are not enough oxidation radicals and active sites on the catalyst to carry out further oxidation, more and more reactants will depart the reactor without engaging in the photo-oxidation reaction [2,70].

4.3. *Effect of Catalyst Loading and Support*

In PCO reactions, a support material is often employed to increase the adsorption capacity and decrease the aggregation of nanoparticles on the photocatalyst surface. PCO efficiency is considerably influenced by the catalyst loading amount and the kind of support material used [2]. Increasing the amount of catalyst loading can improve the reaction rate by reducing the adsorption competition, increasing the surface area, and increasing the availability of the active sites of the catalysts. However, studies have indicated that when catalyst loading exceeds a particular threshold, the degradation rate decreases. This is attributed to the reduced light exposure and agglomeration of particles produced by the large amount of catalyst loading [71].

It is desirable to increase quantum efficiency by immobilizing the catalyst on a support material with suitable properties, like high surface area, porous structure, strong transparency, and stability under UV irradiation. One common support used in many projects is activated carbon (AC), which improves the retention of VOCs on catalytic beds by lowering the competition of water vapor to reactants [51]. Using AC with P25 might nullify the effects of A short residence time and excess humidity on the degradation of pollutants in the oxidation of aromatic VOCs and formaldehyde [68,72]. Li et al. synthesized a novel composite by combining a charcoal skeleton and "chrysanthemum" g-C_3N_4 (C/CN-x). The catalyst was used for the visible-light-induced breakdown of formaldehyde and the most effective photocatalytic material was found to be C/CN-8, with a removal rate of 84.63%, which was 130% more effective than the conventionally used g-C_3N_4. The favorable result is mainly due to the high surface area provided by the catalyst support material [73]. Liu et al. performed PCO of benzene with Ti^{3+} self-doped TiO_2 and found that the photocatalytic activity of benzene was largely improved with doping. The insertion of Ti^{3+} into TiO_2 microspheres decreased the electron–hole recombination and enhanced the light absorption ability, thereby increasing the photocatalytic efficiency of the catalyst. Figure 5a shows the rate of benzene oxidation with Ti-doped TiO_2 catalyst at various calcination temperatures. The possible mechanism of the degradation reaction is given in Figure 5b. CO_2 and CO are the major products formed during the reaction (Figure 5c,d) [74].

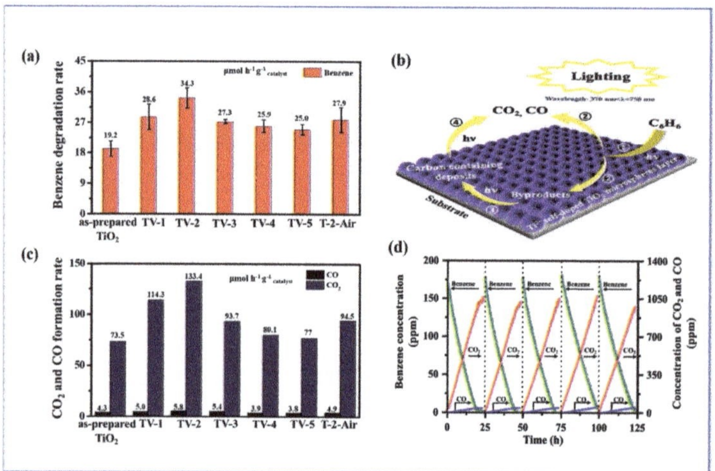

Figure 5. (**a**) Benzene oxidation rate at various calcination temperatures. (**b**) Probable mechanism of benzene oxidation. (**c**) Product formation rate at different calcination temperatures. (**d**) Repeated cycles of PCO of benzene on TV-2 catalyst. Reproduced with permission from [74]. Copyright 2020 Elsevier Ltd.

Using pure aerolyst 7710 pellets coated on TiO_2 glass beads as photocatalysts, Verbruggen et al. examined the oxidation reaction of ethylene. A significant increase in ethylene removal efficiency (up to 70%) was observed for TiO_2-coated glass beads when compared with the parent catalyst. The support provided enhanced light scattering and increased the interaction time between the reactants and photons, thereby increasing the reaction rate [75]. To increase the stability and efficiency of the photocatalyst, a binder is often used while depositing the photocatalyst on the support material. Thevenet et al. used a silica binder to attach P25 on organic fibers and demonstrated that the conversion rate was substantially lower with the binder present, and vice versa [76]. Mo et al. synthesized TiO_2 with silica and mordenite as the support and observed a higher adsorption constant for P25/silica composite because of the strong adsorption capability of silica. The enhanced surface area improves mass transfer; therefore, the activity of the catalyst can be improved by adding a suitable support material [77].

4.4. Effect of Light Intensity

The energy needed to initiate the PCO reaction is provided by UV-visible light, and it is the first step in the activation of the photocatalyst. Theoretically, wavelengths less than 380 nm can activate the TiO_2 photocatalyst, which is equivalent to the bandwidth of TiO_2 (3.2 eV) [71]. Electrons in the valence band of TiO_2 are excited by a UV source, and the frequently used light sources in PCO reactions are fluorescent black light lamps (300–400 nm) and germicidal lamps (UVC, 254 nm). Another common light source used is UV light-emitting diodes (365 nm) due to their long lifespan and excellent performance. Also, a comparison study of photo oxidation of several VOCs revealed that the two light sources, black light and germicidal lamp, with the same power showed no significant performance difference between them [4,51]. When the light intensity is increased, more photons will be formed and, as a result, more e^-h^+ pairs will be generated, which will increase the oxidation rate. For example, the PCO of TCE and 1-propanol both moved from the first-order to half-order rate when the light intensity is increased from 1.5 to 4 W/m^2. Although increasing the light intensity can enhance the pollutant removal rate, excessive light reduces the quantum efficiency and creates unwanted energy loss [51,60,78].

According to Kibanova et al., toluene oxidation over a P25 catalyst was finished nearly twice as quickly by utilizing an ozone-producing lamp as opposed to a UVA-generating lamp [79]. Studies on the photo-oxidation of formaldehyde, toluene, and benzene over P25 showed that the degradation rate follows the order of UVC/O_3 > UVC > UVA. That is, the low-energy photons and ozone of UVA play a positive role in PCO reactions of VOCs [53,62,80]. The same has been confirmed by Krichevskaya et al. by the PCO of a mixture of VOCs. Thus, the amount of CO_2 produced during the reaction increased from 7 to 20 ppm when the O_3 concentration in the chamber was increased from 40 to 80 $\mu g\ L^{-1}$ [81]. Increasing the light intensity at the catalyst surface or utilizing a shorter wavelength light source generally improves the removal efficiency of VOCs. The generation of intermediates or by-products, as well as catalyst deactivation, can also be controlled using the appropriate light sources with optimal intensity [51].

Table 2 highlights how various reaction conditions affect the removal efficiency of photocatalysts in VOC oxidation reactions.

Table 2. Effect of operating parameters on VOC oxidation.

Catalyst	VOC Type	Concentration	RH (%)	Light Source	Removal Efficiency (%)	Ref.
TiO_2/diatomite composite	Acetone, MEK	10 ppm	15	UVA lamp	~42–61	[82]
TiO_2 thin film	Acetone, toluene, *p*-xylene	0.1–0.3 mol/m^3	35	UV lamp	~55–77	[83]
C-doped TiO_2	Toluene	150 mg/m^3	60	Visible light	~60	[84]
TiO_2/Mg-Al LDH	Toluene	100 ppm	50	UV lamp	~74	[85]
Pt-TiO_2	MEK	1 ppm	100	UV lamp	~73	[86]
CNT-TiO_2	Gaseous styrene	25 ppm	0	UV-LED light	~50	[87]
Pt-TiO_2-R	*m*-xylene	1 ppm	50	UV lamp	~73	[88]
Ce-TiO_2	Toluene	150–600 ppb	<3	Visible light	~22	[89]
Pd/WO_3	Acetaldehyde	5 ppm	50	Fluorescent-visible light	~99	[90]
Pt-TiO_2	BuAc	1 ppm	100	UV lamp	~98	[86]
TiO_2/diatomite composite	2-Heptanone	10 ppm	15	UVA lamp	~38	[82]
MOF(Ti)	Acetaldehyde	200 ppm	80	Visible light	~98	[91]
TiO_2	Ethanol	200–2000 ppb	9–60	UVC lamp	~4–44	[92]
SiO_2 coated TiO_2	Isoflurane	500 ppb	50	UVC fluorescent lamp	~100	[93]
N-doped TiO_2	Ethyl benzene	20–140 ppm	50	TUV lamp	~25–100	[94]
F-TiO_2	Toluene	30 ppm	50	VUV lamp	~80	[95]
Zn-Ti-LDH	Toluene	500 ppm	50	UV lamp	~75	[96]
MOF(Fe)/Fe_2O_3	*o*-xylene	25 ppm	50	Xenon lamp	~100	[97]
GO/ZnO	Benzene	100 mg/L	35–45	UV lamp	~87	[98]
Zn_2SO_4/LDH	Toluene	500 ppm	60	UV lamp	~90	[99]

5. Catalyst Deactivation and Possible Solutions

5.1. Main Reasons for Catalyst Deactivation

Prolonged irradiation and periodic uses cause the rapid loss of activity of the photocatalytic materials used for VOC destruction. In contrast to the liquid–solid interface catalytic reactions, photocatalytic processes occurring at the gas–solid interface normally show a steady decline in the activity over time. The wetting ability of liquid can keep the catalyst surfaces fresh, whereas the air medium is unable to offer the same cleaning property. Hence, photocatalysts operating in the air are more vulnerable to deactivation [100]. Intermediates are formed on the surface of the catalysts during VOC oxidation, and many of them interact more strongly with the catalyst than the initial reactant molecules. The deposition of these intermediates on the catalyst surface blocks the photons and prevents the active transfer of O_2 and other reactive species onto the surface of the catalyst. Einaga et al. showed that the effectiveness of VOC removal is drastically decreased after 2 h of photoirradiation of organic compounds like toluene, benzene, cyclohexene, and cyclohexane [101]. According to Taranto et al., the methanol degradation rate was largely reduced over the P25 catalyst immobilized on cellulose fibers at high irradiance (108 W/m^2). They proposed that since cellulose is susceptible to UV radiation, it is likely that the by-products of the reaction will compete with the reactant for the active sites and eventually slow down the reaction rate [102].

One of the most common reasons for the photocatalyst deactivation is the adsorption of chemical intermediates on the catalyst surface. Thevenet et al. used a liquid-phase acidic extraction technique and identified that formic acid, oxalic acid, and acetic acid were adsorbed on the TiO_2 surface during the photo-oxidation of acetylene. They concluded that the accumulation of these carboxylic acids was the primary cause of catalyst deactivation [103]. In a similar work, Einaga et al. observed the evolution of CO_x from the P25 catalyst regeneration process during the oxidation of cyclohexane, cyclohexene benzene, and toluene. They summarized that carbon residues formed from the aromatic compounds were more resistant to oxidation than those from the non-aromatic compounds [101]. Another important finding in VOC oxidation is that the degradation of hydrophobic molecules can be severely hampered by the binding of interfering molecules with hydrophilic properties (such as acetone or benzaldehyde) on the TiO_2 surface [36]. Hay et al. showed that the deactivation of the photocatalyst was caused by the production of a silica layer from photo-oxidation of siloxanes, which clog the pores of P25 in ambient air at low concentrations [104].

Cha et al. showed that half-oxidized organic compounds with low volatility bind to the TiO_2 catalyst surface, hindering the efficiency of air purification in the PCO of toluene [105]. Benzyl radicals are produced during toluene degradation and are further oxidized to produce benzyl alcohol or benzaldehyde. In the presence of superoxide radicals, benzoic acid can be produced selectively, which can then dimerize into benzyl benzoate. High molecular weight polymeric compounds could be produced during this process, which could effectively close the active center of the catalyst, thereby impeding photocatalytic reactions [106]. Relative humidity also plays a significant role in photocatalytic deactivation, and it can affect the extent of deactivation by (1) replacement of hydroxyl groups on the photocatalyst surface by water molecules and (2) absorption of water molecules on the photocatalyst surface, preventing VOC molecules from accessing the active sites [107].

5.2. Properties of Deactivated Photocatalysts

The nature and characteristics of a deactivated photocatalyst often differ from those of a fresh catalyst. For example, when a TiO_2 catalyst undergoes fouling during VOC oxidation, the primary change is that the color of the catalyst turns from white to pale brown. Hay et al. confirmed the formation of silica on the catalyst surface with TEM and electron microprobe techniques for VOC oxidation reactions [18,104]. Analysis using electron microscopy can validate the extent of photocatalyst deactivation. Elemental mapping using TEM electron energy loss spectroscopy (EELS) can be used for the detection of carbon remaining on deactivated TiO_2 after various cycles of VOC oxidation. Carbon deposits on the photocatalyst surface can be calculated by measuring the concentration of CO_2 formed during the photodegradation reaction in clean airflow. Einaga et al. performed photoreaction of several VOCs with constant UV irradiation for 2 h in a continuous flow reactor. The amount of CO_2 produced from the oxidation of cyclohexane and cyclohexene was much less when compared to the CO_2 formed from the oxidation of benzene and toluene. That is, the photodegradation of benzene and toluene deposits more carbon on the TiO_2 surface, thereby resulting in the faster deactivation of the catalyst [101]. In a similar study, the amount of CO_2 that evolved from the F-TiO_2/Pt catalyst was significantly lower than that of pure TiO_2, indicating that the former was more deactivation resistant during the breakdown of toluene. The amount of CO_2 formed from the used photocatalyst surface can be considered as an indicator of the extent of photocatalyst deactivation [107].

In a different work, Chen et al. investigated the intrinsic causes of VOC-mediated intermediates deposition on TiO_2 (P25), in combination with the findings of experiments and DFT calculations. They used β-Ga_2O_3 as a model catalyst with enhanced reactant adsorption to prevent the deactivation of the catalyst during VOC oxidation reactions. The efficient ring-opening abilities of the intermediates on the surface of the β-Ga_2O_3 catalyst increased the durability and achieved excellent resistance to deactivation during the prolonged VOC oxidation process (Figure 6) [108]. In contrast to these findings, several studies have asserted that there has not been any significant photocatalyst deactivation after various VOC oxidation cycles. For example, Verbruggen et al. reported that throughout the 30 h of oxidation, there was no real reduction in the conversion rate of acetylene over P25 and PC500 catalysts. Similarly, Taranto et al. proved that the degradation efficiency of methanol and *n*-octane over titanium catalysts remained the same for an entire day's operation [109,110].

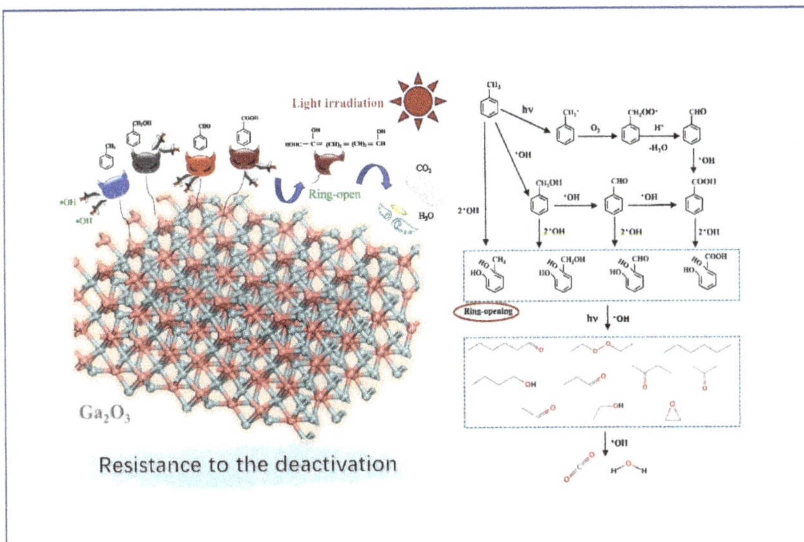

Figure 6. Schematic representation of intermediate formation and reaction mechanism of PCO of toluene on β-Ga_2O_3 catalyst. Reproduced with permission from [108]. Copyright 2020 Elsevier Ltd.

5.3. Solutions for Reducing the Catalyst Deactivation

There are two possibilities once a photocatalyst reaches its limiting point: replacement and regeneration. Since catalyst replacement is not an economically viable option, regeneration techniques have been studied most often. Long-term exposure to air, use of UV radiation, treatment with vaporized H_2O_2 solution, plasma treatment, and ozone purging in the presence of moisture are some of the methods used for catalyst regeneration [51].

UV cleaning has been employed as a post-treatment method in some studies to overcome catalyst deactivation caused by heteroatoms containing VOCs. However, after UV cleaning, the photocatalytic activity was not entirely restored, and the regeneration of the catalyst was influenced by light intensity and illumination duration. For example, a UV cleaning process involving a low-power (8 W) UV lamp was unable to completely restore the activity of the catalyst used to degrade benzene, whereas a UV cleaning experiment using a 16 W light was successful in doing so (Figure 7) [111]. Numerous studies have used heat treatment as a post-treatment method to entirely convert carbonaceous deposits on the catalyst surface into CO_2. Here, the catalyst surface must be heated above the typical vaporization temperature of VOC molecules to break the strong bonds between the catalyst and the organic deposits. However, severe heat treatment can cause the phase transition from anatase to rutile phase in the TiO_2 catalyst and reduce the surface hydroxyl groups, which in turn will cause a decrease in the photocatalytic activity [101].

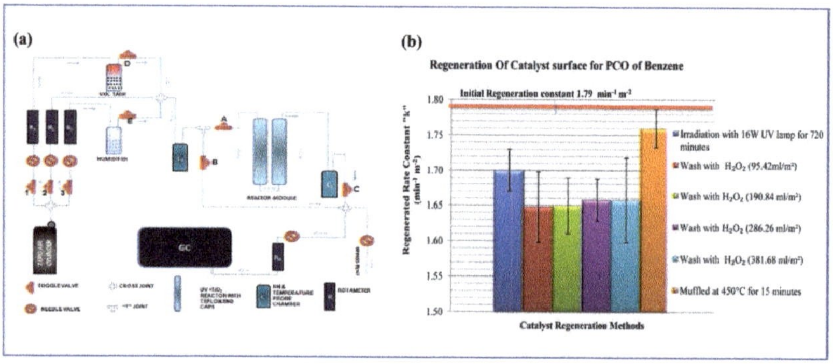

Figure 7. (**a**) Schematic diagram of a photocatalytic reactor used for VOC oxidation. (**b**) Catalyst regeneration rate with different post-catalytic treatments. Reproduced with permission from [111]. Copyright 2015 American Chemical Society.

Another common solution for reducing photocatalyst deactivation involves cleaning the surface of the used photocatalyst with an aqueous solution, and the effectiveness of the regeneration depends on the washing conditions. For example, washing the catalyst with NaOH is a useful technique to get rid of the surface SiO_x residues created during the oxidation of siloxane [112,113]. One potential technique for the prevention of catalyst deactivation is the combined use of adsorbents with photocatalysts. In this way, pollutants are first absorbed by the adsorbent molecules and, additionally, the adsorbent may hold onto the intermediate species generated during the photocatalytic oxidation so that they can be further oxidized without affecting the active sites of the catalyst [114,115]. Arana et al. evaluated the oxidation of gaseous alcohols on TiO_2 and AC-TiO_2 composites. While AC-TiO_2 showed no signs of deactivation, pure TiO_2 lost its activity during mineralization of methanol and ethanol. But the opacity of activated carbon restricts its applications [116]. By the direct incorporation of TiO_2 in NH_2-UiO-66, Yao et al. created a TiO_2/MOF nanocomposite with enhanced degradability and durability. The catalyst was used in the degradation of styrene and the residence time of the reactant was extended by the highly porous MOF, which reduced the deactivation of the catalyst. In a similar work, Suárez et al. used sepiolite, a common clay mineral to combine with TiO_2 to create a hybrid composite material. The composite exhibited excellent mineralization efficiency by absorbing the undesirable TCE degradation intermediates such as dichloroacetyl chloride and phosgene during the PCO reaction [117,118].

In another work, Krichevskaya et al. considerably reduced the time needed to restore P25 activity from hours of post-treatment (15 h) to a relatively shorter time (20 min) with the utilization of UVA + O_3 treatment at high temperature (180 °C) [66]. Additionally, TiO_2 in the form of dispersed powder and thin films showed enhanced VOC degrading activity compared to pellets and honeycomb structures. Thus, a high TiO_2 surface area and enhanced light absorption were achieved with powdered and film forms of TiO_2 [119,120]. Due to its distinctive ordered open-channel structure, TiO_2 nanotubes (TNTs) have shown a great deal of promise in the making of the deactivation-resistant photocatalyst. TNT surfaces are largely exposed to reactants compared with other forms of TiO_2 catalysts which make it easier for the gas molecules to diffuse through the surface. Because the intermediates are immediately destroyed in the presence of oxygen, the development of carbonaceous intermediates on the surface of TNTs can be prevented during the oxidation of VOCs [18,107]. Weon et al. studied the catalytic efficiency of TNT and TiO_2 nanoparticles (TNP) during the PCO of toluene and acetaldehyde. The color change of fresh and used catalyst was visible (Figure 8a) and during post-catalytic treatment, CO_2 was released from the catalyst surface (Figure 8b). The durability of TNT was excellent even after five repeated cycles of toluene oxidation compared to TNP, and the results are shown in Figure 8c,d [18].

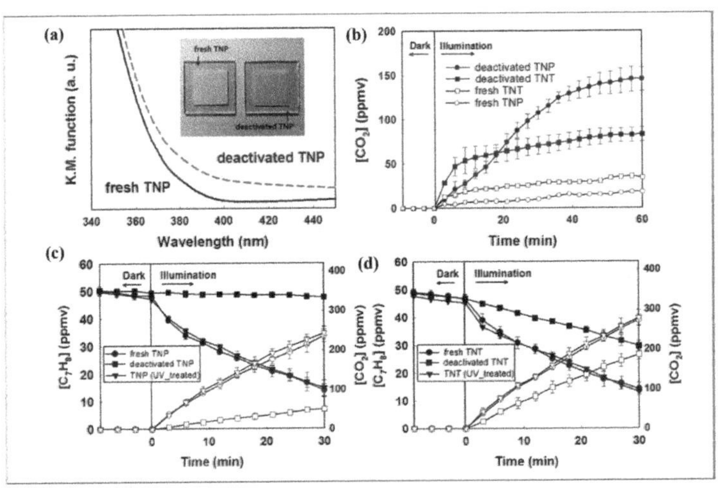

Figure 8. (**a**) Diffuse reflectance UV-visible spectra of fresh and deactivated TNP catalyst. (**b**) CO_2 formation rate from fresh and deactivated TNP and TNT catalysts. (**c**) Profile of PCO of toluene with TNP catalyst. (**d**) Photocatalytic oxidation rate of toluene using TNT catalyst. Reproduced with permission from [18]. Copyright 2016 American Chemical Society.

6. Conclusions and Outlook

The recent developments in the PCO technology for the removal of atmospheric pollutants are assessed in detail in this review. Compared with various other kinds of pollution, air pollution has more sources, and it spreads very quickly, making it difficult to control. Thus, in a polluted atmosphere, especially in indoor environments, air pollution control and prevention present numerous problems. Photocatalysis is actively being researched as a potential air cleansing method. It has been particularly well suited for indoor air purification since it breaks and converts harmful pollutants into CO_2 and H_2O under ambient operating parameters. TiO_2, the most investigated photocatalyst for air purification, does not absorb visible light, and the efficiency of the photocatalyst depends on various external operating parameters. In addition, leaching and deactivation of the photocatalyst are major issues faced during the practical application of PCO but have not received much research interest. Although considerable progress has been made in the use of nano-catalysts for the photocatalytic degradation of VOCs, several issues must be resolved before this technology can be put into practice in various fields. Novel catalytic materials have found applications in photocatalytic reactions, but to replace TiO_2 based catalysts with novel materials in practical applications still has a long way to go.

Based on the recent developments in PCO technology, the following conclusions were drawn and several directions for future development are provided. Some of the important challenges and areas of improvement are as follows:

(1) The photocatalytic efficiency of VOC oxidation depends on various operational parameters (airflow, reactor type, residence time, wavelength and intensity of light, humidity, temperature, etc.). To understand how the degradation efficiency depends on various external factors and to achieve the maximum conversion rate, it is always recommended to perform the PCO reaction in a standard reactor system with different operational parameters.

(2) The stability of the investigated photocatalyst is demonstrated by how long it lasts until the activity starts reducing during continuous operation. The important questions are why the deactivation occurs and how we can reduce the deactivation.

In brief, optimizing reaction conditions and preventing catalyst deactivation could maximize the efficiency of the photocatalyst used for air purification. Photocatalytic oxidation techniques can be a potential solution for effectively eliminating indoor air pollutants. However, with the current progress in VOC oxidation, the commercialization of this technology would be difficult. Considering this, photocatalysis should be an upgraded technique and the catalytic nanomaterials synthesized for the removal of atmospheric VOCs under realistic conditions should be the focus of substantial scientific research in the near future.

Author Contributions: Writing—original draft preparation and editing, A.J.; writing—review and editing, A.M.; review editing and project administration, Y.-C.L. All authors have read and agreed to the published version of the manuscript.

Funding: This work was supported by the Basic Science Research Program through the National Research Foundation of Korea (NRF) funded by the Ministry of Education (RS-2022-00166502) and a grant from the Subway Fine Dust Reduction Technology Development Project of the Ministry of Land Infrastructure and Transport, Republic of Korea (21QPPW-B152306-03).

Data Availability Statement: Not applicable.

Conflicts of Interest: The authors declare that they have no known financial or interpersonal conflicts that would have appeared to have an impact on the research presented in this study.

References

1. Huang, G.; Liu, L.; Chen, L.; Gao, L.; Zhu, J.; Fu, H. Unique Insights into Photocatalytic VOCs Oxidation over WO_3/Carbon Dots Nanohybrids Assisted by Water Activation and Electron Transfer at Interfaces. *J. Hazard. Mater.* **2022**, *423*, 127134. [CrossRef]
2. Li, Y.W.; Ma, W.L. Photocatalytic Oxidation Technology for Indoor Air Pollutants Elimination: A Review. *Chemosphere* **2021**, *280*, 130667. [CrossRef]
3. Mamaghani, A.H.; Haghighat, F.; Lee, C.S. Effect of Titanium Dioxide Properties and Support Material on Photocatalytic Oxidation of Indoor Air Pollutants. *Build. Environ.* **2021**, *189*, 107518. [CrossRef]
4. Shah, K.W.; Li, W. A Review on Catalytic Nanomaterials for Volatile Organic Compounds VOC Removal and Their Applications for Healthy Buildings. *Nanomaterials* **2019**, *9*, 910. [CrossRef]
5. Schiffman, S.S.; Williams, C.M. Science of Odor as a Potential Health Issue. *J. Environ. Qual.* **2005**, *34*, 129–138. [PubMed]
6. Guo, Y.; Wen, M.; Li, G.; An, T. Recent Advances in VOC Elimination by Catalytic Oxidation Technology onto Various Nanoparticles Catalysts: A Critical Review. *Appl. Catal. B Environ.* **2021**, *281*, 119447. [CrossRef]
7. Louis, J.; Padmanabhan, N.T.; John, H. Commercial Applications and Future Trends of Photocatalytic Materials. In *Photocatalysis*; Walter de Gruyter GmbH: Berlin, Germany, 2021; pp. 343–380. [CrossRef]
8. Geng, Q.; Wang, H.; Chen, R.; Chen, L.; Li, K.; Dong, F. Advances and challenges of photocatalytic technology for air purification. *Natl. Sci. Open* **2022**, *1*, 20220025. [CrossRef]
9. Liu, B.; Zhan, Y.; Xie, R.; Huang, H.; Li, K.; Zeng, Y.; Shrestha, R.P.; Kim Oanh, N.T.; Winijkul, E. Efficient Photocatalytic Oxidation of Gaseous Toluene in a Bubbling Reactor of Water. *Chemosphere* **2019**, *233*, 754–761. [CrossRef]

10. Zou, W.; Gao, B.; Ok, Y.S.; Dong, L. Integrated Adsorption and Photocatalytic Degradation of Volatile Organic Compounds (VOCs) Using Carbon-Based Nanocomposites: A Critical Review. *Chemosphere* **2019**, *218*, 845–859. [CrossRef] [PubMed]
11. Zhao, W.; Adeel, M.; Zhang, P.; Zhou, P.; Huang, L.; Zhao, Y.; Ahmad, M.A.; Shakoor, N.; Lou, B.; Jiang, Y.; et al. A Critical Review on Surface-Modified Nano-Catalyst Application for the Photocatalytic Degradation of Volatile Organic Compounds. *Environ. Sci. Nano* **2022**, *9*, 61–80. [CrossRef]
12. Chen, J.; Chen, L.; Wang, X.; Sun, J.; Chen, A.; Xie, X. Er Single Atoms Decorated TiO_2 and Er^{3+} Ions Modified TiO_2 for Photocatalytic Oxidation of Mixed VOCs. *Appl. Surf. Sci.* **2022**, *596*, 153655. [CrossRef]
13. Mo, J.; Zhang, Y.; Xu, Q.; Zhu, Y.; Lamson, J.J.; Zhao, R. Determination and Risk Assessment of By-Products Resulting from Photocatalytic Oxidation of Toluene. *Appl. Catal. B Environ.* **2009**, *89*, 570–576. [CrossRef]
14. Hodgson, A.T.; Destaillats, H.; Sullivan, D.P.; Fisk, W.J. Performance of Ultraviolet Photocatalytic Oxidation for Indoor Air Cleaning Applications. *Indoor Air* **2007**, *17*, 305–316. [CrossRef]
15. Asahi, R.; Morikawa, T.; Ohwaki, T.; Aoki, K.; Taga, Y. Visible-Light Photocatalysis in Nitrogen-Doped Titanium Oxides. *Science* **2001**, *293*, 269–271. [CrossRef] [PubMed]
16. Li, Y.; Ouyang, S.; Xu, H.; Wang, X.; Bi, Y.; Zhang, Y.; Ye, J. Constructing Solid-Gas-Interfacial Fenton Reaction over Alkalinized-C_3N_4 Photocatalyst to Achieve Apparent Quantum Yield of 49% at 420 Nm. *J. Am. Chem. Soc.* **2016**, *138*, 13289–13297. [CrossRef] [PubMed]
17. Mo, J.; Zhang, Y.; Xu, Q.; Lamson, J.J.; Zhao, R. Photocatalytic Purification of Volatile Organic Compounds in Indoor Air: A Literature Review. *Atmos. Environ.* **2009**, *43*, 2229–2246. [CrossRef]
18. Weon, S.; Choi, W. TiO_2 Nanotubes with Open Channels as Deactivation-Resistant Photocatalyst for the Degradation of Volatile Organic Compounds. *Environ. Sci. Technol.* **2016**, *50*, 2556–2563. [CrossRef]
19. Mamaghani, A.H.; Haghighat, F.; Lee, C.S. Photocatalytic Degradation of VOCs on Various Commercial Titanium Dioxides: Impact of Operating Parameters on Removal Efficiency and by-Products Generation. *Build. Environ.* **2018**, *138*, 275–282. [CrossRef]
20. Saoud, W.A.; Kane, A.; Le Cann, P.; Gerard, A.; Lamaa, L.; Peruchon, L.; Brochier, C.; Bouzaza, A.; Wolbert, D.; Assadi, A.A. Innovative Photocatalytic Reactor for the Degradation of VOCs and Microorganism under Simulated Indoor Air Conditions: Cu-Ag/TiO_2-Based Optical Fibers at a Pilot Scale. *Chem. Eng. J.* **2021**, *411*, 128622. [CrossRef]
21. Guo, D.; Feng, D.; Zhang, Y.; Zhang, Z.; Wu, J.; Zhao, Y.; Sun, S. Synergistic Mechanism of Biochar-Nano TiO_2 Adsorption-Photocatalytic Oxidation of Toluene. *Fuel Process. Technol.* **2022**, *229*, 107200. [CrossRef]
22. Weitekamp, C.A.; Stevens, T.; Stewart, M.J.; Bhave, P.; Gilmour, M.I. Health Effects from Freshly Emitted versus Oxidatively or Photochemically Aged Air Pollutants. *Sci. Total Environ.* **2020**, *704*, 135772. [CrossRef]
23. Ajmal, Z.; ul Haq, M.; Naciri, Y.; Djellabi, R.; Hassan, N.; Zaman, S.; Murtaza, A.; Kumar, A.; Al-Sehemi, A.G.; Algarni, H.; et al. Recent Advancement in Conjugated Polymers Based Photocatalytic Technology for Air Pollutants Abatement: Cases of CO_2, NO_x, and VOCs. *Chemosphere* **2022**, *308*, 136358. [CrossRef]
24. Tsai, W.T. An Overview of Health Hazards of Volatile Organic Compounds Regulated as Indoor Air Pollutants. *Rev. Environ. Health* **2019**, *34*, 81–89. [CrossRef] [PubMed]
25. Wolkoff, P.; Nielsen, G.D. Organic Compounds in Indoor Air-Their Relevance for Perceived Indoor Air Quality? *Atmos. Environ.* **2001**, *35*, 4407–4417. [CrossRef]
26. Guieysse, B.; Hort, C.; Platel, V.; Munoz, R.; Ondarts, M.; Revah, S. Biological Treatment of Indoor Air for VOC Removal: Potential and Challenges. *Biotechnol. Adv.* **2008**, *26*, 398–410. [CrossRef]
27. David, E.; Niculescu, V.C. Volatile Organic Compounds (Vocs) as Environmental Pollutants: Occurrence and Mitigation Using Nanomaterials. *Int. J. Environ. Res. Public Health* **2021**, *18*, 13147. [CrossRef]
28. Wang, H.; Lou, S.; Huang, C.; Qiao, L.; Tang, X.; Chen, C.; Zeng, L.; Wang, Q.; Zhou, M.; Lu, S.; et al. Source Profiles of Volatile Organic Compounds from Biomass Burning in Yangtze River Delta, China. *Aerosol Air Qual. Res.* **2014**, *14*, 818–828. [CrossRef]
29. Kamal, M.S.; Razzak, S.A.; Hossain, M.M. Catalytic Oxidation of Volatile Organic Compounds (VOCs)—A Review. *Atmos. Environ.* **2016**, *140*, 117–134. [CrossRef]
30. Özçelik, Z.; Soylu, G.S.P.; Boz, I. Catalytic Combustion of Toluene over Mn, Fe and Co-Exchanged Clinoptilolite Support. *Chem. Eng. J.* **2009**, *155*, 94–100. [CrossRef]
31. Patdhanagul, N.; Rangsriwatananon, K.; Siriwong, K.; Hengrasmee, S. Combined Modification of Zeolite NaY by Phenyl Trimethyl Ammonium Bromide and Potassium for Ethylene Gas Adsorption. *Microporous Mesoporous Mater.* **2012**, *153*, 30–34. [CrossRef]
32. Bakheet, A.A.; Mohd Zain, M.F.; Kadhum, A.A.; Abdalla, Z. Photocatalytic Oxidation Performance to Removal of Volatile Organic Compounds in Indoor Environment. *Environ. Res. Eng. Manag.* **2011**, *58*, 27–33. [CrossRef]
33. Halios, C.H.; Landeg-Cox, C.; Lowther, S.D.; Middleton, A.; Marczylo, T.; Dimitroulopoulou, S. Chemicals in European Residences–Part I: A Review of Emissions, Concentrations and Health Effects of Volatile Organic Compounds (VOCs). *Sci. Total Environ.* **2022**, *839*, 156201. [CrossRef]
34. Wan, Y.; Zhao, D. On the Controllable Soft-Templating Approach to Mesoporous Silicates. *Chem. Rev.* **2007**, *107*, 2821–2860. [CrossRef] [PubMed]
35. Tong, H.; Ouyang, S.; Bi, Y.; Umezawa, N.; Oshikiri, M.; Ye, J. Nano-Photocatalytic Materials: Possibilities and Challenges. *Adv. Mater.* **2012**, *24*, 229–251. [CrossRef] [PubMed]

36. Hu, Z.; Yang, C.; Lv, K.; Li, X.; Li, Q.; Fan, J. Single Atomic Au Induced Dramatic Promotion of the Photocatalytic Activity of TiO_2 Hollow Microspheres. *Chem. Commun.* **2020**, *56*, 1745–1748. [CrossRef]
37. He, F.; Weon, S.; Jeon, W.; Chung, M.W.; Choi, W. Self-Wetting Triphase Photocatalysis for Effective and Selective Removal of Hydrophilic Volatile Organic Compounds in Air. *Nat. Commun.* **2021**, *12*, 6259. [CrossRef] [PubMed]
38. Sydorenko, J.; Mere, A.; Krunks, M.; Krichevskaya, M.; Acik, I.O. Transparent TiO_2 Thin Films with High Photocatalytic Activity for Indoor Air Purification. *RSC Adv.* **2022**, *12*, 35531–35542. [CrossRef]
39. Yamashita, H.; Mori, K.; Kuwahara, Y.; Kamegawa, T.; Wen, M.; Verma, P.; Che, M. Single-Site and Nano-Confined Photocatalysts Designed in Porous Materials for Environmental Uses and Solar Fuels. *Chem. Soc. Rev.* **2018**, *47*, 8072–8096. [CrossRef]
40. Verma, P.; Kondo, Y.; Kuwahara, Y.; Kamegawa, T.; Mori, K.; Raja, R.; Yamashita, H. Design and Application of Photocatalysts Using Porous Materials. *Catal. Rev. Sci. Eng.* **2021**, *63*, 165–233. [CrossRef]
41. Chu, S.; Wang, Y.; Guo, Y.; Feng, J.; Wang, C.; Luo, W.; Fan, X.; Zou, Z. Band Structure Engineering of Carbon Nitride: In Search of a Polymer Photocatalyst with High Photooxidation Property. *ACS Catal.* **2013**, *3*, 912–919. [CrossRef]
42. Sun, R.; Shi, Q.; Zhang, M.; Xie, L.; Chen, J.; Yang, X.; Chen, M.; Zhao, W. Enhanced Photocatalytic Oxidation of Toluene with a Coral-like Direct Z-Scheme $BiVO_4$/g-C_3N_4 Photocatalyst. *J. Alloys Compd.* **2017**, *714*, 619–626. [CrossRef]
43. Zhong, D.K.; Sun, J.; Inumaru, H.; Gamelin, D.R. Solar Water Oxidation by Composite Catalyst/α-Fe_2O_3 Photoanodes. *J. Am. Chem. Soc.* **2009**, *131*, 6086–6087. [CrossRef] [PubMed]
44. Liu, Y.; Mallouk, K.; Emamipour, H.; Rood, M.J.; Liu, X.; Yan, Z. Isobutane Adsorption with Carrier Gas Recirculation at Different Relative Humidities Using Activated Carbon Fiber Cloth and Electrothermal Regeneration. *Chem. Eng. J.* **2019**, *360*, 1011–1019. [CrossRef]
45. Yue, Z.; Vakili, A.; Wang, J. Activated Carbon Fibers from Meltblown Isotropic Pitch Fiber Webs for Vapor Phase Adsorption of Volatile Organic Compounds. *Chem. Eng. J.* **2017**, *330*, 183–190. [CrossRef]
46. Ahmad, A.; Ali, M.; Al-Sehemi, A.G.; Al-Ghamdi, A.A.; Park, J.W.; Algarni, H.; Anwer, H. Carbon-Integrated Semiconductor Photocatalysts for Removal of Volatile Organic Compounds in Indoor Environments. *Chem. Eng. J.* **2023**, *452*, 139436. [CrossRef]
47. Yan, S.; Song, H.; Li, Y.; Yang, J.; Jia, X.; Wang, S.; Yang, X. Integrated Reduced Graphene Oxide/Polypyrrole Hybrid Aerogels for Simultaneous Photocatalytic Decontamination and Water Evaporation. *Appl. Catal. B Environ.* **2022**, *301*, 120820. [CrossRef]
48. Enesca, A.; Cazan, C. Volatile Organic Compounds (VOCS) Removal from Indoor Air by Heterostructures/Composites/Doped Photocatalysts: A Mini-Review. *Nanomaterials* **2020**, *10*, 1965. [CrossRef] [PubMed]
49. Akerdi, A.G.; Bahrami, S.H. Application of Heterogeneous Nano-Semiconductors for Photocatalytic Advanced Oxidation of Organic Compounds: A Review. *J. Environ. Chem. Eng.* **2019**, *7*, 103283. [CrossRef]
50. Chen, H.; Nanayakkara, C.E.; Grassian, V.H. Titanium Dioxide Photocatalysis in Atmospheric Chemistry. *Chem. Rev.* **2012**, *112*, 5919–5948. [CrossRef]
51. Mamaghani, A.H.; Haghighat, F.; Lee, C.S. Photocatalytic Oxidation Technology for Indoor Environment Air Purification: The State-of-the-Art. *Appl. Catal. B Environ.* **2017**, *203*, 247–269. [CrossRef]
52. Zhao, J.; Yang, X. Photocatalytic Oxidation for Indoor Air Purification: A Literature Review. *Build. Environ.* **2003**, *38*, 645–654. [CrossRef]
53. Quici, N.; Vera, M.L.; Choi, H.; Puma, G.L.; Dionysiou, D.D.; Litter, M.I.; Destaillats, H. Effect of Key Parameters on the Photocatalytic Oxidation of Toluene at Low Concentrations in Air under 254 + 185 nm UV Irradiation. *Appl. Catal. B Environ.* **2010**, *95*, 312–319. [CrossRef]
54. Van Durme, J.; Dewulf, J.; Sysmans, W.; Leys, C.; Van Langenhove, H. Abatement and Degradation Pathways of Toluene in Indoor Air by Positive Corona Discharge. *Chemosphere* **2007**, *68*, 1821–1829. [CrossRef] [PubMed]
55. Korologos, C.A.; Philippopoulos, C.J.; Poulopoulos, S.G. The Effect of Water Presence on the Photocatalytic Oxidation of Benzene, Toluene, Ethylbenzene and m-Xylene in the Gas-Phase. *Atmos. Environ.* **2011**, *45*, 7089–7095. [CrossRef]
56. Park, E.J.; Seo, H.O.; Kim, Y.D. Influence of Humidity on the Removal of Volatile Organic Compounds Using Solid Surfaces. *Catal. Today* **2017**, *295*, 3–13. [CrossRef]
57. Talaiekhozani, A.; Rezania, S.; Kim, K.; Sanaye, R.; Mohammad, A. Recent Advances in Photocatalytic Removal of Organic and Inorganic Pollutants in Air. *J. Clean. Prod.* **2021**, *278*, 123895. [CrossRef]
58. Assadi, A.A.; Bouzaza, A.; Wolbert, D. Photocatalytic Oxidation of Trimethylamine and Isovaleraldehyde in an Annular Reactor: Influence of the Mass Transfer and the Relative Humidity. *J. Photochem. Photobiol. A Chem.* **2012**, *236*, 61–69. [CrossRef]
59. Vildozo, D.; Portela, R.; Ferronato, C.; Chovelon, J.M. Photocatalytic Oxidation of 2-Propanol/Toluene Binary Mixtures at Indoor Air Concentration Levels. *Appl. Catal. B Environ.* **2011**, *107*, 347–354. [CrossRef]
60. Vincent, G.; Marquaire, P.M.; Zahraa, O. Photocatalytic Degradation of Gaseous 1-Propanol Using an Annular Reactor: Kinetic Modelling and Pathways. *J. Hazard. Mater.* **2009**, *161*, 1173–1181. [CrossRef]
61. Debono, O.; Thévenet, F.; Gravejat, P.; Héquet, V.; Raillard, C.; Le Coq, L.; Locoge, N. Gas Phase Photocatalytic Oxidation of Decane at ppb Levels: Removal Kinetics, Reaction Intermediates and Carbon Mass Balance. *J. Photochem. Photobiol. A Chem.* **2013**, *258*, 17–29. [CrossRef]
62. Kibanova, D.; Sleiman, M.; Cervini-Silva, J.; Destaillats, H. Adsorption and Photocatalytic Oxidation of Formaldehyde on a Clay-TiO_2 Composite. *J. Hazard. Mater.* **2012**, *211–212*, 233–239. [CrossRef]

63. Yu, K.P.; Whei-May Lee, G.; Huang, W.M.; Wu, C.C.; Lou, C.I.; Yang, S. Effectiveness of Photocatalytic Filter for Removing Volatile Organic Compounds in the Heating, Ventilation, and Air Conditioning System. *J. Air Waste Manag. Assoc.* **2006**, *56*, 666–674. [CrossRef]
64. Mamaghani, A.H.; Haghighat, F.; Lee, C.S. Photocatalytic Oxidation of MEK over Hierarchical TiO_2 Catalysts: Effect of Photocatalyst Features and Operating Conditions. *Appl. Catal. B Environ.* **2019**, *251*, 1–16. [CrossRef]
65. Vildozo, D.; Ferronato, C.; Sleiman, M.; Chovelon, J.M. Photocatalytic Treatment of Indoor Air: Optimization of 2-Propanol Removal Using a Response Surface Methodology (RSM). *Appl. Catal. B Environ.* **2010**, *94*, 303–310. [CrossRef]
66. Boulamanti, A.K.; Philippopoulos, C.J. Photocatalytic Degradation of C5-C7 Alkanes in the Gas-Phase. *Atmos. Environ.* **2009**, *43*, 3168–3174. [CrossRef]
67. Shaw, J.T.; Rickard, A.R.; Newland, M.J.; Dillon, T.J. Rate Coefficients for Reactions of OH with Aromatic and Aliphatic Volatile Organic Compounds Determined by the Multivariate Relative Rate Technique. *Atmos. Chem. Phys.* **2020**, *20*, 9725–9736. [CrossRef]
68. Ao, C.H.; Lee, S.C. Combination Effect of Activated Carbon with TiO_2 for the Photodegradation of Binary Pollutants at Typical Indoor Air Level. *J. Photochem. Photobiol. A Chem.* **2004**, *161*, 131–140. [CrossRef]
69. Li, Y.; Liu, F.; Li, M.; Li, W.; Qi, X.; Xue, M.; Wang, Y.; Han, F. Study on Adsorption Coupling Photodegradation on Hierarchical Nanostructured G-C_3N_4/TiO_2/Activated Carbon Fiber Composites for Toluene Removal. *J. Sol-Gel Sci. Technol.* **2020**, *93*, 402–418. [CrossRef]
70. Huang, H.; Xu, Y.; Feng, Q.; Leung, D.Y.C. Low Temperature Catalytic Oxidation of Volatile Organic Compounds: A Review. *Catal. Sci. Technol.* **2015**, *5*, 2649–2669. [CrossRef]
71. Malayeri, M.; Haghighat, F.; Lee, C.S. Modeling of Volatile Organic Compounds Degradation by Photocatalytic Oxidation Reactor in Indoor Air: A Review. *Build. Environ.* **2019**, *154*, 309–323. [CrossRef]
72. Lu, Y.; Wang, D.; Ma, C.; Yang, H. The Effect of Activated Carbon Adsorption on the Photocatalytic Removal of Formaldehyde. *Build. Environ.* **2010**, *45*, 615–621. [CrossRef]
73. Li, X.; Qian, X.; An, X.; Huang, J. Preparation of a Novel Composite Comprising Biochar Skeleton and "Chrysanthemum" g-C_3N_4 for Enhanced Visible Light Photocatalytic Degradation of Formaldehyde. *Appl. Surf. Sci.* **2019**, *487*, 1262–1270. [CrossRef]
74. Liu, X.; Zhang, Y.; Matsushima, S.; Hojo, H.; Einaga, H. Photocatalytic Oxidation Process for Treatment of Gas Phase Benzene Using Ti^{3+} Self-Doped TiO_2 Microsphere with Sea Urchin-like Structure. *Chem. Eng. J.* **2020**, *402*, 126220. [CrossRef]
75. Verbruggen, S.W.; Ribbens, S.; Tytgat, T.; Hauchecorne, B.; Smits, M.; Meynen, V.; Cool, P.; Martens, J.A.; Lenaerts, S. The Benefit of Glass Bead Supports for Efficient Gas Phase Photocatalysis: Case Study of a Commercial and a Synthesised Photocatalyst. *Chem. Eng. J.* **2011**, *174*, 318–325. [CrossRef]
76. Thevenet, F.; Guaïtella, O.; Herrmann, J.M.; Rousseau, A.; Guillard, C. Photocatalytic Degradation of Acetylene over Various Titanium Dioxide-Based Photocatalysts. *Appl. Catal. B Environ.* **2005**, *61*, 58–68. [CrossRef]
77. Mo, J.; Zhang, Y.; Xu, Q.; Yang, R. Effect of TiO_2/Adsorbent Hybrid Photocatalysts for Toluene Decomposition in Gas Phase. *J. Hazard. Mater.* **2009**, *168*, 276–281. [CrossRef] [PubMed]
78. Yang, L.; Liu, Z. Study on Light Intensity in the Process of Photocatalytic Degradation of Indoor Gaseous Formaldehyde for Saving Energy. *Energy Convers. Manag.* **2007**, *48*, 882–889. [CrossRef]
79. Kibanova, D.; Cervini-Silva, J.; Destaillats, H. Efficiency of Clay-TiO_2 Nanocomposites on the Photocatalytic Elimination of a Model Hydrophobic Air Pollutant. *Environ. Sci. Technol.* **2009**, *43*, 1500–1506. [CrossRef]
80. Jeong, J.; Sekiguchi, K.; Lee, W.; Sakamoto, K. Photodegradation of Gaseous Volatile Organic Compounds (VOCs) Using TiO_2 Photoirradiated by an Ozone-Producing UV Lamp: Decomposition Characteristics, Identification of by-Products and Water-Soluble Organic Intermediates. *J. Photochem. Photobiol. A Chem.* **2005**, *169*, 279–287. [CrossRef]
81. Krichevskaya, M.; Preis, S.; Moiseev, A.; Pronina, N.; Deubener, J. Gas-Phase Photocatalytic Oxidation of Refractory VOCs Mixtures: Through the Net of Process Limitations. *Catal. Today* **2017**, *280*, 93–98. [CrossRef]
82. Zhang, G.; Peyravi, A.; Hashisho, Z.; Sun, Z.; Liu, Y.; Zheng, S.; Zhong, L. Integrated Adsorption and Photocatalytic Degradation of VOCs Using a TiO_2/Diatomite Composite: Effects of Relative Humidity and Reaction Atmosphere. *Catal. Sci. Technol.* **2020**, *10*, 2378–2388. [CrossRef]
83. Liang, W.J.; Li, J.; Jin, Y.Q. Photocatalytic Degradation of Gaseous Acetone, Toluene, and p-Xylene Using a TiO_2 Thin Film. *J. Environ. Sci. Health Part A Toxic/Hazardous Subst. Environ. Eng.* **2010**, *45*, 1384–1390. [CrossRef]
84. Dong, F.; Guo, S.; Wang, H.; Li, X.; Wu, Z. Enhancement of the Visible Light Photocatalytic Activity of C-Doped TiO_2 Nanomaterials Prepared by a Green Synthetic Approach. *J. Phys. Chem. C* **2011**, *115*, 13285–13292. [CrossRef]
85. Dong, X.; Cui, Z.; Sun, Y.; Dong, F. Humidity-Independent Photocatalytic Toluene Mineralization Benefits from the Utilization of Edge Hydroxyls in Layered Double Hydroxides (LDHs): A Combined Operando and Theoretical Investigation. *ACS Catal.* **2021**, *11*, 8132–8139. [CrossRef]
86. You, J.-W.; Vikrant, K.; Lim, D.-H.; Verma, S.; Kukkar, D.; Anwer, H.; Kim, K.-H. Photocatalytic Potential of a Titanium Dioxide–Supported Platinum Catalyst against VOCs with Complicated Composition under Varying Humidity Conditions. *J. Clean. Prod.* **2022**, *371*, 133487. [CrossRef]
87. An, T.; Chen, J.; Nie, X.; Li, G.; Zhang, H.; Liu, X.; Zhao, H. Synthesis of Carbon Nanotube-Anatase TiO_2 Sub-Micrometer-Sized Sphere Composite Photocatalyst for Synergistic Degradation of Gaseous Styrene. *ACS Appl. Mater. Interfaces* **2012**, *4*, 5988–5996. [CrossRef]

88. Zhang, J.; Vikrant, K.; Kim, K.H.; Dong, F.; Won Chung, M.; Weon, S. Unveiling the Collective Effects of Moisture and Oxygen on the Photocatalytic Degradation of M-Xylene Using a Titanium Dioxide Supported Platinum Catalyst. *Chem. Eng. J.* **2022**, *439*, 135747. [CrossRef]
89. Sidheswaran, M.; Tavlarides, L.L. Visible Light Photocatalytic Oxidation of Toluene Using a Cerium-Doped Titania Catalyst. *Ind. Eng. Chem. Res.* **2008**, *47*, 3346–3357. [CrossRef]
90. Arai, T.; Horiguchi, M.; Yanagida, M.; Gunji, T.; Sugihara, H.; Sayama, K. Complete Oxidation of Acetaldehyde and Toluene over a Pd/WO_3 Photocatalyst under Fluorescent or Visible-Light Irradiation. *Chem. Commun.* **2008**, 5565–5567. [CrossRef]
91. Gao, Z.; Wang, J.; Muhammad, Y.; Zhang, Y.; Jalil, S.; Hu, Y. Enhanced Moisture-Resistance and Excellent Photocatalytic Performance of Synchronous N/Zn-Decorated MIL-125(Ti) for Vaporous Acetaldehyde Degradation. *Chem. Eng. J.* **2020**, *388*, 124389. [CrossRef]
92. Zhong, L.; Haghighat, F. Modeling of By-Products from Photocatalytic Oxidation (PCO) Indoor Air Purifiers: A Case Study of Ethanol. *Build. Environ.* **2018**, *144*, 427–436. [CrossRef]
93. Whyte, H.E.; Raillard, C.; Subrenat, A.; Héquet, V. Photocatalytic Oxidation of Isoflurane, an Anesthetic Gas: The Influence of Operating Parameters. *Chem. Eng. J.* **2018**, *352*, 441–449. [CrossRef]
94. Kamaei, M.; Rashedi, H.; Dastgheib, S.M.M.; Tasharrofi, S. Comparing Photocatalytic Degradation of Gaseous Ethylbenzene Using N-Doped and Pure TiO_2 Nano-Catalysts Coated on Glass Beads under Both UV and Visible Light Irradiation. *Catalysts* **2018**, *8*, 466. [CrossRef]
95. Xie, R.; Lei, D.; Zhan, Y.; Liu, B.; Tsang, C.H.A.; Zeng, Y.; Li, K.; Leung, D.Y.C.; Huang, H. Efficient Photocatalytic Oxidation of Gaseous Toluene over F-Doped TiO_2 in a Wet Scrubbing Process. *Chem. Eng. J.* **2020**, *386*, 121025. [CrossRef]
96. Liu, Y.; Chen, S.; Li, K.; Wang, J.; Chen, P.; Wang, H.; Li, J.; Dong, F. Promote the Activation and Ring Opening of Intermediates for Stable Photocatalytic Toluene Degradation over Zn-Ti-LDH. *J. Colloid Interface Sci.* **2022**, *606*, 1435–1444. [CrossRef]
97. Chen, L.; Wang, X.; Rao, Z.; Tang, Z.; Wang, Y.; Shi, G.; Lu, G.; Xie, X.; Chen, D.; Sun, J. In-Situ Synthesis of Z-Scheme MIL-100(Fe)/α-Fe_2O_3 Heterojunction for Enhanced Adsorption and Visible-Light Photocatalytic Oxidation of O-Xylene. *Chem. Eng. J.* **2021**, *416*, 129112. [CrossRef]
98. Jafari, A.J.; Kalantary, R.R.; Esrafili, A.; Arfaeinia, H. Synthesis of Silica-Functionalized Graphene Oxide/ZnO Coated on Fiberglass and Its Application in Photocatalytic Removal of Gaseous Benzene. *Process Saf. Environ. Prot.* **2018**, *116*, 377–387. [CrossRef]
99. Lei, B.; Cui, W.; Chen, P.; Chen, R.; Sun, Y.; Kim, K.H.; Dong, F. Rational Design of LDH/Zn_2SnO_4 Heterostructures for Efficient Mineralization of Toluene Through Boosted Interfacial Charge Separation. *Energy Environ. Mater.* **2022**, *6*, e12291. [CrossRef]
100. Liqiang, J.; Baifu, X.; Fulong, Y.; Baiqi, W.; Keying, S.; Weimin, C.; Honggang, F. Deactivation and Regeneration of ZnO and TiO_2 Nanoparticles in the Gas Phase Photocatalytic Oxidation of N-C_7H_{16} or SO_2. *Appl. Catal. A Gen.* **2004**, *275*, 49–54. [CrossRef]
101. Einaga, H.; Futamura, S.; Ibusuki, T. Heterogeneous Photocatalytic Oxidation of Benzene, Toluene, Cyclohexene and Cyclohexane in Humidified Air: Comparison of Decomposition Behavior on Photoirradiated TiO_2 Catalyst. *Appl. Catal. B Environ.* **2002**, *38*, 215–225. [CrossRef]
102. Taranto, J.; Frochot, D.; Pichat, P. Photocatalytic Air Purification: Comparative Efficacy and Pressure Drop of a TiO_2-Coated Thin Mesh and a Honeycomb Monolith at High Air Velocities Using a 0.4 M3 Close-Loop Reactor. *Sep. Purif. Technol.* **2009**, *67*, 187–193. [CrossRef]
103. Thevenet, F.; Guillard, C.; Rousseau, A. Acetylene Photocatalytic Oxidation Using Continuous Flow Reactor: Gas Phase and Adsorbed Phase Investigation, Assessment of the Photocatalyst Deactivation. *Chem. Eng. J.* **2014**, *244*, 50–58. [CrossRef]
104. Hay, S.O.; Obee, T.N.; Thibaud-Erkey, C. The Deactivation of Photocatalytic Based Air Purifiers by Ambient Siloxanes. *Appl. Catal. B Environ.* **2010**, *99*, 435–441. [CrossRef]
105. Cha, B.J.; Woo, T.G.; Han, S.W.; Saqlain, S.; Seo, H.O.; Cho, H.K.; Kim, J.Y.; Kim, Y.D. Surface Modification of TiO_2 for Obtaining High Resistance against Poisoning during Photocatalytic Decomposition of Toluene. *Catalysts* **2018**, *8*, 500. [CrossRef]
106. Tripathy, J.; Lee, K.; Schmuki, P. Tuning the Selectivity of Photocatalytic Synthetic Reactions Using Modified TiO_2 Nanotubes. *Angew. Chemie* **2014**, *126*, 12813–12816. [CrossRef]
107. Weon, S.; He, F.; Choi, W. Status and Challenges in Photocatalytic Nanotechnology for Cleaning Air Polluted with Volatile Organic Compounds: Visible Light Utilization and Catalyst Deactivation. *Environ. Sci. Nano* **2019**, *6*, 3185–3214. [CrossRef]
108. Chen, P.; Cui, W.; Wang, H.; Dong, X.; Li, J.; Sun, Y.; Zhou, Y.; Zhang, Y.; Dong, F. The Importance of Intermediates Ring-Opening in Preventing Photocatalyst Deactivation during Toluene Decomposition. *Appl. Catal. B Environ.* **2020**, *272*, 118977. [CrossRef]
109. Verbruggen, S.W.; Masschaele, K.; Moortgat, E.; Korany, T.E.; Hauchecorne, B.; Martens, J.A.; Lenaerts, S. Factors Driving the Activity of Commercial Titanium Dioxide Powders towards Gas Phase Photocatalytic Oxidation of Acetaldehyde. *Catal. Sci. Technol.* **2012**, *2*, 2311–2318. [CrossRef]
110. Taranto, J.; Frochot, D.; Pichat, P. Photocatalytic Treatment of Air: Comparison of Various TiO_2, Coating Methods, and Supports Using Methanol or n-Octane as Test Pollutant. *Ind. Eng. Chem. Res.* **2009**, *48*, 6229–6236. [CrossRef]
111. Dhada, I.; Nagar, P.K.; Sharma, M. Challenges of TiO_2-Based Photooxidation of Volatile Organic Compounds: Designing, Coating, and Regenerating Catalyst. *Ind. Eng. Chem. Res.* **2015**, *54*, 5381–5387. [CrossRef]
112. Chemweno, M.K.; Cernohlavek, L.G.; Jacoby, W.A. Deactivation of Titanium Dioxide Photocatalyst by Oxidation of Polydimethylsiloxane and Silicon Sealant Off-Gas in a Recirculating Batch Reactor. *J. Air Waste Manag. Assoc.* **2008**, *58*, 12–18. [CrossRef]

113. De Sun, R.; Nakajima, A.; Watanabe, T.; Hashimoto, K. Decomposition of Gas-Phase Octamethyltrisiloxane on TiO_2 Thin Film Photocatalysts-Catalytic Activity, Deactivation, and Regeneration. *J. Photochem. Photobiol. A Chem.* **2003**, *154*, 203–209. [CrossRef]
114. Ajmal, A.; Majeed, I.; Malik, R.N.; Idriss, H.; Nadeem, M.A. Principles and Mechanisms of Photocatalytic Dye Degradation on TiO_2 Based Photocatalysts: A Comparative Overview. *RSC Adv.* **2014**, *4*, 37003–37026. [CrossRef]
115. Zhang, X.; Zhou, M.; Lei, L. Preparation of Photocatalytic TiO_2 Coatings of Nanosized Particles on Activated Carbon by AP-MOCVD. *Carbon N. Y.* **2005**, *43*, 1700–1708. [CrossRef]
116. Araña, J.; Doña-Rodríguez, J.M.; Cabo, C.G.I.; González-Díaz, O.; Herrera-Melián, J.A.; Pérez-Peña, J. FTIR Study of Gas-Phase Alcohols Photocatalytic Degradation with TiO_2 and AC-TiO_2. *Appl. Catal. B Environ.* **2004**, *53*, 221–232. [CrossRef]
117. Yao, P.; Liu, H.; Wang, D.; Chen, J.; Li, G.; An, T. Enhanced Visible-Light Photocatalytic Activity to Volatile Organic Compounds Degradation and Deactivation Resistance Mechanism of Titania Confined inside a Metal-Organic Framework. *J. Colloid Interface Sci.* **2018**, *522*, 174–182. [CrossRef]
118. Suárez, S.; Coronado, J.M.; Portela, R.; Martín, J.C.; Yates, M.; Avila, P.; Sánchez, B. On the Preparation of TiO_2-Sepiolite Hybrid Materials for the Photocatalytic Degradation of TCE: Influence of TiO_2 Distribution in the Mineralization. *Environ. Sci. Technol.* **2008**, *42*, 5892–5896. [CrossRef]
119. Qian, X.; Ren, M.; Yue, D.; Zhu, Y.; Han, Y.; Bian, Z.; Zhao, Y. Mesoporous TiO_2 Films Coated on Carbon Foam Based on Waste Polyurethane for Enhanced Photocatalytic Oxidation of VOCs. *Appl. Catal. B Environ.* **2017**, *212*, 1–6. [CrossRef]
120. Yan, T.; Long, J.; Shi, X.; Wang, D.; Li, Z.; Wang, X. Efficient Photocatalytic Degradation of Volatile Organic Compounds by Porous Indium Hydroxide Nanocrystals. *Environ. Sci. Technol.* **2010**, *44*, 1380–1385. [CrossRef]

catalysts

MDPI

Article

Preparation of S-Scheme g-C_3N_4/ZnO Heterojunction Composite for Highly Efficient Photocatalytic Destruction of Refractory Organic Pollutant

Buse Sert [1], Zeynep Bilici [2], Kasim Ocakoglu [1,*], Nadir Dizge [2], Tannaz Sadeghi Rad [3] and Alireza Khataee [3,4,*]

1 Department of Engineering Fundamental Sciences, Faculty of Engineering, Tarsus University, 33400 Tarsus, Turkey
2 Department of Environmental Engineering, Mersin University, 33343 Mersin, Turkey
3 Department of Environmental Engineering, Faculty of Engineering, Gebze Technical University, 41400 Gebze, Turkey
4 Research Laboratory of Advanced Water and Wastewater Treatment Processes, Department of Applied Chemistry, Faculty of Chemistry, University of Tabriz, Tabriz 51666-16471, Iran
* Correspondence: kasim.ocakoglu@tarsus.edu.tr (K.O.); akhataee@gtu.edu.tr (A.K.)

Abstract: In this study, graphitic carbon nitride (g-C_3N_4)-based ZnO heterostructure was synthesized using a facile calcination method with urea and zinc nitrate hexahydrate as the initiators. According to the scanning electron microscopic (SEM) images, spherical ZnO particles can be seen along the g-C_3N_4 nanosheets. Additionally, the X-ray diffraction (XRD) analysis reveals the successful synthesis of the g-C_3N_4/ZnO. The photocatalytic activity of the synthesized catalyst was tested for the decolorization of crystal violet (CV) as an organic refractory contaminant. The impacts of ZnO molar ratio, catalyst amount, CV concentration, and H_2O_2 concentration on CV degradation efficiency were investigated. The obtained outcomes conveyed that the ZnO molar ratio in the g-C_3N_4 played a prominent role in the degradation efficiency, in which the degradation efficiency reached 95.9% in the presence of 0.05 mmol of ZnO and 0.10 g/L of the catalyst in 10 mg/L of CV through 120 min under UV irradiation. Bare g-C_3N_4 was also tested for dye decolorization, and a 76.4% dye removal efficiency was obtained. The g-C_3N_4/ZnO was also tested for adsorption, and a 32.3% adsorption efficiency was obtained. Photocatalysis, in comparison to adsorption, had a dominant role in the decolorization of CV. Lastly, the results depicted no significant decrement in the CV degradation efficiency in the presence of the g-C_3N_4/ZnO photocatalyst after five consecutive runs.

Keywords: S-scheme photocatalyst; g-C_3N_4/ZnO; advanced oxidation processes; crystal violet

1. Introduction

Wastewaters from the textile industry have adverse impacts on aquatic life due to the high concentration of hazardous chemicals released into the environment [1]. Diverse salts, chlorinated chemicals, and surfactants are commonly found in such wastewaters [2]. Since most dyes own high reactivity, even a trace amount of dye in water affects the gas solubility in water bodies [3,4]. Hence, researchers have devised several treatment techniques for the disintegration of hazardous contaminants [5]. Amongst these techniques, the photocatalytic process as a commonly used method for wastewater treatment has gained high interest [6].

The photocatalytic process is considered an eco-friendly, low-cost, and effective alternative method for decreasing water pollution when compared to other chemical, physical, and biological approaches. [7,8]. By using light irradiation, electrons in the valence band (VB) of a semiconductor are stimulated to the conduction band (CB), generating electron–hole pairs [9,10]. Afterward, the formed electron–hole pairs are separated to trigger the desired redox reaction on the active catalyst surfaces [11]. Metal oxides, especially titanium dioxide (TiO_2) and zinc oxide (ZnO), are frequently utilized in photocatalytic investigations owing to their physical stability and strong catalytic activity [12,13]. Moreover, ZnO has unique features, such as high abundance, low toxicity, ecofriendliness and cost-effectiveness [14]. ZnO is an n-type semiconductor oxide comparable to TiO_2. As ZnO has the same band gap energy as TiO_2 but exhibits greater absorption efficiency across a significant portion of the solar spectrum, ZnO has been suggested as a potential replacement photocatalyst [15–17]. Mirzaeifard et al. [18] reported that in the presence of 0.1 g of S-doped ZnO and 5 mg/L of RhB at a pH = 5, a degradation efficiency of 100% was achieved through 90 min. In addition, Ahmad et al. [19] prepared Au-ZnO which could degrade RhB (95%) within 60 min under the basic pH of RhB. Nevertheless, a high recombination ratio of electron hole pairs and photo corrosion have constrained the usage of ZnO as a photocatalyst [20].

Direct Z-scheme heterojunction and S-scheme heterojunction both use the same basic carrier separation mechanism. The first two generations of direct Z-scheme heterojunctions, conventional and all-solid-state, both have significant drawbacks. As a result, S-scheme heterojunction with a clear identification and a practical charge-transfer channel can successfully prevent needless misunderstandings (Figure 1). In addition, due to their distinctive structure and photoelectronic properties, such as greater specific surface area and more active sites, 2D-layered heterojunction photocatalysts have drawn significant interest and exhibit considerable potential in photocatalysis. Theoretically, by combining the S-scheme heterojunction and gradual 2D hybrid interfaces, one can maximize each one's benefits and improve charge separation and utilization in photocatalysis. The development of sophisticated S-scheme heterojunctions with gradual 2D coupling interfaces for significantly accelerated photocatalytic destruction of refractory organic pollutants, however, has received very little attention to date. ZnO-based heterostructures have been identified as an effective way of increasing photocatalytic efficiency [21–24]. Recently, graphitic carbon nitride (g-C_3N_4)-based ZnO nanocomposites have attracted great attention due to their low band gap, high physicochemical/photochemical stability, and electron–hole separation ability [25,26]. In general, semiconductor-mediated photocatalysis begins with the absorption of proper photon energy equal to or greater than the energy gap of the desired photocatalyst to produce e^-/h^+ pairs, followed by the production of reactive species that leads to various redox reactions on the heterogeneous photocatalysts' surface [27,28].

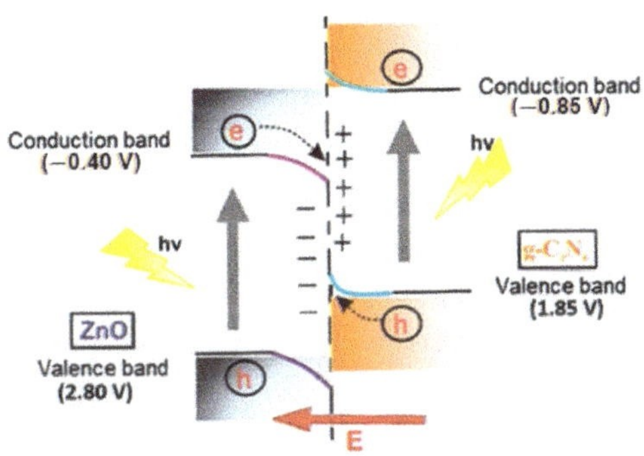

Figure 1. A representative diagram for charge-transfer processes in an S-scheme heterojunction under light irradiation.

The latest achievements in the photocatalytic synthesis of organic chemicals using g-C_3N_4-based composites have been described by Akhundi et al. [29]. Owing to the intriguing properties of g-C_3N_4, including an appropriate bandgap, biocompatibility, cost-effectiveness, and proper stability, g-C_3N_4 has attracted widespread attention [30,31]. As a polymeric semiconductor, g-C_3N_4 is used in photovoltaic solar cells, chemical sensors, and in separation, detection, bioimaging, photocatalytic, and membrane applications [32]. Accordingly, electron–hole pairs can be generated by light irradiation, which is responsible for initiating the photocatalytic reactions. However, electron–hole pairs can rapidly recombine, which can hamper the degradation efficiency. Therefore, attempts to enhance the photocatalytic activity of g-C_3N_4-based photocatalysts have propelled them to the frontline of investigations [33]. In this way, various adjustments, such as doping with efficient elements and preparing nanocomposites, can be performed to decrease the electron and hole recombination by capturing the electrons [34]. Various nanocomposites have been designed by combining g-C_3N_4 with inorganic materials, including TiO_2, CdS, and Fe_2O_3 [35–38]. Recently, modified g-C_3N_4 by ZnO has demonstrated remarkable catalytic properties [39]. As an instance, Liu et al. [40] reported a g-C_3N_4/ZnO nanocomposite structure derived from melamine and zinc chloride with improved photoactivity. As a result, the nanocomposite demonstrated better photocatalytic results compared to bare g-C_3N_4 or ZnO. In addition, Sun et al. [41] deduced that a g-C_3N_4/ZnO composite obtained by heating melamine and zinc acetate had a remarkable impact on the photocatalytic degradation of methyl orange and p-nitrophenol under visible light in comparison to the single-phase g-C_3N_4.

Figure 2 provides the energy band structure of the g-C_3N_4/ZnO nanocomposite, which demonstrates the lower CB potential energy of g-C_3N_4 rather than ZnO (ECB = −0.85 eV vs. −0.40 eV). Therefore, by exposing it to visible light, the excited electrons from the CB of g-C_3N_4 can readily inject into the CB of ZnO. The potential energy match in the g-C_3N_4/ZnO is a driving force for transferring electrons, which effectively reduces the recombination of e^-/h^+ pairs. Afterward, the electrons on the CB of ZnO can react with O_2 to form superoxide radicals ($O_2^{\bullet-}$), which are able to generate hydroxyl radicals (·OH) through further reactions. On the other hand, OH radicals are also produced by the oxidation of water molecules by h+ in the VB of g-C_3N_4 [42].

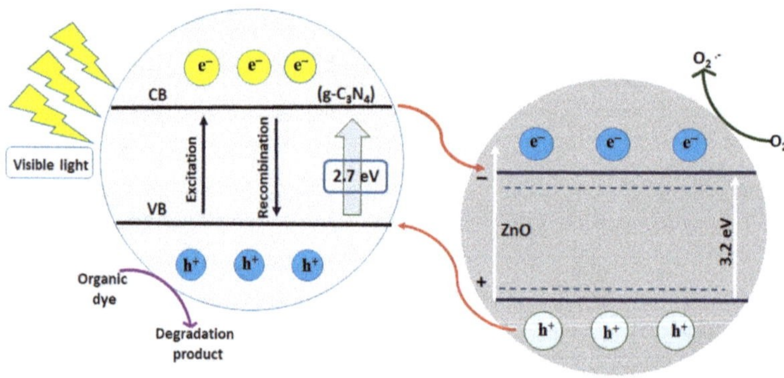

Figure 2. A representative diagram for the energy band structure of g-C_3N_4/ZnO heterostructure.

In this study, g-C_3N_4/ZnO nanocomposites with different ZnO molar ratios were synthesized using a simple method that is different from the literature. However, the present work is focused on the photocatalytic activity of the g-C_3N_4/ZnO nanocomposites with different ZnO molar ratios under UV light, which can help understand their performance under real ambient test conditions. To our knowledge, there are no reports on the application of g-C_3N_4/ZnO nanocomposites for the photocatalytic degradation of crystal violet under UV light. It is observed that the photodegradation of crystal violet is faster and more efficient when the g-C_3N_4/ZnO nanocomposite is used as a photocatalyst. This is because the recombination rate of photogenerated electron–hole pairs decreases, and the g-C_3N_4/ZnO nanocomposite increases visible light absorption compared to individual ZnO and g-C_3N_4.

2. Results and Discussion

2.1. Characterization of g-C_3N_4/ZnO Heterogeneous Photocatalyst

The powder X-ray diffraction patterns of the as-prepared bare g-C_3N_4 and different molar ratios of ZnO-decorated g-C_3N_4 nanocomposites are shown in Figure 3A. The measured powder XRD patterns of the g-C_3N_4/ZnO nanocomposites replicate Joint Committee on Powder Diffraction Standards (JCPDS) card no. 36-1451 and show peaks that are linked to the hexagonal wurtzite structure of ZnO. The XRD patterns of the bare g-C_3N_4 (JCPDS card no. 87 1526), ZnO, g C_3N_4/ZnO (0.25), g-C_3N_4/ZnO (0.50), and g-C_3N_4/ZnO (1.00) demonstrate a strong peak at 27.5°, which corresponds to the (002) plane, revealing graphitic stacking of g-C_3N_4 in the bare g-C_3N_4 and g-C_3N_4/ZnO nanocomposites (Figure 3A). By further increasing the molar ratio of ZnO from 0.50 mmol to 1.00 mmol, it can be observed that the crystal growth of the bare g-C_3N_4 is limited [41]. As the amount of ZnO increases in the samples, the peaks also increase. In addition, the g-C_3N_4 peaks decrease, which is just opposite of ZnO. This is because the density of g-C_3N_4 is lower than ZnO (Figure 3A).

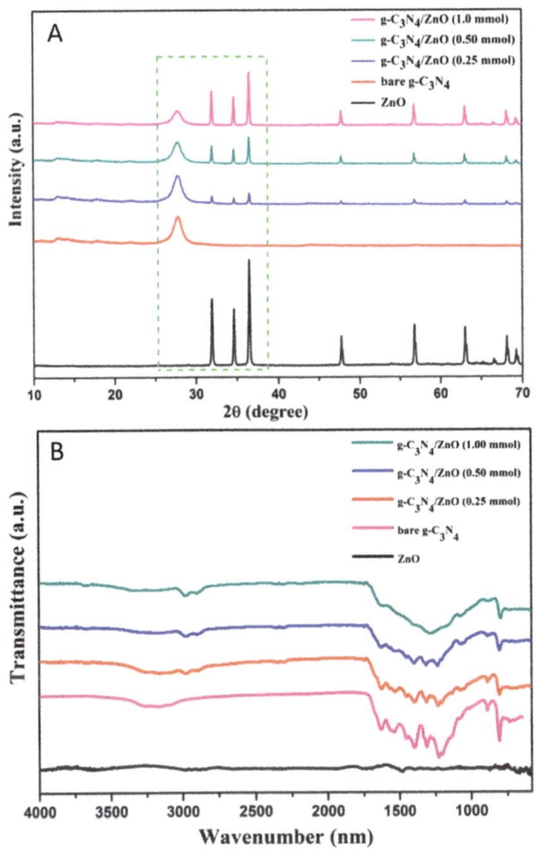

Figure 3. (**A**) XRD patterns of the bare g-C_3N_4, ZnO, and synthesized different molar ratios g-C_3N_4/ZnO samples, and (**B**) FTIR spectra of the g-C_3N_4/ZnO samples.

Figure 3B displays the Fourier-transform infrared spectra of the g-C_3N_4 and g-C_3N_4/ZnO samples. The N–H and hydroxyl groups of adsorbed H_2O molecules give g-C_3N_4 a broad peak at 3180 cm^{-1}. Aromatic C–N stretching modes generated by the g-C_3N_4 peaks can be assigned to several peaks ranging from 1285 to 1367 cm^{-1}. The out-of-plane bending vibration of triazine rings is revealed by the intense band at 806 cm^{-1}. The ZnO sample shows characteristic bands in the region of 593–458 cm^{-1}, which correspond to the ZnO bond vibration, and another band of about 3160 cm^{-1}, which is related to the hydroxyl group stretching vibrations. For the g-C_3N_4/ZnO hybrid structure, all the characteristic peaks related to g-C_3N_4 emerge, and the structure is proven to be successfully synthesized. The typical bands of the ZnO bond vibration appear at about 458, 473, and 455 cm^{-1} for the g-C_3N_4/ZnO (0.25 mmol), g-C_3N_4/ZnO (0.50 mmol), and g-C_3N_4/ZnO (1.00 mmol) samples, respectively. Through ZnO crystallization, the intensities of the peaks at 1626–1206 cm^{-1} drop, and the peaks are merged into a broad absorption band with increasing ZnO content. Hence, the creation of a conjugated material consisting of g-C_3N_4 and ZnO is vivid in the FT-IR spectra of the g-C_3N_4/ZnO composites, offering a channel for charge carrier transfer, resulting in photocatalytic activity enhancement [39].

SEM analysis was used to investigate the surface morphology and microstructure of the prepared samples. The SEM images of the bare g-C_3N_4 (Figure 4A) are in good consistency with the morphology of g-C_3N_4, which has been previously reported in the literature [42,43]. As illustrated in Figure 4A, the bare g-C_3N_4 appears as agglomerated pieces containing irregular smaller crystals. Then, a comparison of the surface morphologies of the g-C_3N_4/ZnO composite structure was performed (Figure 4B–D). The SEM images shown in Figure 4 reveal the formation of ZnO nanoparticles on the g-C_3N_4 nanosheets. The presented ZnO nanoparticles are not in a globular shape due to the nearness of g-C_3N_4 sheets. Moreover, Figure 4A shows that the uneven and projecting boundaries of the overlapping graphitic carbon nitride layers are what have created the g-C_3N_4 multilayer roughness. In Figure 4B–D, we can see the morphological changes of the coating around the g-C_3N_4 sample as the amount of ZnO increases.

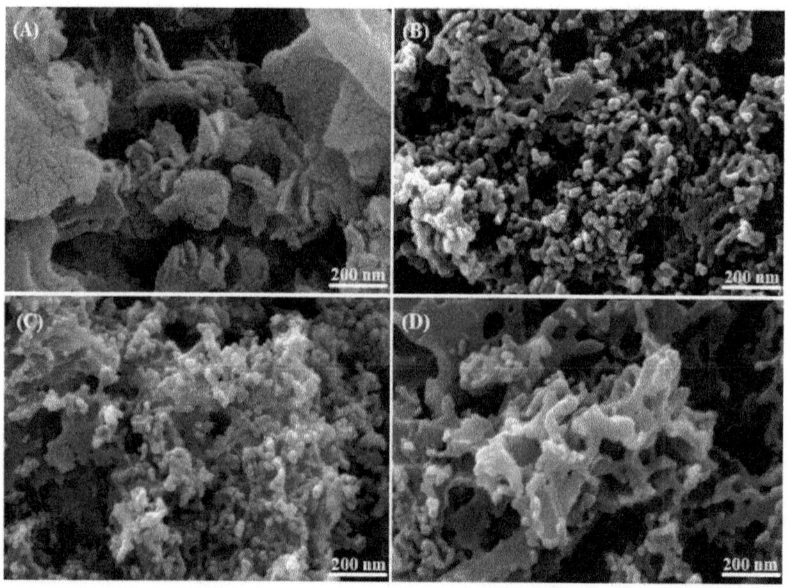

Figure 4. SEM images of (**A**) bare g-C_3N_4, (**B**) g-C_3N_4/ZnO (0.25), (**C**) g-C_3N_4/ZnO (0.50), and (**D**) g-C_3N_4/ZnO (1.00) samples.

The optical properties of the samples were analyzed using a UV–Vis absorption spectroscopy. The normalized absorption spectra of the prepared g-C_3N_4 and g-C_3N_4/ZnO nanocomposites are depicted in Figure 5A. The bare g-C_3N_4 sample has a characteristic absorption peak at 410 nm. We also observe that the absorbance of the g-C_3N_4/ZnO nanocomposite sample is slightly increased compared to the g-C_3N_4 sample in the UV–Vis wavelength region. The absorption nature of the g-C_3N_4/ZnO nanocomposites increases over the entire visible spectrum compared to the bare g-C_3N_4 and ZnO samples. This improved light absorption intensity in the visible range will be beneficial for the photocatalytic degradation process. The energy bandgap was characterized using the Kubelka–Munk equation for direct band-gap semiconductors, which was derived from a Tauc plot of $(\alpha h\upsilon)^2$

vs. hυ, as shown in Figure 5B. According to Figure 5B, the intersection of the line extrapolation and the x-axis reveals the energy band gaps. The equation is $\alpha = A(h\upsilon\text{-}Eg)^n/h\upsilon$, where α is the absorption coefficient, hυ is the energy of the incident photon, Eg is the band gap of the material, A is the constant, and n is a factor that depends on the nature of the electron transition [44]. The band gap energies of the ZnO, bare g-C_3N_4, g-C_3N_4/ZnO (0.25), g-C_3N_4/ZnO (0.50), and g-C_3N_4/ZnO (1.00) samples are determined to be 3.10, 2.70, 2.73, 2.66, and 2.61 eV, respectively, as demonstrated in Figure 5B.

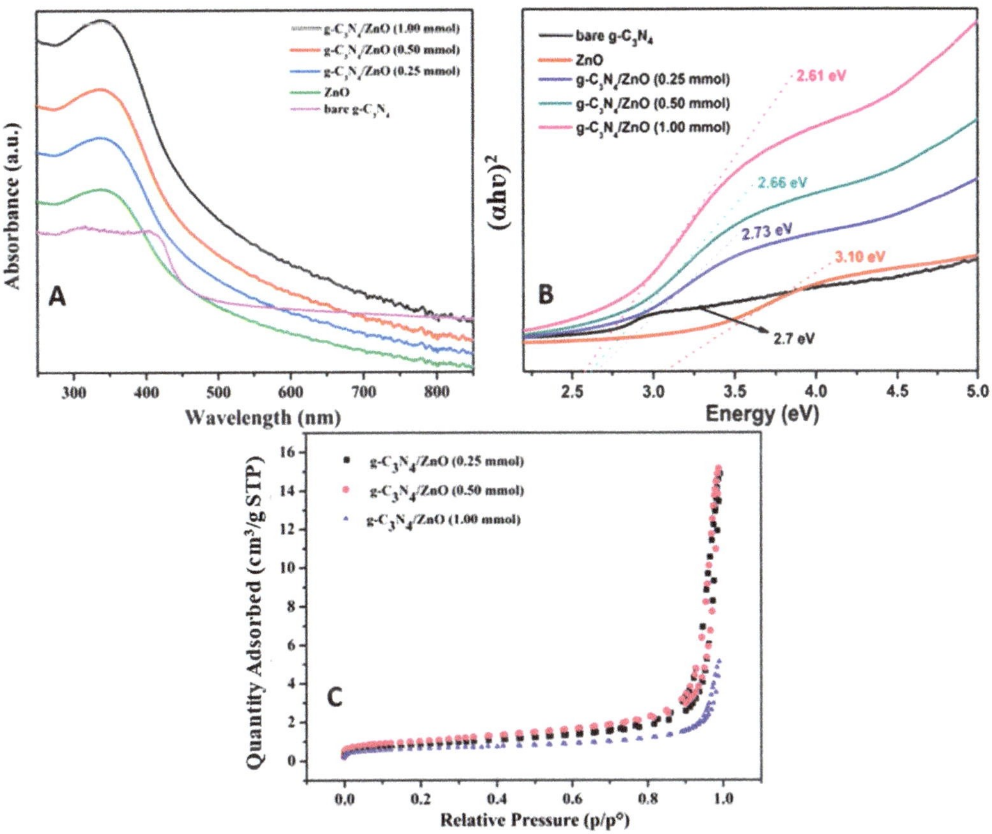

Figure 5. (**A**) UV–Vis absorption spectra, (**B**) Tauc plot of the g-C_3N_4, ZnO, and different molar ratios g-C_3N_4/ZnO nanocomposites, and (**C**) nitrogen adsorption–desorption isotherms of the g-C_3N_4/ZnO nanocomposites.

Figure 5C displays the adsorption–desorption N_2 isotherms for the g-C_3N_4/ZnO hybrid structures. The N_2 adsorption–desorption isotherms produced at 77 K were used to determine the surface area and pore volumes of the g-C_3N_4/ZnO (0.25 mmol), g-C_3N_4/ZnO (0.50 mmol) and g-C_3N_4/ZnO (1.00 mmol) hybrid structures (Table 1). The surface areas obtained for the g-C_3N_4, g-C_3N_4/ZnO (0.25 mmol), g-C_3N_4/ZnO (0.50 mmol), and g-C_3N_4/ZnO (1.00 mmol) nanocomposites are, respectively, 47.01 m^2/g, 67.15 m^2/g, 80.77 m^2/g, and 49.97 m^2/g. The highest surface area and pore volume are obtained for the g-C_3N_4/ZnO (0.50 mmol) sample as 80.77 m^2/g and 0.38 cm^3/g, respectively. The low surface area and pore volume of the hybrid structures can be predicted owing to the changes in the bond structures.

Table 1. The surface area and pore volume of the prepared samples.

Sample	Surface Area (m^2/g)	Pore Volume (cm^3/g)
g-C_3N_4/ZnO (0.25 mmol)	67.15	0.41
g-C_3N_4/ZnO (0.50 mmol)	80.77	0.38
g-C_3N_4/ZnO (1.00 mmol)	49.97	0.12
g-C_3N_4	47.01	0.17

2.2. *The Effect of Different Amounts of ZnO*

To determine the impact of diverse amounts of ZnO on CV degradation efficiency, different amounts of ZnO in the g-C_3N_4/ZnO photocatalyst were provided. The removal of CV was also investigated for the bare g-C_3N_4 catalyst to observe the photocatalytic performances of the designed nanocomposites. Initially, the amounts of ZnO selected to explore the degradation efficiency were 0.25, 0.50, and 1.00 mmol. Afterward, 0.10 g/L of the prepared photocatalyst was inserted into the CV aqueous solution, and the photocatalytic decolorization was investigated under UV light irradiation (Figure 6A). The results depict that 0.50 mmol of ZnO can degrade the CV solution up to 95.9% within 120 min. Lower removal efficiencies are obtained for 0.25 mmol and 1.00 mmol of ZnO. The reason could be explained by the fact that lower amounts of ZnO are insufficient to degrade the dye molecules. On the other hand, the active centers of the catalyst might have been inactivated because 1.00 mmol of ZnO was agglomerated due to overdose. Therefore, the 0.50 mmol ZnO-decorated g-C_3N_4 catalyst was chosen for further evaluation.

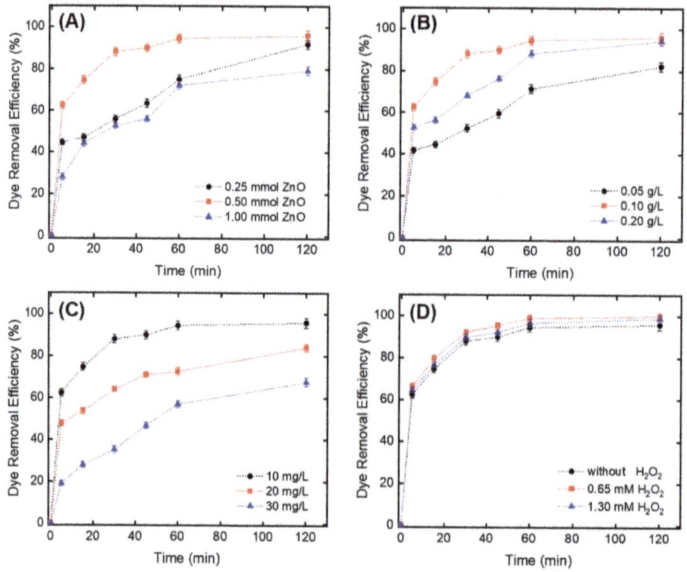

Figure 6. (**A**) The effect of ZnO on dye removal efficiency (experimental conditions: [catalyst] = 0.10 g/L). (**B**) The effect of catalyst amount on dye removal efficiency (experimental conditions: ZnO ratio = 0.5 mmol and [dye] = 10 mg/L). (**C**) The effect of dye concentration on dye removal efficiency (experimental conditions: ZnO ratio = 0.5 mmol and [catalyst] = 0.10 g/L). (**D**) The effect of H_2O_2 concentration on dye removal efficiency (experimental conditions: ZnO ratio = 0.5 mmol, [catalyst] = 0.10 g/L, and [dye] = 10 mg/L).

2.3. *The Effect of g-C_3N_4/ZnO Catalyst Concentration*

The effect of g-C_3N_4/ZnO catalyst concentration on dye removal efficiency was examined. Three different catalyst amounts (50, 100, and 200 mg/L) were used to examine the impact on CV removal efficiency (Figure 6B). The outcomes reveal that by increasing the amount of the catalyst from 0.05 g/L to 0.10 g/L, the degradation efficiency can be increased from 82.2% to 95.9%, respectively. By increasing the catalyst amount, the provided surface area as a platform for the reaction and generation of diverse reactive oxygen species (ROSs) is incremented. However, enhancing the catalyst amount up to 0.20 g/L can decrease the removal efficiency to 94.2%, which might be due to the light penetration prevention.

2.4. The Effect of CV Concentration

The influence of initial CV concentration (10, 20, and 30 mg/L) on the removal efficiency is presented in Figure 6C. Increasing the dye concentration from 10 to 30 mg/L causes competition as there are more dye molecules on the constant active sites of the catalyst and the generated ROS [45]. Moreover, high contaminant concentrations cause more intense color and adversely affect dye removal efficiency due to decreasing UV light penetration into the pollutant solution. It seems that most of the UV light might be absorbed by the dye molecules rather than the catalyst [46].

2.5. The Effect of H_2O_2 Concentration

The effect of H_2O_2 concentration on CV removal efficiency is demonstrated in Figure 6D. As it is well known, the addition of H_2O_2 promotes the formation of more OH• radicals by reacting with UV light irradiation, and these OH• radicals are able to improve the photocatalytic oxidation [47]. A complete decolorization (100%) of the CV dye was obtained at 0.65 mM of H_2O_2 within 120 min. However, an increase in H_2O_2 concentration cannot increase the dye removal efficiency due to the prevention of OH• formation by excessive H_2O_2 molecules [48].

The efficiency of catalyst reuse was studied by recycling the same catalyst slurry for the degradation of 10 mg/L of CV under similar batch conditions. The reuse experiments were performed for three cycles and by maintaining the catalyst slurry concentration at constant values for each run.

2.6. Kinetic Study

The pseudo-first- and second-order reaction kinetics were calculated for the present study. The kinetic constants of the reaction k_1 (1/min) and k_2 (L/mg.min) were calculated from the slope of the line of the kinetic models of pseudo-first order ($\ln C_0/C = k_1 t$) and second order ($1/C - 1/C_0 = k_2 t$). The kinetic constants k_1 and k_2 and the regression coefficients (R^2) are compared in Table 2. The results show that the kinetic constants decrease with an increase in the initial concentration of CV from 10 to 30 mg/L. The pseudo-second-order kinetic model supplies the best fit to the experimental data for photocatalytic decolorization of CV using the g-C_3N_4/ZnO catalyst (Figure S1).

Table 2. The pseudo-first- and second-order kinetic rate constants with regression coefficients for decolorization of CV at different dye concentration conditions (experimental conditions: solution pH = 6.8, and [catalyst] = 0.10 g/L).

C_{CV} (mg/L)	First Order		Second Order	
	k_1 (1/min)	R^2	k_2 (L/mg min)	R^2
10	0.0235	0.7594	0.0202	0.9497
20	0.0127	0.8297	0.0020	0.9780
30	0.009	0.9290	0.0006	0.9805

2.7. The Effect of Different Processes

Adsorption experiments were also performed to compare the efficiency of adsorption and the photocatalytic potential of the g-C_3N_4/ZnO catalyst on the removal of the CV dye solution using 0.1 g/L of catalyst loading. The results in Figure 7A indicate that photocatalysis is the dominant mechanism for the decolorization of the CV dye. It is reported in the literature that the pH of zero-point charge (pH_{PZC}) of ZnO nanoparticles is around 9.5–10, which means that for pH values lower than 10, the surface of ZnO nanoparticles is positively charged [49,50]. Additionally, the g-C_3N_4 sample has positive zeta potential values [51]. In an aqueous solution, the methyl groups of CV dissociate and then CV molecules are converted to cationic dye ions. Therefore, the surface of the g-C_3N_4 and ZnO nanoparticles is positively charged at the original pH (6.8) of the solution, which is disadvantageous for the reach of positively charged CV dye ions due to electrostatic repulsive force. This causes less adsorption of the CV dye on the g-C_3N_4/ZnO catalyst. Moreover, the bare g-C_3N_4 was also tested for dye decolorization, and a 76.4% dye removal efficiency was obtained. Moreover, a comparative study with the literature is presented in Table 3. The results depict that our heterogeneous catalyst supplies effective decolorization with a lower catalyst amount compared to other studies.

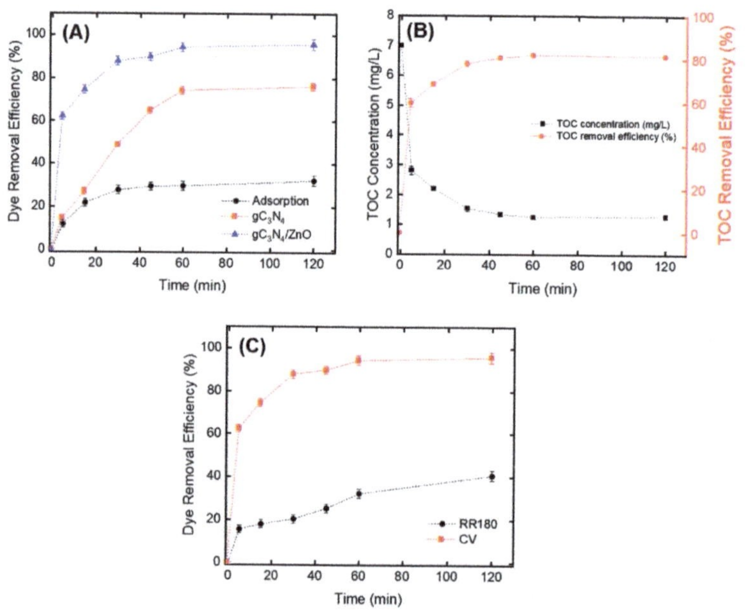

Figure 7. (**A**) The effect of different processes on dye removal efficiency, (**B**) TOC alterations during photocatalysis of CV, and (**C**) comparison of the degradation efficiency of RR180 and CV during photocatalysis. (Experimental conditions: ZnO ratio = 0.5 mmol, [catalyst] = 0.10 g/L, and [dye] = 10 mg/L).

Table 3. Comparison of g-C_3N_4-based heterogeneous catalysts for dye decolorization.

Catalysts	Dye Concentration (mg/L)	Pollutants	Degradation Efficiency	References
g-C_3N_4/ZnO	10	Crystal Violet	95.9% for 120 min	This study
g-C_3N_4/Ag	10	Methyl Orange	98.7% for 120 min	[36]
g-C_3N_4/CdS	5	Rhodamine B	95% for 120 min	[37]
g-C_3N_4/BiOBr	10	Rhodamine B	95% for 30 min	[52]
g-C_3N_4/$TiOF_2$	5	Rhodamine B	78.3% for 80 min	[53]
g-C_3N_4/MoS_2	10	Rhodamine B	97.6% for 50 min	[54]
g-C_3N_4/$LaFeO_3$	15	Rhodamine B	70% for 160 min	[55]
g-C_3N_4/Cu_2O/Cu	10	Methyl Orange	99% for 40 min	[56]

The TOC of the solution was monitored during the degradation reactions to ensure the mineralization of the pollutants. It can be clearly seen that the degradation reaches steady-state conditions after 45 min of the process, with 80.12% efficiency (Figure 7B), which indicates that the g-C_3N_4/ZnO nanocomposite can effectively mineralize CV. After 45 min, degradation is almost not observed. The result show that when the catalyst is new (virgin), the rate of TOC removal is faster, and later its performance remains constant.

2.8. The Comparison of Removal Efficiency in the Presence of an Anionic Dye

Additionally, the photocatalytic degradation efficiency of a negatively charged reactive red 180 (RR180) dye was compared with the positively charged CV dye (Figure 7C). Removal efficiencies of 40.8% and 95.9% were obtained for the RR180 and CV dyes within 120 min, respectively. This could be attributed to the molecular structure of the RR180 dye (linear formula: $C_{29}H_{19}N_3Na_4O_{17}S_5$ with molecular weight: 933.76) being more complex than the CV dye (linear formula: $C_{25}H_{30}N_3Cl$ with molecular weight: 407.98).

3. Materials and Methods

3.1. Materials

Zinc nitrate hexahydrate ($Zn(NO_3)_2{\cdot}6H_2O$, >99%), urea (CH_4N_2O >99.5%), and ammonium hydroxide solution (NH_4OH, 26%) were purchased from Sigma-Aldrich (USA). All reagents used in this work were of analytical grade and were used without further purifications. Crystal violet (CV), which was used as a model pollutant, was procured from Sigma-Aldrich (USA).

3.2. Synthesis of g-C_3N_4

Graphite-like g-C_3N_4 nanosheets were synthesized according to the procedure reported previously in the literature [39]. A total of 50 g of urea was first calcinated at 580 °C for 3 h at a heating rate of 5 °C/min under argon gas flow using a one-step synthesis technique. The material was cooled to room temperature, washed with HNO_3 solution (0.1 mol/L) and water, and dried at 70 °C overnight.

3.3. The Synthesis of Different Molar Ratios of ZnO-Decorated g-C_3N_4

The ZnO-decorated g-C_3N_4 (named g-C_3N_4/ZnO) nanohybrid was obtained in the presence of urea and different proportions of $Zn(NO_3)_2{\cdot}6H_2O$. Firstly, 10 mL of distilled water was adjusted to a pH of 10 with the ammonium solution. A total of 166 mmol of urea was added to the distilled water/ammonia solution mixture. The urea was completely dissolved, and different proportions of $[Zn(NO_3)_2{\cdot}6H_2O]_x$ (x; 0.25 mmol, 0.50 mmol, and 1.00 mmol) were added to form the hybrid structure. Then, the mixture was vigorously stirred at 70 °C for 2 h, and the solvent was evaporated at 80 °C. The obtained precipitation was heated at 550 °C for 3 h in a furnace (heating rate of 5 °C/min) and then allowed to cool down to room temperature. A schematic illustration of the g-C_3N_4/ZnO synthesis procedure is shown in Figure 8.

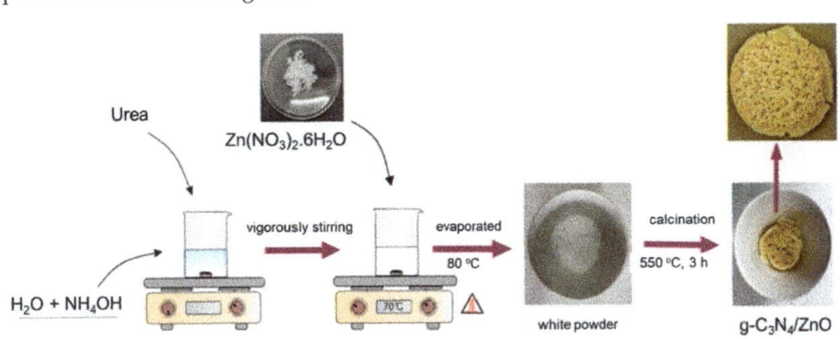

Figure 8. The synthesis procedure of different molar ratios of ZnO-decorated g-C_3N_4.

3.4. Photocatalytic Degradation Experiments

The experimental studies were performed in a batch Pyrex photo reactor (column-shaped) with a 500 mL capacity. A total of 100 mL of dye solution was prepared. The photocatalytic reactor was blown from the bottom at a flow rate of 150 mL/min to accelerate the mixing of the catalyst. Cooled air was used to keep a constant temperature of 25 ± 1 °C. Six UVA lamps (Philips TL8 W Actinic BL, emitting wavelengths at 365 nm) were placed in a hexagonal position around the reactor, which was surrounded by an aluminum-coated film, to obtain a uniform reflection. The samples were taken from the photocatalytic reactor at predetermined times (5, 15, 30, 45, 60, and 120 min) during the 120 min reaction and centrifuged at 6000 rpm for 5 min to separate the catalyst from the aqueous solution. Maximum absorbance was measured by using a Hach DR3900 UV–Vis spectrophotometer (Düsseldorf, Germany) to calculate dye removal efficiency. The dye removal efficiency was calculated by subtracting the final concentration from the initial concentration and dividing it by the initial concentration.

The reusability behavior of the catalyst was examined through five consecutive runs. After each run, the treated solution was discharged from the reactor after the catalyst was precipitated. Then, a new solution was filled into the reactor and the catalyst was utilized again. The catalyst was washed with 50 mL of deionized water, centrifuged, and dried in an oven.

3.5. Characterizations Techniques

The crystal structures of the as-prepared samples were characterized using an X-ray diffraction (XRD, Bruker AXS D8 Advance, Billerica, Massachusetts, USA) diffractometer. The absorption spectra of the g-C_3N_4/ZnO nanocomposites were recorded using a UV–Vis spectrophotometer (Thermo scientific-Genesys 150, Menlo Park, CA, USA). A Quanta 650 FEG field emission scanning electron microscope manufactured by FEI (USA) was used to evaluate the surface and section morphology of the nanocomposites. The Fourier-transform infrared spectroscopy (FTIR) spectra of the nanocomposites were recorded using a PerkinElmer FTIR spectrometer (Waltham, MA, USA). The surface area and total pore volume were measured by using the Brunauer Emmett-Teller analysis (BET, MicroActive for TriStar II Plus 2.00, Norcross, GA, USA).

4. Conclusions

In summary, the ZnO-decorated g-C_3N_4 heterogeneous catalyst was successfully synthesized via a one-step calcination method. It was characterized using FTIR, XRD, UV–Vis, SEM, and BET analyses. The XRD and FTIR measurements demonstrated the strong coordination of ZnO with bare g-C_3N_4 and the partial breakdown of the g-C_3N_4 catalyst's crystalline structure upon the addition of ZnO. Furthermore, the ZnO-decorated g-C_3N_4 composite exhibited high photocatalytic activity for the degradation of CV dye under UV light. The results displayed no decrease in the CV dye removal efficiency up to five cycles for the g-C_3N_4 /ZnO photocatalyst. The synergistic interactions between the ZnO and g-C_3N_4 provide a forceful method for combining two visible-light photocatalysts to expand their UV light absorption. This study provides some new insights into the fabrication of advanced catalysts for highly efficient photocatalytic applications.

Supplementary Materials: The following supporting information can be downloaded at: https://www.mdpi.com/article/10.3390/catal13030485/s1, Figure S1: Kinetic study for CV decoloriaztion.

Author Contributions: Conceptualization, B.S. and Z.B.; methodology, T.S.R.; investigation, N.D., K.O. and A.K.; writing—original draft preparation, K.O. and A.K.; writing—review and editing, N.D., K.O. and AK. All authors have read and agreed to the published version of the manuscript.

Funding: This research received no external funding.

Data Availability Statement: Not applicable.

Conflicts of Interest: The authors declare no conflict of interest.

References

1. Hachem, F.C.; Bocquillon, O.; Zahraa, M. Decolourization of textile industry wastewater by the photocatalytic degradation process. *Dye. Pigment.* **2001**, *49*, 117–125. [CrossRef]
2. Natarajan, S.; Bajaj, H.C.; Tayade, R.J. Recent advances based on the synergetic effect of adsorption for removal of dyes from wastewater using photocatalytic process. *J. Environ. Sci.* **2017**, *65*, 201–222. [CrossRef] [PubMed]
3. Anisuzzaman, S.M.; Joseph, C.G.; Affandi, N.A.; Maruja, S.N.; Vijayan, V. Current Trends in the Utilization of Photolysis and Photocatalysis Treatment Processes for the Remediation of Dye Wastewater: A Short Review. *Chem. Eng.* **2022**, *6*, 58. [CrossRef]
4. Ambaye, T.G.; Hagos, K. Photocatalytic and biological oxidation treatment of real textile wastewater. *Nanotechnol. Environ. Eng.* **2020**, *5*, 28. [CrossRef]
5. Souza, R.P.; Freitas, T.K.F.S.; Domingues, F.S.; Pezoti, O.; Ambrosio, E.; Ferrari, L.A.M.; Garcia, J.C. Photocatalytic activity of TiO_2, ZnO and Nb_2O_5 applied to degradation of textile wastewater. *J. Photochem. Photobiol. A Chem* **2016**, *329*, 9–17. [CrossRef]
6. Eskikaya, O.; Ozdemir, S.; Tollu, G.; Dizge, N.; Ramaraj, R.; Manivannan, A.; Balakrishnan, D. Synthesis of two different zinc oxide nanoflowers and comparison of antioxidant and photocatalytic activity. *Chemosphere* **2022**, *306*, 153389. [CrossRef] [PubMed]
7. Lv, Y.; Lin, J.; Peng, S.; Zhang, L.; Yu, L. Effective ways to enhance the photocatalytic activity of ZnO nanopowders: High crystalline degree, more oxygen vacancies, and preferential growth. *New J. Chem.* **2019**, *43*, 19223–19231. [CrossRef]
8. Vattikuti, S.V.P.; Reddy, P.A.K.; Shim, J.; Chan, B. Visible-Light-Driven Photocatalytic Activity of SnO_2−ZnO Quantum Dots Anchored on g-C_3N_4 Nanosheets for Photocatalytic Pollutant Degradation and H_2 Production. *ACS Omega* **2018**, *3*, 7587–7602. [CrossRef]
9. Qamar, M.A.; Shahid, S.; Javed, M.; Shariq, M.; Fadhali, M.M.; Madkhali, O.; Ali, S.K.; Syed, I.S.; Awaji, M.Y.; Shakir Khan, M.; et al. Accelerated Decoloration of Organic Dyes from Wastewater Using Ternary Metal/g-C_3N_4/ZnO Nanocomposites: An Investigation of Impact of g-C_3N_4 Concentration and Ni and Mn Doping. *Catalysts* **2022**, *12*, 1388. [CrossRef]
10. Mohammed, I. The photocatalytic performance of the g-C_3N_4/ZnOcomposite photocatalyst toward degradation of organic pollutants and its inactivity toward hydrogen evolution: The influence of light irradiation and charge transfer. *Chem. Phys. Lett.* **2019**, *739*, 136992. [CrossRef]
11. Li, J.; Zhou, M.; Ye, Z.; Ma, C.; Wang, H.; Huo, P.; Yan, Y. Enhanced photocatalytic activity of g-C_3N_4-ZnO/HNTs composited heterostructure photocatalysts for degradation of tetracycline under visible light irradiation. *RSC Adv.* **2015**, *5*, 91177–91189. [CrossRef]
12. Manikanika; Chopra, L. Photocatalytic activity of zinc oxide for dye and drug degradation: A review. *Mater. Today Proc.* **2022**, *52*, 1653–1656. [CrossRef]
13. Nam, Y.; Lim, J.H.; Ko, K.C.; Lee, J.Y. Photocatalytic activities of TiO_2 nanoparticles: A theoretical aspect. *J. Mater. Chem. A* **2019**, *7*, 13833–13859. [CrossRef]
14. Jadoun, S.; Yáñez, J.; Mansilla, H.D.; Riaz, U.; Chauhan, N.P.S. Conducting polymers/zinc oxide-based photocatalysts for environmental remediation: A review. *Environ. Chem. Lett.* **2022**, *20*, 2063–2083. [CrossRef]
15. Nagaraju, G.; Shivaraju, G.C.; Banuprakash, G.; Rangappa, D. Photocatalytic Activity of ZnO Nanoparticles: Synthesis via Solution Combustion Method. *Mater. Today Proc.* **2017**, *4*, 11700–11705. [CrossRef]
16. Forouzandeh, M.M.; Ganjali, F.; Zamiri, E.; Zarei, S.S.; Jalali, F.; Padervand, M.; Taheri, L.R.; Maleki, A. Efficient Photodegradation of Eriochrome Black-T by a Trimetallic Magnetic Self-Synthesized Nanophotocatalyst Based on Zn/Au/Fe-Embedded Poly(vinyl alcohol). *Langmuir* **2022**, *38*, 13728–13743. [CrossRef] [PubMed]
17. Mclaren, A.; Valdes-Solis, T.; Li, G.; Tsang, S.C. Shape and Size Effects of ZnO Nanocrystals on Photocatalytic Activity. *J. Am. Chem. Soc.* **2009**, *131*, 12540–12541. [CrossRef]
18. Mirzaeifard, Z.; Shariatinia, Z.; Jourshabani, M.; Darvishi, S.M.R. ZnO Photocatalyst Revisited: Effective Photocatalytic Degradation of Emerging Contaminants Using S-Doped ZnO Nanoparticles under Visible Light Radiation. *Ind. Eng. Chem. Res.* **2020**, *59*, 15894–15911. [CrossRef]
19. Ahmad, M.R.; Khan, W.; Qureshi, M.M.; Gul, M.T.; Haq, A.; Ullah, S.R.; Rab, A.; Menaa, F. Phytogenic fabrication of ZnO and gold decorated ZnO nanoparticles for photocatalytic degradation of Rhodamine B. *J. Environ. Chem. Eng.* **2021**, *9*, 104725. [CrossRef]
20. Chen, D.; Wang, K.; Xiang, D.; Zong, R.; Yao, W.; Zhu, Y. Significantly enhancement of photocatalytic performances via core–shell structure of ZnO@mpg-C_3N_4. *Appl. Catal. B Environ.* **2014**, *147*, 554–561. [CrossRef]
21. Safaei, J.; Mohamed, N.A.; Mohamad, N.; Mohamad, F.; Soh, M.F.; Ludin, N.A.; Ibrahim, M.A.; Roslam, W.I.; Wan, N.; Mat, T.; et al. Graphitic carbon nitride (g-C_3N_4) electrodes for energy conversion and storage: A review on photoelectrochemical water splitting, solar cells and supercapacitors. *J. Mater. Chem. A* **2018**, *6*, 22346–22380. [CrossRef]
22. Padervand, M.; Heidarpour, H.; Goshadehzehn, M.; Hajiahmadi, S. Photocatalytic degradation of 3-methyl-4-nitrophenol over Ag/AgCl-decorated/[MOYI]-coated/ZnO nanostructures: Material characterization, photocatalytic performance, and in-vivo toxicity assessment of the photoproducts. *Environ. Technol. Innov.* **2021**, *21*, 101212. [CrossRef]
23. Deng, X.; Wang, D.; Li, H.; Jiang, W.; Zhou, T.; Wen, Y.; Yu, B.; Che, G.; Wang, L. Boosting interfacial charge separation and photocatalytic activity of 2D/2D g-C_3N_4/$ZnIn_2S_4$ S-scheme heterojunction under visible light irradiation. *J. Alloy. Compd.* **2020**, *894*, 162209. [CrossRef]
24. Huu, H.T.; Thi, M.D.N.; Nguyen, V.P.; Thi, L.N.; Phan, T.T.T.; Hoang, Q.D.; Vo, V. One-pot synthesis of S-scheme MoS_2/g-C_3N_4 heterojunction as effective visible light photocatalyst. *Sci. Rep.* **2021**, *11*, 14787. [CrossRef]

25. Mohamed, R.M.; Mkhalid, I.A.; Alhaddad, M.; Basaleh, A.; Alzahrani, K.A.; Ismail, A.A. Enhanced CO_2 photocatalytic conversion into CH_3OH over visible-light-driven Pt nanoparticle-decorated mesoporous ZnO–ZnS S-scheme heterostructures. *Ceram. Int.* **2021**, *47*, 26779–26788. [CrossRef]
26. Kang, S.; Li, Z.; Xu, Z.; Zhang, Z.; Sun, J.; Bian, J.; Bai, L.; Qu, Y.; Jing, L. Synthesis of mixed-valence Cu phthalocyanine/graphene/g-C_3N_4 ultrathin heterojunctions as efficient photocatalysts for CO_2 reduction. *Catal. Sci. Technol.* **2022**, *12*, 4817–4825. [CrossRef]
27. Swamy, C.K.; Hezam, A.; Ramesh, A.M.; Ramakrishnegowda, D.H.; Purushothama, D.K.; Krishnegowda, J.; Shivanna, S. Microwave hydrothermal synthesis of copper induced ZnO/g-C_3N_4 heterostructure with efficient photocatalytic degradation through S-scheme mechanism. *J. Photochem. Photobiol. A Chem.* **2021**, *418*, 113394. [CrossRef]
28. Chao, Y.; Liu, B.; Zhou, B.; Tian, S.; Zhang, L. Structure and properties of graphene oxide during the synthesis process at fixed temperatures. *Ceram. Int.* **2021**, *47*, 1748–17493. [CrossRef]
29. Akhundi, A.; Habibi-Yangjeh, A. High performance magnetically recoverable g-C_3N_4/Fe_3O_4/Ag/Ag_2SO_3 plasmonic photocatalyst for enhanced photocatalytic degradation of water pollutants. *Adv. Powder Technol.* **2017**, *28*, 565–574. [CrossRef]
30. Fu, J.; Yu, J.; Jiang, C.; Cheng, B. g-C_3N_4-Based Heterostructured Photocatalysts. *Energy Mater.* **2018**, *8*, 1701503. [CrossRef]
31. Padervand, M.; Rhimi, B.; Wang, C. One-pot synthesis of novel ternary Fe_3N/Fe_2O_3/C_3N_4 photocatalyst for efficient removal of rhodamine B and CO_2 reduction. *J. Alloy. Compd.* **2021**, *852*, 156955. [CrossRef]
32. Pomilla, F.R.; Cortes, M.A.L.R.M.; Hamilton, J.W.J.; Molinari, R.; Barbieri, G.; Marci, G.; Palmisano, L.; Sharma, P.K.; Brown, A.; Byrne, J.A. An Investigation into the Stability of Graphitic CN as a Photocatalyst for CO_2 Reduction. *J. Phys. Chem. C* **2018**, *122*, 28727–28738. [CrossRef]
33. Dong, G.; Zhang, L. Porous structure dependent photoreactivity of graphitic carbon nitride under visible light. *J. Mater. Chem.* **2012**, *22*, 1160–1166. [CrossRef]
34. Akhundi, A.; Badiei, A.; Ziarani, G.M.; Habibi-Yangjeh, A.; Muñoz-Batista, M.J.; Luque, R. Graphitic carbon nitride-based photocatalysts: Toward efficient organic transformation for value-added chemicals production. *Mol. Catal.* **2020**, *488*, 110902. [CrossRef]
35. Wu, Y.; Tao, L.; Zhao, J.; Yue, X.; Deng, W.; Li, Y.; Wang, C. TiO_2/g-C_3N_4 nanosheets hybrid photocatalyst with enhanced photocatalytic activity under visible light irradiation. *Res. Chem. Intermed.* **2016**, *42*, 3609–3624. [CrossRef]
36. Liu, R.; Yang, W.; He, G.; Zheng, W.; Li, M.; Tao, W.; Tian, M. Ag-Modified g-C_3N_4 Prepared by a One-Step Calcination Method for Enhanced Catalytic Efficiency and Stability. *ACS Omega* **2020**, *5*, 19615–19624. [CrossRef]
37. Xu, Y.; Zhang, W.D. CdS/ g-C_3N_4 Hybrids with Improved Photostability and Visible Light Photocatalytic Activity. *Eur. J. Inorg. Chem.* **2015**, *10*, 1744–1751. [CrossRef]
38. Li, Y.; Zhu, S.; Liang, Y.; Li, Z.; Wu, S.; Chang, C.; Luo, S.; Cui, Z. Synthesis of α-Fe_2O_3/g-C_3N_4 Photocatalyst for high-efficiency water splitting under full light. *Mater. Des.* **2020**, *196*, 109191. [CrossRef]
39. Paul, D.R.; Gautam, S.; Panchal, P.; Nehra, S.P.; Choudhary, P.; Sharma, A. ZnO-Modified g-C_3N_4: A Potential Photocatalyst for Environmental Application. *ACS Omega* **2020**, *5*, 3828–3838. [CrossRef] [PubMed]
40. Liu, W.; Wang, M.; Xu, C.; Chen, S. Facile synthesis of g-C_3N_4/ZnO composite with enhanced visible light photooxidation and photoreduction properties. *Chem. Eng. J.* **2012**, *209*, 386–393. [CrossRef]
41. Sun, J.-X.; Yuan, Y.-P.; Qiu, L.-G.; Jiang, X.; Xie, A.-J.; Shen, Y.-H.; Zhu, J.F. Fabrication of composite photocatalyst g-C_3N_4–ZnO and enhancement of photocatalytic activity under visible light. *Dalton Trans.* **2012**, *41*, 6756–6763. [CrossRef] [PubMed]
42. Fageria, P.; Nazir, R.; Gangopadhyay, S.; Barshilia, H.C.; Pande, S. Graphitic-carbon nitride support for the synthesis of shape-dependent ZnO and their application in visible light photocatalysts. *RSC Adv.* **2015**, *5*, 80397–80409. [CrossRef]
43. Tauc, J.; Grigorovici, R.; Vancu, A. Optical Properties and Electronic Structure of Amorphous Germanium. *Phys. Stat. Sol.* **1966**, *15*, 627–637. [CrossRef]
44. Kumaresan, N.; Sinthiya, M.M.A.; Kumar, M.P.; Ravichandran, S.; Babu, R.R.; Sethurman, K.; Ramamurthi, K. Investigation on the g-C_3N_4 encapsulated ZnO nanorods heterojunction coupled with GO for effective photocatalytic activity under visible light irradiation. *Arab. J. Chem.* **2018**, *13*, 2826–2843. [CrossRef]
45. Ngullie, R.C.; Alaswad, S.O.; Bhuvaneswari, K.; Shanmugam, P.; Pazhanivel, T.; Arunachalam, P. Synthesis and Characterization of Efficient g-C_3N_4/ZnONanocomposites Photocatalyst for Photocatalytic Degradation of Methylene Blue. *Coatings* **2020**, *10*, 500. [CrossRef]
46. Wahab, H.S.; Hussain, A.A. Photocatalytic oxidation of phenol red onto nanocrystalline TiO_2 particles. *J. Nanostruct. Chem.* **2016**, *6*, 261–274. [CrossRef]
47. Bilici, Z.; Bouchareb, R.; Sacak, T.; Yatmaz, H.C.; Dizge, N. Recycling of TiO_2-containing waste and utilization by photocatalytic degradation of a reactive dye solution. *Water Sci. Technol.* **2021**, *83*, 1242–1249. [CrossRef]
48. Vahid, V.; Somayeh, F.; Ramazan, K.; Alireza, K. Enhancing the permeability and antifouling properties of cellulose acetate ultrafiltration membrane by incorporation of ZnO@graphitic carbon nitride nanocomposite. *Carbohydr. Polym.* **2021**, *256*, 117413. [CrossRef]
49. Daneshvar, N.; Rasoulifard, M.H.; Khataee, A.R.; Hosseinzadeh, F. Removal of C.I. Acid Orange 7 from aqueous solution by UV irradiation in the presence of ZnO nanopowder. *J. Hazard. Mater.* **2007**, *143*, 95–101. [CrossRef]
50. Miao, J.; Jia, Z.; Lu, H.B.; Habibi, D.; Zhang, J.C. Heterogeneous photocatalytic degradation of mordant black 11 with ZnO nanoparticles under UV–Vis light. *J. Taiwan Inst. Chem. Eng.* **2014**, *45*, 1636–1641. [CrossRef]

51. Kumar, R.; Barakat, M.A.; Alseroury, F.A. Oxidized g-C_3N_4/polyaniline nanofiber composite for the selective removal of hexavalent chromium. *Sci. Rep.* **2017**, *7*, 12850. [CrossRef] [PubMed]
52. Ye, L.; Liu, J.; Jiang, Z.; Peng, T.; Zan, L. Facets coupling of BiOBr-g-C_3N_4 composite photocatalyst for enhanced visible-light-driven photocatalytic activity. *Appl. Catal. B Environ.* **2013**, *142–143*, 1–7. [CrossRef]
53. Liu, Y.; Ma, Z. $TiOF_2$/ g-C_3N_4 composite for visible-light driven photocatalysis. *Colloids Surf. A Physicochem. Eng. Asp.* **2021**, *618*, 126471. [CrossRef]
54. Tian, J.; Chen, Z.; Jing, J.; Feng, C.; Sun, M.; Li, W. Enhanced photocatalytic performance of the MoS_2/g-C_3N_4 heterojunction composite prepared by vacuum freeze drying method. *J. Photochem. Photobiol. A Chem.* **2020**, *390*, 112260. [CrossRef]
55. Qian, L.; Jie, J.; Changhai, L.; Song, X.; Zhongyu, L. Constructing a novel p-n heterojunction photocatalyst $LaFeO_3$/g-C_3N_4 with enhanced visible-light-driven photocatalytic activity. *J. Alloy. Compd.* **2017**, *709*, 542–548. [CrossRef]
56. Yongchao, B.; Kezheng, C. A novel Z-scheme visible light driven Cu_2O/Cu/g-C_3N_4 photocatalyst using metallic copper as a charge transfer mediator. *Mol. Catal.* **2017**, *432*, 187–195. [CrossRef]

catalysts

MDPI

Article

Assembled Porphyrin Nanofiber on the Surface of g-C_3N_4 Nanomaterials for Enhanced Photocatalytic Degradation of Organic Dyes

Hoan Thi Lai [1], Giang Thi Nguyen [2], Nga Thuy Tran [1], Thanh Tung Nguyen [2], Chinh Van Tran [3,*], Duy Khiem Nguyen [4], S. W. Chang [5], W. Jin Chung [5], Dinh Duc Nguyen [5,*], Hoai Phuong Nguyen Thi [3] and Duong Duc La [3,*]

1 Department of Fundamental Science, University of Transport and Communications, Cau Giay, Dong Da, Hanoi 100000, Vietnam
2 Institute of Materials Science, Vietnam Academy of Science and Technology, 18-Hoang Quoc Viet, Hanoi 100000, Vietnam
3 Institute of Chemistry and Materials, 17 Hoang Sam, Nghia Do, Cau Giay, Hanoi 100000, Vietnam
4 VN-UK Institute for Research and Executive Education, The University of Danang, Danang 550000, Vietnam
5 Department of Environmental Energy Engineering, Kyonggi University, Goyang 10285, Republic of Korea
* Correspondence: chinhpkkq@gmail.com (C.V.T.); nguyensyduc@gmail.com (D.D.N.); duc.duong.la@gmail.com (D.D.L.)

Abstract: In this work, a g-C_3N_4/porphyrin nanocomposite was fabricated through the self-assembling of monomeric Tetrakis (4-carboxyphenyl) porphyrin (TCPP) molecules with g-C_3N_4 nanomaterials. The characterizing results showed a good distribution of TCPP nanofibers with a diameter of < 100 nm and several micrometers in length on the g-C_3N_4 nanoflakes' surfaces. The prepared g-C_3N_4/porphyrin nanocomposite had two bandgap energies of 2.38 and 2.7 eV, which could harvest a wide range of photon energy in the light spectrum, particularly in visible light. The obtained C_3N_4/TCPP nanocomposite revealed a remarkable photodegradation efficiency toward rhodamine B dyes, with a RhB removing rate of 3.3×10^{-2} min^{-1}. The plausible mechanism for the photocatalytic performance of the g-C_3N_4/porphyrin photocatalyst for the RhB dye's degradation was also studied and discussed.

Keywords: porphyrin-based nanomaterials; C_3N_4/TCPP hybrid material; dye photodegradation; enhanced photocatalysis

1. Introduction

The development of global economics has caused many environmental issues d the discarding of pollutants in wastewater, posing serious threats to human and nature's health [1–3]. Of the various pollutants disposed of in wastewater, organic dyes, originally derived from paper, food, mining, printing leather, and pharmaceutical industries, are highly toxic and contain carcinogenic agents and non-biodegradables. Most common organic dyes found in industrial wastewater can be listed as azo, phenothiazine, and triphenylmethane derivatives [4]. Among these, a commonly used fluorescent label or colorant in industries is rhodamine B (RhB). This organic dye is highly soluble in water and possesses extreme toxicity, being harmful to human health and the environment and posing risks such as teratogenicity, carcinogenicity, and mutagenicity [5–12]. Many strategies have been effectively utilized thus far for removing RhB dye from wastewater, such as redox degradation, photodegradation, coagulation, adsorption, and mixed approaches [13–15]. Each method has various advantages as well as disadvantages. Among them, the photodegradation process has been considered a promising pathway for RhB removal in aqueous media, using photon energy from the solar spectrum [16,17]. However, most photocatalysts for dye degradation can only harvest photon energy in ultraviolet light due to their large bandgap energy [18]. Therefore, there is an urgency to develop a photocatalyst that is able to absorb photon energy in the visible light region (approximately 42.3% in the solar spectrum).

Organic semiconductor nanostructures with low bandgap energy activated in the visible light region and high absorbing efficiency for photon energy have been extensively employed in photocatalysis. Among organic-based semiconductor photocatalysts, soft solid-state materials constructed from basic organic building blocks of π-conjugated porphyrin derivatives exhibit promising photocatalytic activity [19–25]. Many approaches have been effectively employed to prepare the π-conjugated organic semiconductors for photocatalysis applications [26,27]. Among these, the self-assembly of supramolecules is one of the most used methodologies to design and prepare soft nanostructured materials from monomeric organic molecules, including porphyrin derivatives. Porphyrin-based nanomaterials have been considered a new advanced nanomaterial for many applications, especially in photocatalysis. Porphyrin supramolecular nanostructures are commonly obtained via self-assembly from monomeric molecules and employed as photocatalysts. In order to enhance the photocatalytic activity, porphyrin nanostructures could be incorporated with other inorganic semiconductors to enable photocatalytic efficiency in a wide solar spectrum range and to improve the charge separation. Ionic self-assembly, reprecipitation, and surfactant-assisted self-assembly (SAS) are common self-assembly protocols used for the preparation of porphyrin nanostructures with controlled morphologies [28–31]. Based on each self-assembly technique, porphyrin monomers were designed and synthesized to obtain desired nanostructures. Self-assembled porphyrin nanostructures have been evidenced to be novel photocatalysts for application in environmental treatments [32]. For example, nanostructured porphyrin photocatalysts successfully obtained via self-assembly in the shape of microspheres, octahedra, and nanosheets exhibited remarkable performance for the photodegradation of several organic dyes [33]. However, free-standing nanostructured porphyrin photocatalysts have been hindered from practical application because of their low stability and quick recombining of electrons and holes. Thus, the incorporation of free-standing aggregates with other inorganic semiconductors could be a promising strategy to enhance charge separation, photon energy absorption, and stability. Our group also produced a range of porphyrin-based nanomaterials in regard to incorporate with several inorganic semiconductors, which revealed the enhanced photodegradation efficiency of organic dyes [30,34–37].

With low bandgap energy activated in the visible light region, high chemical and thermal stabilities, graphitic carbon nitride (g-C_3N_4) is an inorganic semiconductor, which has been extensively studied for photocatalytic applications such as water splitting, pollutant removal, photovoltaic solar cells, and chemical sensors [38–40]. Many approaches are available for the preparation of the g-C_3N_4 material, including, but not limited to, solution solvothermal methods, microwave heating-assisted approaches, mixing processes, hydrolysis, the sol-gel process, and mechanical grinding [41]. The g-C_3N_4-based nanocomposite showed high removal performance toward organic dyes in an aqueous solution [42]. However, a low surface area, fast recombining of generated electrons and holes, and a low photon energy-harvesting capability have limited the use of the g-C_3N_4 photocatalyst for pollutant degradation in practice [43]. To tackle these disadvantages, g-C_3N_4 could be doped (metal or nonmetal doping) or incorporated with other semiconductors to improve the photocatalytic efficiency. For example, the g-C_3N_4/$NiSO_4$ nanocomposite was facilely fabricated with a solvent evaporation method, which revealed a rapid removal of rhodamine B (RhB), Congo red (CR), methylene blue (MB), and indigo carmine (IC) under sunlight irradiation [44]. Durga et al. also incorporated g-C_3N_4 with $Ca_2Fe_2O_5$ to obtain the g-C_3N_4/$Ca_2Fe_2O_5$ nanocomposite [45]. The resultant nanocomposite exhibited an improved removal efficiency of up to approximately 96% toward methylene blue. However, most of the semiconductors incorporated with g-C_3N_4 were inorganic semiconductors; no organic semiconductors have been reported for incorporation with g-C_3N_4 for enhancing photocatalytic performance.

Herein, we propose a simple strategy to synthesize a g-C_3N_4/porphyrin hybrid material via the self-assembling of TCPP monomers with g-C_3N_4 nanomaterials. The resultant nanocomposite is well-characterized using scanning electron microscopy (SEM), FTIR spectroscopy, X-ray diffraction (XRD), and a UV-Vis spectrophotometer. The photodegradation evaluation of the synthesized nanomaterial was investigated for RhB removal. This work also proposes and discusses the plausible photodegradation mechanism of RhB using the g-C_3N_4/porphyrin photocatalyst.

2. Results and Discussion

The UV-Vis spectrophotometer is a reliable tool for investigating optical characteristics of materials. The optical properties of the TCPP monomers, free-standing assembled TCPP nanofibers, and g-C_3N_4/porphyrin materials were studied with UV–Vis spectra, and the result is shown in Figure 1. The UV-Vis spectra of the monomeric TCPP molecules revealed a sharp peak at approximately 412 nm, which was due to the transition of a_{1u} (π) to $e^*_g(\pi)$, which is the absorption of the Soret band. The a_{2u} (π) to $e^*_g(\pi)$ transition in porphyrin molecules was also presented, with the appearance of four weak Q-absorption bands ranging from 500 to 700 nm [46,47]. Interestingly, after the self-assembly, a strong peak at approximately 418 nm in the UV-Vis spectra of the TCPP aggregates was observed, which was assigned to the Soret absorption peaks, with a significant decrease in the peak's intensity, which was primarily evident for the self-assembly of the TCPP monomer. The distinct bathochromic and hypsochromic shift in the Soret absorption band from 412 nm to 418 nm (blue shift) before and after the self-assembly indicated the supramolecular assembly of the monomeric porphyrin molecules following the *J*-type aggregates [31,48,49]. Furthermore, four Q-absorption bands in the monomeric porphyrin molecule transformed into one band at approximately 664 nm after aggregation, which was considered further evidence for the *J*-type aggregates. After the TCPP monomers self-assembled on the surface of g-C_3N_4 the UV-Vis spectrum of the hybrid material showed similar characteristic peaks, as evidenced in the TCPP aggregates' UV-Vis spectra. The lower intensity of absorption peaks of the hybrid material's UV-Vis spectrum in comparison to that of the TCPP aggregates demonstrated that the TCPP aggregates were well-integrated with the g-C_3N_4 nanomaterials.

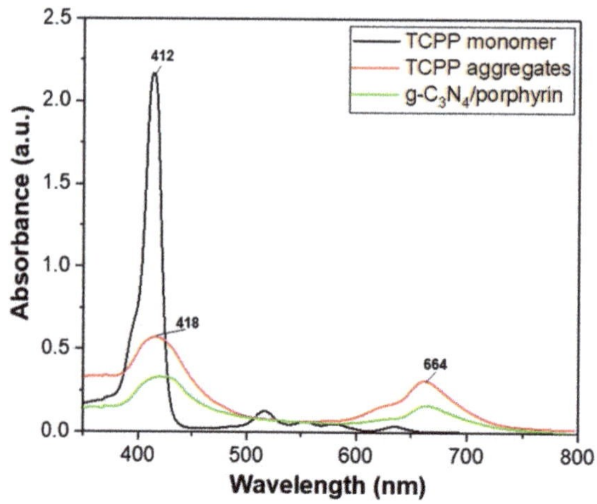

Figure 1. UV-Vis spectra of porphyrin monomer, free-standing self-assembled TCPP, and g-C_3N_4/porphyrin hybrid material.

The morphologies of the free-standing TCPP aggregates, g-C_3N_4, and g-C_3N_4/porphyrin hybrid material were studied using the SEM images (Figure 2). Figure 2a shows the fiber nanostructure of the free-standing TCPP aggregates of less than 100 nm in diameter and several micrometers in length. The $g\text{-}C_3N_4$ had a flake shape with a nanoscale thickness and a diameter of approximately 0.5 to 2 μm (Figure 2b). Figure 2c,d show the low and high SEM images of the $g\text{-}C_3N_4$/porphyrin. It could be observed that the assembled TCPP nanofibers were uniformly distributed on the surface of the $g\text{-}C_3N_4$ nanoflakes while retaining the morphologies of the free standing porphyrin nanofibers (Figure 2a) and $g\text{-}C_3N_4$ (Figure 2b). The result was in good agreement with the UV-Vis results, and confirmed the successful synthesis of a well-defined $g\text{-}C_3N_4$/porphyrin hybrid material via self-assembly. The EDS analysis of $g\text{-}C_3N_4$ revealed the atomic percentages of C and N to be 36.62 and 63.38%, respectively, (Figure S1). The self-assembly of $g\text{-}C_3N_4$ with porphyrin witnessed an increase in the C atomic percentage (44.06%) and a decrease in the N atomic percentage (46.22%), as evidenced for the presence of porphyrin in the composite (Figure S2). The O atomic percentage of approximately 8.6% was further evidence of the presence of porphyrin, as it was the main element in the TCPP porphyrin molecules.

Figure 2. SEM images of (**a**) free-standing porphyrin aggregates, (**b**) $g\text{-}C_3N_4$, and (**c**,**d**) $g\text{-}C_3N_4$/porphyrin nanofibers.

The bonding nature of the $g\text{-}C_3N_4$/porphyrin hybrid material was investigated with FTIR spectroscopy, and the result is shown in Figure 3. The FTIR spectrum exhibited a broad absorption band in the range of 3000 to 3500 cm^{-1}, which was attributed to the characteristic vibrations of the -OH and -COOH groups in porphyrin molecules and to the moisture. The intense vibration peaks at 1410 and 1640 cm^{-1} were attributed to the C-O and C=O bonds in the carbonyl groups in the porphyrin molecules, respectively [29]. Firm peaks ranging from 1200 to 1650 cm^{-1} were ascribed to the vibrations of the C-N heterocycles, as well as heptazine-derived repeating units [50]. The presence of vibration bands at 1321 and 1242 cm^{-1} were attributed to the stretching vibration modes of the C-NH-C bridge or C-N(-C) trigonal units in $g\text{-}C_3N_4$ [51]. The characteristic vibration of the tri-s-triazine modes was also observed, with the appearance of a stretching band at 812 cm^{-1}. These results further demonstrated that the assembled TCPP porphyrin nanofibers were well-incorporated with the $g\text{-}C_3N_4$ nanoparticles to form a hybrid material. The FTIR spectrum of the monomeric TCPP porphyrin monomers showed similar characteristic vibration bands of porphyrin aggregates in the $g\text{-}C_3N_4$/porphyrin hybrid material, with a small shift, which demonstrated the successful self-assembly of porphyrin monomers into the porphyrin nanofibers.

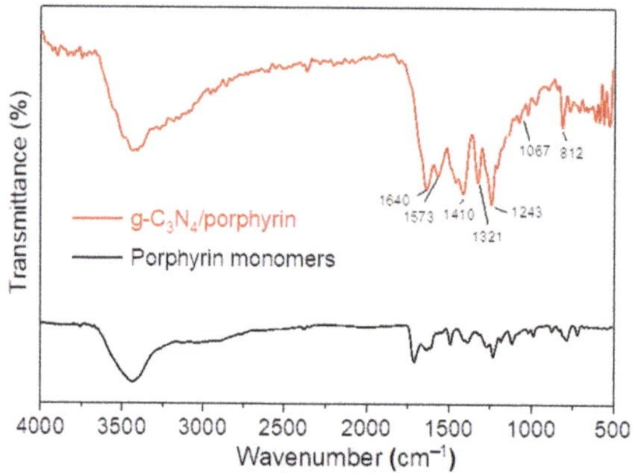

Figure 3. FTIR spectrum of TCPP porphyrin monomer and g-C_3N_4/porphyrin hybrid material.

The phase's characteristics and structure of the hybrid material were studied using the X-ray diffraction (XRD) technique. Illustrated in Figure 4 are the XRD patterns of the prepared samples. The XRD of the monomeric porphyrin molecule revealed no diffraction peaks, conveying the amorphous nature of the porphyrin monomer. The XRD pattern of g-C_3N_4 (black line) depicted diffraction peaks at approximately 13 and 27.8°, which were in good agreement with the g-C_3N_4 material reported previously [52–54]. The strong peak at 27.8° attributed to the plane of (002) in g-C_3N_4 resulted from the layer stacking of the aromatic system. The weak peak at approximately 13° was assigned to the (100) plane (JCPDS NO. 87-1526), which resulted from the tri-s-triazine unit stacking. The XRD pattern of the g-C_3N_4/porphyrin hybrid material also showed characteristic peaks of g-C_3N_4 at approximately 13 and 27.8°. The XRD pattern of the hybrid material showed a small shift in the diffraction peaks of g-C_3N_4, which might be attributed to the electronic interaction between porphyrin and g-C_3N_4. Intriguingly, several additional peaks with an asterisk at two theta values of 19 and 22° were also observed in the hybrid material's XRD pattern, which belonged to the porphyrin aggregates. This result demonstrated the crystalline nature of the porphyrin aggregates [30,36], which was additional confirmation for the successful self-assembling of the TCPP monomer on the surface of g-C_3N_4.

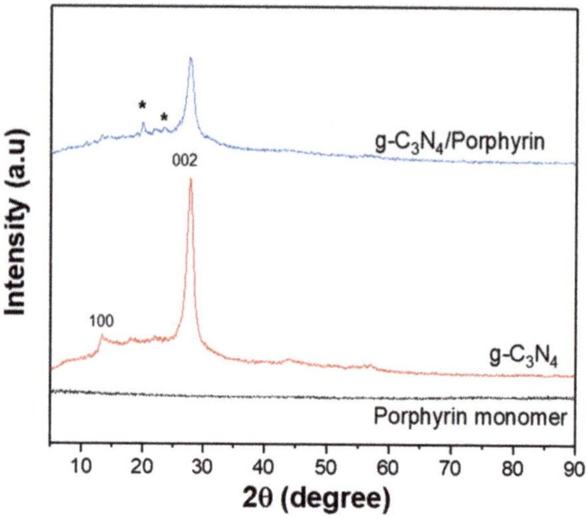

Figure 4. XRD patterns of TCPP porphyrin monomer, g-C_3N_4, and g-C_3N_4/porphyrin material.

The optical properties of the samples were further studied with the diffuse reflectance UV-Vis spectrophotometer. The result is depicted in Figure 5a. The spectrum of the free-standing g-C_3N_4 showed strong absorption in the ultraviolet region (300–400 nm). Interestingly, the diffuse reflectance UV-Vis spectrum of the g-C_3N_4/porphyrin hybrid material showed two broad peaks at approximately 450 nm and 650 nm, indicating that the hybrid material could favorably absorb light in the visible light and near-UV light region. This broad wavelength absorption in the light region was significantly favorable for photocatalytic applications. The difference between the photon energy and bandgaps was responsible for the optical absorption strength of the material, following the below equation:

$$(\alpha h\upsilon)^{1/n} = A(h\upsilon - E_g) \quad (1)$$

where E_g is the bandgap energy, α is the coefficient of the absorption band, h is Planck's constant (6626.10^{-34} Js), υ is the speed of light (3.108 ms^{-1}), and A is the absorption intensity. The Tauc plot was drawn based on $(\alpha h\upsilon)^2$ versus $h\upsilon$, and, then, the bandgap energy of the material was determined using the Tauc plot. Figure 5b shows the Tauc plots of the free-standing g-C_3N_4 and g-C_3N_4/porphyrin hybrid material. The bandgap energies of g-C_3N_4 were calculated from the Tauc plot to be approximately 2.91 eV, only able to harvest the energy of photon energy from ultraviolet light. From the Tauc plot, two bandgap energies of the g-C_3N_4/porphyrin hybrid material were determined to be 2.38 and 2.7 eV, which were favorable for harvesting photon energy from visible and near-ultraviolet light. This result implied that the successful integration of the g-C_3N_4 and TCPP porphyrin aggregates could create a novel photocatalyst with bandgap energies in a wide range of light regions, particularly in visible light.

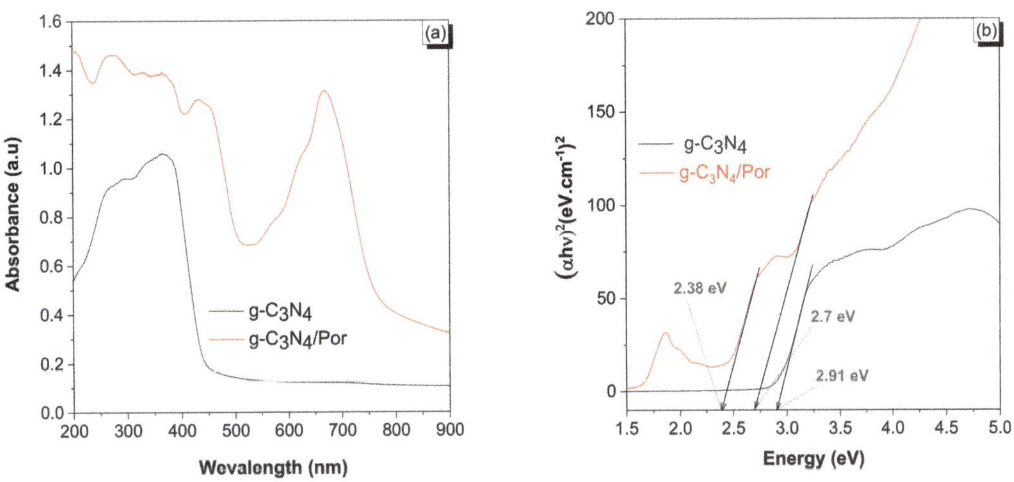

Figure 5. (**a**) The diffuse reflectance UV-Vis and (**b**) the Tauc plot of the free-standing g-C_3N_4 and g-C_3N_4/porphyrin hybrid material.

The photodegradation performance of the prepared g-C_3N_4/porphyrin hybrid material was investigated for RhB removal under simulated sunlight irradiation. Figure 6 reveals the effect of the g-C_3N_4 loading on the photodegradation behavior of the hybrid material. It was evident that the photodegradation performance of the g-C_3N_4/porphyrin hybrid material toward RhB was remarkably higher than that of the control, free-standing g-C_3N_4, and TCPP porphyrin aggregates. The RhB removal percentages of the free-standing g-C_3N_4 and TCPP aggregates were determined to be 38 and 42%, respectively, under photocatalytic experimental conditions. When g-C_3N_4 proportions in the g-C_3N_4/porphyrin increased, the photocatalytic degrading efficiency of RhB also increased. At a TCPP:g-C_3N_4 weight ratio of 1:1, the RhB removal percentage was approximately 71%, and enhanced significantly at a ratio of 1:2, with a degrading efficiency of 89%. The RhB degrading efficiency reached a maximum at a TCPP:g-C_3N_4 ratio of 1:3, with a removal percentage of 98%. When the g-C_3N_4 contents further increased, the removal efficiency of the RhB slightly decreased. Thus, the TCPP:g-C_3N_4 ratio of 1:3 was considered to be the optimal proportion of g-C_3N_4 in the hybrid material, and this ratio was selected for the preparation of a hybrid material for further photocatalytic studies.

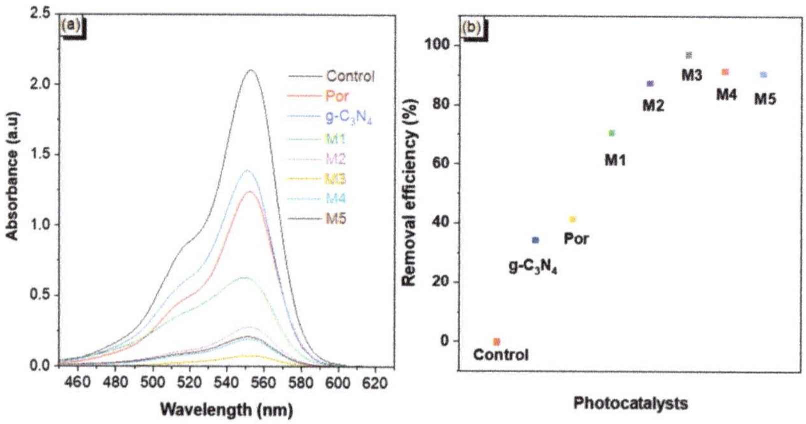

Figure 6. Effect of g-C_3N_4 contents on the photodegradation efficiency of the hybrid material with various TCPP aggregates: g-C_3N_4 weight ratios of 1:1, 1:2, 1:3, 1:4, and 1:6 labeled as M1, M2, M3, M4, and M5, respectively, under simulated sunlight irradiation for 100 min. (**a**) UV-vis spectra of remaining RhB concentration and (**b**) removal efficiencies with various photocatalysts.

The g-C_3N_4 with bandgap energies in the range of 2–3 eV has been extensively used as an effective photocatalyst under near-UV and visible light irradiation [55,56]. It has been evidenced that assembled porphyrin has a molecular structure similar to chlorophyll (photoactive molecules), which is a major component participating in many processes of biological energy transduction in algae and plants [57,58]. The bandgap energies of the hybrid material were determined to be 2.38 and 2.7 eV, indicating the effective harvesting of photon energy from a wide range of light regions, especially from the light in the visible light spectrum. Therefore, in this work, photocatalytic experiments were performed for the degradation of RhB using the prepared g-C_3N_4/porphyrin photocatalyst under simulated sunlight irradiation. The result is depicted in Figure 7. It can be seen from Figure 7a that the hybrid material exhibited remarkable photodegradation performance, with a RhB removal efficiency of almost 100% after irradiating with the simulated sunlight for 120 min. Illustrated in Figure 7b is the ln (C_t/C_0) vs. time plot for studying the photocatalytic reaction's kinetics, in which C_t is the absorption peak's intensity at time t, and C_0 is the absorption peak's intensity at time zero. The photocatalytic performance was evaluated through the changes in the RhB absorption peak at a wavelength of 553 nm as a function of time. Under experimental conditions, the negligible self-sensitized degradation of RhB was observed with no decrease in the concentration of RhB. When free-standing g-C_3N_4 and TCPP porphyrin aggregates were used as photocatalysts, the RhB concentrations remarkably decreased, with a rate constant of cal. 3.75×10^{-3} and 5×10^{-3} min^{-1}, respectively. Intriguingly, when the g-C_3N_4/porphyrin hybrid material was employed as a photocatalyst, the RhB removal percentage remarkably increased, with a rate constant calculated to be approximately 3.3×10^{-2} min^{-1}, which was approximately six- and ten-fold higher compared to the g-C_3N_4 and TCPP aggregates, respectively. In comparison to the other assembled porphyrin-based nanomaterials reported in the literature, this rate constant for the degradation of RhB using the g-C_3N_4/porphyrin hybrid material was higher, and much higher than the commercial photocatalyst of TiO_2 [30,34,36,59].

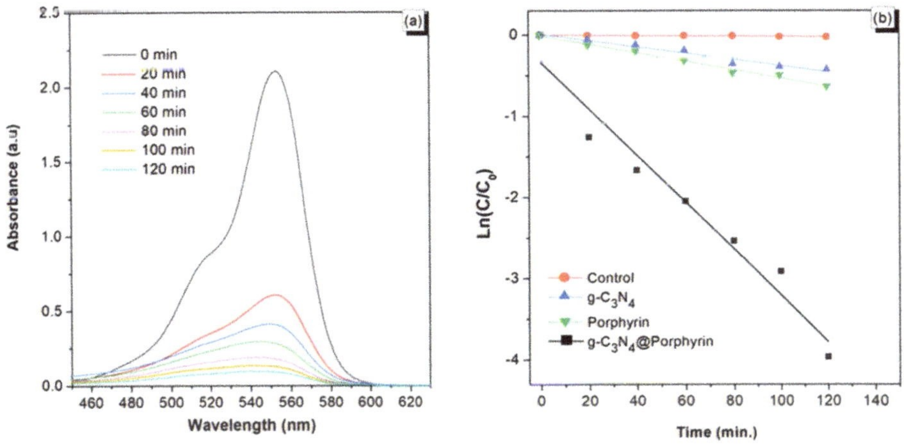

Figure 7. Effect of photocatalytic time on the removal efficiency of RhB using (**a**) the g-C_3N_4/porphyrin hybrid material prepared with a TCPP:g-C_3N_4 ratio of 1:3 (M3) as the photocatalyst and (**b**) the kinetics study of the photocatalysts under irradiation of simulated sunlight for 120 min.

It is well-known that J-type porphyrin aggregates are of a photo-semiconductor property, because of their robust π–π intermolecular interactions in the π- conjugate of the porphyrin molecules [20,31,37]. Due to the photo-semiconductor property, porphyrin aggregates via self-assembly can receive energy from photons to enable electrons to move to conduction bands from valence bands and form electron and hole pairs for the photocatalytic reaction [29,49,60]. Additionally, the photocatalytic performance of hybrid materials remarkably increased due to the improvement of charge separation induced through processes of an exciton-coupled charge transfer between the g-C_3N_4 and porphyrin aggregates [61,62]. The g-C_3N_4/porphyrin hybrid material had two bandgap energies of 2.38 and 2.7 eV, which enabled the photocatalyst to receive energy from photons in a wide range of the light spectrum, especially with visible light. Figure 8 proposes the possible photodegrading mechanism of the g-C_3N_4/porphyrin hybrid material for the removal of the RhB organic dye. When g-C_3N_4/porphyrin was irradiated with the sunlight region, the g-C_3N_4 and TCPP nanofibers harvested photon energy to enable the jumping of electrons from the valence band to the conduction band, generating electron/hole pairs [29]. Thanks to processes of the exciton-coupled charge transfer, newly generated electrons diffused from the conduction band of the g-C_3N_4 to the TCPP aggregates, and holes formed the valence band of the TCPP nanofibers to the g-C_3N_4 through the interface between the two materials. These transfer processes significantly avoided the electron and hole recombination; as a result, more electrons and holes could participate in the photocatalysis. Remarkably, the electrons reduced O_2 and H_2O to form free radicals, such as $^{\bullet}O_2^-$, $OH^{\bullet}$ … ; these free radicals oxidized the RhB dye to fewer toxic compounds. The holes also migrated and directly oxidized the RhB molecules to unharmed products.

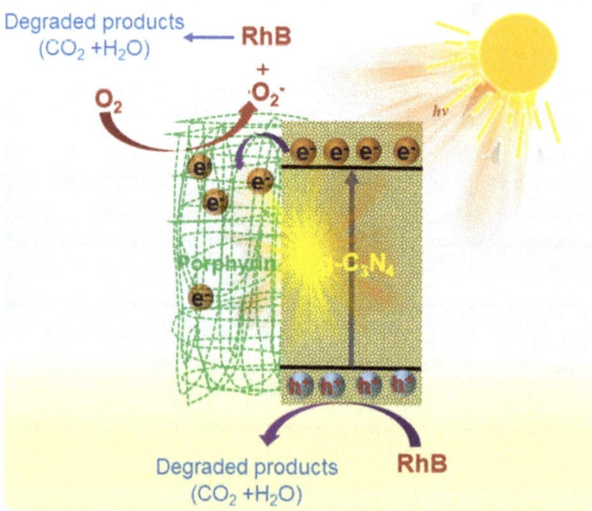

Figure 8. Plausible photodegradation mechanism of the g-C_3N_4/porphyrin for the rhodamine B removal.

3. Materials and Methods

3.1. Chemicals

Shanghai Macklin Biochemical Company provided monomeric tetrakis (4-carboxyphenyl) porphyrin (TCPP) molecules. Hydrochloric acid (HCl, 99%), ethanol (99%), sodium hydroxide (NaOH, 99%), and urea (CH_4N_2O, 99%) were purchased from Xilong Chemical Co., Ltd., Guangdong, China. All the chemicals without further purification and double-distilled water were employed for all experiments.

3.2. Preparation of g-C_3N_4 Nanomaterial

The g-C_3N_4 nanomaterial was prepared through the direct annealing of the urea compound. As in a typical process, a crucible with a cover containing 10 g of urea was heated at 550 °C for 3 h. The crucible was then naturally cooled to room temperature, after which a yellow powder was obtained as raw pristine. The raw g-C_3N_4 was washed with distilled water several times and ethanol to remove impurities. After that, the sample was completely dried at 70 °C for five hours.

3.3. Synthesis of g-C_3N_4/Porphyrin

An acid-based neutralization-induced self-assembly approach was employed to synthesize the g-C_3N_4@porphyrin hybrid material. As in a typical process, 8 mg TCPP was dissolved in 2 mL NaOH 0.1M to obtain solution A. Then, different proportions of the g-C_3N_4 nanomaterial were added into solution A, with various TCPP:g-C_3N_4 weight ratios of 0.5:1, 1:1, 1:2, 1:4, and 1:6. The resultant mixture was thoroughly mixed for 20 min through bath sonicating to obtain a homogeneous mixture of the g-C_3N_4 and porphyrin monomers. HCl 0.1M was gradually introduced to the mixture to induce the reprecipitation self-assembly of the porphyrin nanofibers on the surface of g-C_3N_4. The pH value was then adjusted to a pH of 6–7. The reprecipitates were collected with the use of centrifugation, thoroughly washed with double-distilled water, and dried at 60 °C for 12 h. The samples were stored at room temperature, in ambient humidity for other testing.

3.4. Material Characterization Techniques

The X′Pert PRO PANalytical instrument with a 0.15405 nm Cu-Kα radiation source was employed to obtain an X-ray diffraction (XRD) from the investigation of the structure and crystallinity of the C_3N_4@porphyrin hybrid material. Scanning electron microscopy and energy dispersive X-ray spectroscopy (SEM-EDX, Hitachi S-4600, Fukuoka, Japan) were utilized to observe the morphology and the elemental composition of the sample. FTIR spectra were measured in a TENSOR II (Bruker) IR spectrometer by using KBr pellets. A UV-vis spectrophotometer (Jasco V730) in the range of 200–900 nm was used to obtain the optical absorption spectra of the materials. A UV-vis spectrophotometer was also used to study the photodegradation efficiency of the g-C_3N_4@porphyrin hybrid materials to remove RhB dye at an absorption wavelength of 553 nm.

3.5. Photocatalytic Experiments

The photodegradation efficiency of the g-C_3N_4@porphyrin hybrid materials for the removal of a simulated pollutant using RhB organic dye was studied under simulated sunlight irradiation generated from a 350 W xenon lamp (China, 350 W). Following the typical procedure, 0.4 mg of the g-C_3N_4@porphyrin hybrid material was introduced into 20 mL of a 10 ppm RhB solution. The adsorption equilibrium state of the hybrid material was established by placing the mixture in the dark for 2 h, before irradiating with light. The mixture was taken out after each time point and the UV–Vis spectra recorded for each sample. For comparison, the photodegradation performance of the free-standing g-C_3N_4 and TCPP nanofibers in the removal of RhB was obtained following the same photocatalytic protocol with the hybrid material.

4. Conclusions

A facile, self-assembling approach was successfully employed to fabricate the g-C_3N_4/porphyrin hybrid material through self-assembling TCPP porphyrin monomers on g-C_3N_4 nanoflake surfaces. The prepared porphyrin aggregates in a fiber shape of less than 100 nm in diameter and several micrometers in length were shown to be well-integrated in the g-C_3N_4 nanomaterials. The TCPP porphyrin aggregates were determined to form on the g-C_3N_4's via *J*-type self-assembling pathway. The bandgap energies of the g-C_3N_4/porphyrin hybrid material were calculated to be 2.38 and 2.7 eV, which could make the hybrid material an efficient photocatalyst in a broad light spectrum range. Obviously, the hybrid material exhibited an enhanced photodegradation efficiency toward the rhodamine B dye with a degrading speed of up to 3.3×10^{-2} min^{-1}, which was superior to other porphyrin-based photocatalysts reported previously. The enhanced photodegradation efficiency was due to the improved harvesting of photon energy, as well as the enhanced charge separation due to the exciton coupled charge transfer processes between the g-C_3N_4 and TCPP aggregates. With a facile fabricating method and highly photocatalytic performance, the g-C_3N_4/porphyrin hybrid material is considered an advanced photocatalyst for many applications, such as environmental remediation and water splitting.

Supplementary Materials: The following supporting information can be downloaded at: https://www.mdpi.com/article/10.3390/catal12121630/s1, Figure S1: EDS spectrum of the g-C3N4; Figure S2: EDS spectrum of the g-C_3N_4/porphyrin.

Author Contributions: Conceptualization: D.D.N., D.D.L. and C.V.T.; methodology: T.T.N., D.D.L. and H.P.N.T.; software: W.J.C. and H.T.L.; investigation: H.T.L., G.T.N., N.T.T. and D.K.N.; data curation: T.T.N. and S.W.C.; writing—original draft: H.T.L. and D.D.L.; writing—reviewing scientific contents and editing: D.D.N., D.D.L. and C.V.T. All authors have read and agreed to the published version of the manuscript.

Funding: This research was funded by the University of Transport and Communications (UTC) under grant number T2022-CB-005.

Data Availability Statement: Not applicable.

Acknowledgments: The authors thank the staff at the Institute of Chemistry and Materials for the support in characterizing and analyses. We also thank the institute for all facilities we used to carry out the work.

Conflicts of Interest: The authors declare no conflict of interest.

References

1. Liang, L.; Cheng, L.; Zhang, Y.; Wang, Q.; Wu, Q.; Xue, Y.; Meng, X. Efficiency and mechanisms of rhodamine B degradation in Fenton-like systems based on zero-valent iron. *RSC Adv.* **2020**, *10*, 28509–28515. [CrossRef] [PubMed]
2. Sandoval, A.; Hernández-Ventura, C.; Klimova, T.E. Titanate nanotubes for removal of methylene blue dye by combined adsorption and photocatalysis. *Fuel* **2017**, *198*, 22–30. [CrossRef]
3. Kumar, S.S.; Kumar, V.; Malyan, S.K.; Sharma, J.; Mathimani, T.; Maskarenj, M.S.; Ghosh, P.C.; Pugazhendhi, A. Microbial fuel cells (MFCs) for bioelectrochemical treatment of different wastewater streams. *Fuel* **2019**, *254*, 115526. [CrossRef]
4. Manzoor, J.; Sharma, M. Impact of Textile Dyes on Human Health and Environment. In *Impact of Textile Dyes on Public Health and the Environment*; IGI Global: Hershey, PA, USA, 2020; pp. 162–169.
5. Daneshvar, N.; Salari, D.; Khataee, A. Photocatalytic degradation of azo dye acid red 14 in water: Investigation of the effect of operational parameters. *J. Photochem. Photobiol. A Chem.* **2003**, *157*, 111–116. [CrossRef]
6. Ledakowicz, S.; Gonera, M. Optimisation of oxidants dose for combined chemical and biological treatment of textile wastewater. *Water Res.* **1999**, *33*, 2511–2516. [CrossRef]
7. Nestmann, E.R.; Douglas, G.R.; Matula, T.I.; Grant, C.E.; Kowbel, D.J. Mutagenic activity of rhodamine dyes and their impurities as detected by mutation induction in Salmonella and DNA damage in Chinese hamster ovary cells. *Cancer Res.* **1979**, *39*, 4412–4417.
8. Salleh, M.A.M.; Mahmoud, D.K.; Karim, W.A.W.A.; Idris, A. Cationic and anionic dye adsorption by agricultural solid wastes: A comprehensive review. *Desalination* **2011**, *280*, 1–13. [CrossRef]
9. Merouani, S.; Hamdaoui, O.; Saoudi, F.; Chiha, M.; Pétrier, C. Influence of bicarbonate and carbonate ions on sonochemical degradation of Rhodamine B in aqueous phase. *J. Hazard. Mater.* **2010**, *175*, 593–599. [CrossRef]
10. Mostafa Hosseini Asl, S.; Ghadi, A.; Sharifzadeh Baei, M.; Javadian, H.; Maghsudi, M.; Kazemian, H. Porous catalysts fabricated from coal fly ash as cost-effective alternatives for industrial applications: A review. *Fuel* **2018**, *217*, 320–342. [CrossRef]
11. Muhmood, T.; Xia, M.; Lei, W.; Wang, F. Under vacuum synthesis of type-I heterojunction between red phosphorus and graphene like carbon nitride with enhanced catalytic, electrochemical and charge separation ability for photodegradation of an acute toxicity category-III compound. *Appl. Catal. B Environ.* **2018**, *238*, 568–575. [CrossRef]
12. Muhmood, T.; Xia, M.; Lei, W.; Wang, F.; Khan, M.A. Efficient and stable ZrO_2/Fe modified hollow-C_3N_4 for photodegradation of the herbicide MTSM. *RSC Adv.* **2017**, *7*, 3966–3974. [CrossRef]
13. Bulut, E.; Özacar, M.; Şengil, İ.A. Equilibrium and kinetic data and process design for adsorption of Congo Red onto bentonite. *J. Hazard. Mater.* **2008**, *154*, 613–622. [CrossRef]
14. Li, Z.-J.; Wang, L.; Yuan, L.-Y.; Xiao, C.-L.; Mei, L.; Zheng, L.-R.; Zhang, J.; Yang, J.-H.; Zhao, Y.-L.; Zhu, Z.-T. Efficient removal of uranium from aqueous solution by zero-valent iron nanoparticle and its graphene composite. *J. Hazard. Mater.* **2015**, *290*, 26–33. [CrossRef]
15. Chen, Z.; Wang, T.; Jin, X.; Chen, Z.; Megharaj, M.; Naidu, R. Multifunctional kaolinite-supported nanoscale zero-valent iron used for the adsorption and degradation of crystal violet in aqueous solution. *J. Colloid Interface Sci.* **2013**, *398*, 59–66. [CrossRef]
16. Bhosale, S.V.; La, D.D. Nanoscale porphyrin superstructures: Properties, self-assembly and photocatalytic applications. In *Nanoscience*; Royal Society of Chemistry: London, UK, 2018; pp. 57–85.
17. Du, J.; Bao, J.; Liu, Y.; Kim, S.H.; Dionysiou, D.D. Facile preparation of porous Mn/Fe_3O_4 cubes as peroxymonosulfate activating catalyst for effective bisphenol A degradation. *Chem. Eng. J.* **2019**, *376*, 119193. [CrossRef]
18. Zhu, X.; Yuan, W.; Lang, M.; Zhen, G.; Zhang, X.; Lu, X. Novel methods of sewage sludge utilization for photocatalytic degradation of tetracycline-containing wastewater. *Fuel* **2019**, *252*, 148–156. [CrossRef]
19. Sakakibara, K.; Hill, J.P.; Ariga, K. Thin-Film-Based Nanoarchitectures for Soft Matter: Controlled Assemblies into Two-Dimensional Worlds. *Small* **2011**, *7*, 1288–1308. [CrossRef]
20. Würthner, F.; Kaiser, T.E.; Saha-Möller, C.R. J-Aggregates: From Serendipitous Discovery to Supramolecular Engineering of Functional Dye Materials. *Angew. Chem. Int. Ed.* **2011**, *50*, 3376–3410. [CrossRef]
21. Zhang, C.; Chen, P.; Dong, H.; Zhen, Y.; Liu, M.; Hu, W. Porphyrin Supramolecular 1D Structures via Surfactant-Assisted Self-Assembly. *Adv. Mater.* **2015**, *27*, 5379–5387. [CrossRef]
22. Drain, C.M.; Varotto, A.; Radivojevic, I. Self-organized porphyrinic materials. *Chem. Rev.* **2009**, *109*, 1630–1658. [CrossRef]
23. Elemans, J.A.; van Hameren, R.; Nolte, R.J.; Rowan, A.E. Molecular Materials by Self-Assembly of Porphyrins, Phthalocyanines, and Perylenes. *Adv. Mater.* **2006**, *18*, 1251–1266. [CrossRef]
24. Hoeben, F.J.; Jonkheijm, P.; Meijer, E.; Schenning, A.P. About supramolecular assemblies of π-conjugated systems. *Chem. Rev.* **2005**, *105*, 1491–1546. [CrossRef] [PubMed]

25. Lee, D.; Kim, K.-D.; Lee, Y.-K. Conversion of V-porphyrin in asphaltenes into V_2S_3 as an active catalyst for slurry phase hydrocracking of vacuum residue. *Fuel* **2020**, *263*, 116620. [CrossRef]
26. Ngai, J.H.; Chan, C.C.M.; Ho, C.H.Y.; Ho, J.K.W.; Cheung, S.H.; Yin, H.; So, S.K. A facile and robust approach to prepare fluorinated polymer dielectrics for probing the intrinsic transport behavior of organic semiconductors. *Mater. Adv.* **2020**, *1*, 891–898. [CrossRef]
27. Marrocchi, A.; Facchetti, A.; Lanari, D.; Petrucci, C.; Vaccaro, L. Current methodologies for a sustainable approach to π-conjugated organic semiconductors. *Energ. Environ. Sci.* **2016**, *9*, 763–786. [CrossRef]
28. Wang, Z.; Medforth, C.J.; Shelnutt, J.A. Porphyrin nanotubes by ionic self-assembly. *J. Am. Chem. Soc.* **2004**, *126*, 15954–15955. [CrossRef]
29. La, D.D.; Bhosale, S.V.; Jones, L.A.; Revaprasadu, N.; Bhosale, S.V. Fabrication of a Graphene@ TiO_2@ Porphyrin hybrid material and its photocatalytic properties under simulated sunlight irradiation. *ChemistrySelect* **2017**, *2*, 3329–3333. [CrossRef]
30. La, D.D.; Bhosale, S.V.; Jones, L.A.; Bhosale, S.V. Arginine-induced porphyrin-based self-assembled nanostructures for photocatalytic applications under simulated sunlight irradiation. *Photochem. Photobiol. Sci.* **2017**, *16*, 151–154. [CrossRef]
31. Guo, P.; Chen, P.; Ma, W.; Liu, M. Morphology-dependent supramolecular photocatalytic performance of porphyrin nanoassemblies: From molecule to artificial supramolecular nanoantenna. *J. Mater. Chem.* **2012**, *22*, 20243–20249. [CrossRef]
32. La, D.D.; Ngo, H.H.; Nguyen, D.D.; Tran, N.T.; Vo, H.T.; Nguyen, X.H.; Chang, S.W.; Chung, W.J.; Nguyen, M.D.-B. Advances and prospects of porphyrin-based nanomaterials via self-assembly for photocatalytic applications in environmental treatment. *Coord. Chem. Rev.* **2022**, *463*, 214543. [CrossRef]
33. Zhong, Y.; Wang, Z.; Zhang, R.; Bai, F.; Wu, H.; Haddad, R.; Fan, H. Interfacial self-assembly driven formation of hierarchically structured nanocrystals with photocatalytic activity. *ACS Nano* **2014**, *8*, 827–833. [CrossRef]
34. La, D.; Hangarge, R.; Bhosale, S.V.; Ninh, H.; Jones, L.; Bhosale, S. Arginine-mediated self-assembly of porphyrin on graphene: A photocatalyst for degradation of dyes. *Appl. Sci.* **2017**, *7*, 643. [CrossRef]
35. La, D.D.; Rananaware, A.; Salimimarand, M.; Bhosale, S.V. Well–dispersed assembled porphyrin nanorods on graphene for the enhanced photocatalytic performance. *ChemistrySelect* **2016**, *1*, 4430–4434. [CrossRef]
36. La, D.D.; Rananaware, A.; Thi, H.P.N.; Jones, L.; Bhosale, S.V. Fabrication of a TiO_2@ porphyrin nanofiber hybrid material: A highly efficient photocatalyst under simulated sunlight irradiation. *Adv. Nat. Sci. Nanosci. Nanotechnol.* **2017**, *8*, 015009.
37. Aljabri, M.D.; La, D.D.; Jadhav, R.W.; Jones, L.A.; Nguyen, D.D.; Chang, S.W.; Dai Tran, L.; Bhosale, S.V. Supramolecular nanomaterials with photocatalytic activity obtained via self-assembly of a fluorinated porphyrin derivative. *Fuel* **2019**, *254*, 115639. [CrossRef]
38. Safaei, J.; Mohamed, N.A.; Noh, M.F.M.; Soh, M.F.; Ludin, N.A.; Ibrahim, M.A.; Isahak, W.N.R.W.; Teridi, M.A.M. Graphitic carbon nitride (g-C_3N_4) electrodes for energy conversion and storage: A review on photoelectrochemical water splitting, solar cells and supercapacitors. *J. Mater. Chem. A* **2018**, *6*, 22346–22380. [CrossRef]
39. Huang, Z.; Zeng, X.; Li, K.; Gao, S.; Wang, Q.; Lu, J. Z-scheme $NiTiO_3$/g-C_3N_4 heterojunctions with enhanced photoelectrochemical and photocatalytic performances under visible LED light irradiation. *ACS Appl. Mater. Interfaces* **2017**, *9*, 41120–41125. [CrossRef]
40. Malik, R.; Tomer, V.K.; Chaudhary, V.; Dahiya, M.S.; Sharma, A.; Nehra, S.; Duhan, S.; Kailasam, K. An excellent humidity sensor based on In–SnO_2 loaded mesoporous graphitic carbon nitride. *J. Mater. Chem. A* **2017**, *5*, 14134–14143. [CrossRef]
41. Ahmaruzzaman, M.; Mishra, S.R. Photocatalytic performance of g-C_3N_4 based nanocomposites for effective degradation/removal of dyes from water and wastewater. *Mater. Res. Bull.* **2021**, *143*, 111417. [CrossRef]
42. Mohanta, D.; Mahanta, A.; Mishra, S.R.; Jasimuddin, S.; Ahmaruzzaman, M. Novel SnO_2@ ZIF-8/gC_3N_4 nanohybrids for excellent electrochemical performance towards sensing of p-nitrophenol. *Environ. Res.* **2021**, *197*, 111077. [CrossRef]
43. Muhmood, T.; Khan, M.A.; Xia, M.; Lei, W.; Wang, F.; Ouyang, Y. Enhanced photo-electrochemical, photo-degradation and charge separation ability of graphitic carbon nitride (g-C_3N_4) by self-type metal free heterojunction formation for antibiotic degradation. *J. Photochem. Photobiol. A Chem.* **2017**, *348*, 118–124. [CrossRef]
44. Yang, S.; Sun, Q.; Han, W.; Shen, Y.; Ni, Z.; Zhang, S.; Chen, L.; Zhang, L.; Cao, J.; Zheng, H. A simple and highly efficient composite based on gC 3 N 4 for super rapid removal of multiple organic dyes from water under sunlight. *Catal. Sci. Technol.* **2022**, *12*, 786–798. [CrossRef]
45. Vavilapalli, D.S.; Peri, R.G.; Sharma, R.; Goutam, U.; Muthuraaman, B.; Rao, R.; Singh, S. g-C_3N_4/$Ca_2Fe_2O_5$ heterostructures for enhanced photocatalytic degradation of organic effluents under sunlight. *Sci. Rep.* **2021**, *11*, 19639. [CrossRef] [PubMed]
46. Sun, J.; Meng, D.; Jiang, S.; Wu, G.; Yan, S.; Geng, J.; Huang, Y. Multiple-bilayered RGO–porphyrin films: From preparation to application in photoelectrochemical cells. *J. Mater. Chem.* **2012**, *22*, 18879–18886. [CrossRef]
47. Chen, Y.; Zhang, C.; Zhang, X.; Ou, X.; Zhang, X. One-step growth of organic single-crystal p–n nano-heterojunctions with enhanced visible-light photocatalytic activity. *Chem. Commun.* **2013**, *49*, 9200–9202. [CrossRef]
48. Guo, P.; Chen, P.; Liu, M. One-dimensional porphyrin nanoassemblies assisted via graphene oxide: Sheetlike functional surfactant and enhanced photocatalytic behaviors. *ACS Appl. Mater. Interfaces* **2013**, *5*, 5336–5345. [CrossRef]
49. Kano, H.; Kobayashi, T. Time-resolved fluorescence and absorption spectroscopies of porphyrin J-aggregates. *J. Chem. Phys.* **2002**, *116*, 184–195. [CrossRef]

50. Lotsch, B.V.; Döblinger, M.; Sehnert, J.; Seyfarth, L.; Senker, J.; Oeckler, O.; Schnick, W. Unmasking melon by a complementary approach employing electron diffraction, solid-state NMR spectroscopy, and theoretical calculations-structural characterization of a carbon nitride polymer. *Chem. A Eur. J.* **2007**, *13*, 4969–4980. [CrossRef]
51. Liu, J.; Zhang, T.; Wang, Z.; Dawson, G.; Chen, W. Simple pyrolysis of urea into graphitic carbon nitride with recyclable adsorption and photocatalytic activity. *J. Mater. Chem.* **2011**, *21*, 14398–14401. [CrossRef]
52. Zhou, D.; Qiu, C. Study on the effect of Co doping concentration on optical properties of g-C_3N_4. *Chem. Phys. Lett.* **2019**, *728*, 70–73. [CrossRef]
53. Li, Y.; Wu, S.; Huang, L.; Wang, J.; Xu, H.; Li, H. Synthesis of carbon-doped g-C_3N_4 composites with enhanced visible-light photocatalytic activity. *Mater. Lett.* **2014**, *137*, 281–284. [CrossRef]
54. Xu, M.; Han, L.; Dong, S. Facile fabrication of highly efficient g-C_3N_4/Ag_2O heterostructured photocatalysts with enhanced visible-light photocatalytic activity. *ACS Appl. Mater. Interfaces* **2013**, *5*, 12533–12540. [CrossRef]
55. Liu, X.; Ma, R.; Zhuang, L.; Hu, B.; Chen, J.; Liu, X.; Wang, X. Recent developments of doped g-C_3N_4 photocatalysts for the degradation of organic pollutants. *Crit. Rev. Environ. Sci. Technol.* **2021**, *51*, 751–790. [CrossRef]
56. Wen, J.; Xie, J.; Chen, X.; Li, X. A review on g-C_3N_4-based photocatalysts. *Appl. Surf. Sci.* **2017**, *391*, 72–123. [CrossRef]
57. McConnell, I.; Li, G.; Brudvig, G.W. Energy conversion in natural and artificial photosynthesis. *Chem. Biol.* **2010**, *17*, 434–447. [CrossRef]
58. Barber, J. Photosynthetic energy conversion: Natural and artificial. *Chem. Soc. Rev.* **2009**, *38*, 185–196. [CrossRef]
59. La, D.D.; Tran, C.V.; Hoang, N.T.; Ngoc, M.D.D.; Nguyen, T.P.; Vo, H.T.; Ho, P.H.; Nguyen, T.A.; Bhosale, S.V.; Nguyen, X.C. Efficient photocatalysis of organic dyes under simulated sunlight irradiation by a novel magnetic $CuFe_2O_4$@ porphyrin nanofiber hybrid material fabricated via self-assembly. *Fuel* **2020**, *281*, 118655. [CrossRef]
60. Meadows, P.J.; Dujardin, E.; Hall, S.R.; Mann, S. Template-directed synthesis of silica-coated J-aggregate nanotapes. *Chem. Commun.* **2005**, *29*, 3688–3690. [CrossRef]
61. Muhmood, T.; Cai, Z.; Lin, S.; Xiao, J.; Hu, X.; Ahmad, F. Graphene/graphitic carbon nitride decorated with AgBr to boost photoelectrochemical performance with enhanced catalytic ability. *Nanotechnology* **2020**, *31*, 505602. [CrossRef]
62. Muhmood, T.; Xia, M.; Lei, W.; Wang, F.; Khan, M.A. Design of graphene nanoplatelet/graphitic carbon nitride heterojunctions by vacuum tube with enhanced photocatalytic and electrochemical response. *Eur. J. Inorg. Chem.* **2018**, *2018*, 1726–1732. [CrossRef]

catalysts

MDPI

Article

Enhanced Removal of Organic Dyes Using Co-Catalytic Ag-Modified ZnO and TiO_2 Sol-Gel Photocatalysts

Nina Kaneva *, Assya Bojinova and Karolina Papazova

Laboratory of Nanoparticle Science and Technology, Department of General and Inorganic Chemistry, Faculty of Chemistry and Pharmacy, University of Sofia, 1 James Bourchier Blvd., 1164 Sofia, Bulgaria
* Correspondence: nina_k@abv.bg

Abstract: Zinc oxide and titanium dioxide semiconductor photocatalysts have been widely utilized in the last few decades for water treatment because of their high photocatalytic efficiency. Recently, a lot of researchers have focused on the improvement of the photocatalytic properties of catalysts through modifying and co-modifying them with different metals and nonmetals. These co-catalytic ions improve the photocatalytic activity of ZnO and TiO_2 by reducing its energy band gap. This might be useful in wastewater treatment for the photocatalytic degradation of organic contaminants. In this study, we prepared semiconductor films that were surface-modified with Ag co-catalyst layers via the photo-fixation of Ag (I) ions with varied concentrations (10^{-2}–10^{-4} M) in the water phase under UV illumination for the first time. The photocatalytic behavior was evaluated by the degradation of malachite green and methylene blue under UV and visible light irradiation. The ZnO/Ag and TiO_2/Ag samples showed a faster degradation of malachite green compared to methylene blue due to the formation of stable intermediates by the reaction of OH radicals with the triarylmethane dye (C=C bond) during the photocatalysis. The co-catalytic-silver-modified films had a higher photocatalytic efficiency in comparison with the pure nanostructures. The dye photodegradation rate constants increased in the following order: pure films < films modified with Ag, 10^{-4} M < films modified with Ag10^{-3} M < films modified with Ag10^{-2} M. The Ag modification and the heterojunction of the composites contributed to trapping and transfer of the electrons. Therefore, the photogenerated charges had a longer lifetime, resulting in a strengthened photocatalytic ability of the ZnO/Ag and TiO_2/Ag films.

Keywords: ZnO; TiO_2; sol–gel; co-catalytic silver modification; photocatalysis

1. Introduction

Dyes are one of the most dangerous pollutants used in many industrial sectors, such as the textiles, clinical substances, and food and cosmetics industries. They are difficult to degrade by conventional methods [1]. Therefore, the World Health Organization and the European Union recommend maximum doses that should not be exceeded when using these dyes. The reason for this is to prevent damage to human and animal health in the environment [2]. In recent years, scientific researchers have focused on developing new techniques and methods to overcome these problems, which are related to the purification of wastewater from industries, homes and businesses. Water pollution is a major problem considering the increasing demand due to the increasing population, global warming and water scarcity in arid regions [3].

Nowadays, advanced oxidation processes (AOPs) are used much more because they make the degradation of pollutants into easily degradable by-products or their complete mineralization to CO_2 and H_2O possible. These processes are based on the formation of superoxide ions and hydroxyl radicals with an oxidizing power greater than that of traditional oxidants such as Cl_2 or O_3. These radicals can partially or completely mineralize most organic compounds [4,5].

Some of the main water pollutants include pharmaceutical drugs, phenol compounds, detergents, herbicides, pesticides, surfactants and dyes [6–10]. Our goal in this study was to eliminate two organic pollutants, namely malachite green (MG) and methylene blue (MB), using AOPs under UV irradiation. These substances belong to the group of triarylmethane dyes. The two dyes are most commonly used as colorants in various fields such as the textile, paper, pharmaceutical and food industries or in diagnostics, training and research laboratories. Despite all their positives, these dyes are known to have genotoxic and carcinogenic properties [11].

Heterogeneous photocatalysis is a promising method for the removal of many organic materials and the conversion of toxic inorganic materials into harmless compounds [12]. In their nanocrystalline form, titanium dioxide and zinc oxide are well-known semiconductors with photocatalytic activities, and they have great potential for applications such as in environmental purification, pigments, catalyst supports, fillers, coatings, photoconductors, dielectric materials and the generation of hydrogen gas. Their involvement in all these applications is due to their important characteristics such as their non-toxicity, high photosensitivity, low cost, high photostability towards light, suitable electronic band structures, catalytic activity and oxidative power and good transparency as thin films [13]. Therefore, they are effective photocatalysts because their photoinduced electron–hole pairs (e^-/h^+) are strong oxidizing and reducing agents for pollutant decomposition. However, TiO_2 and ZnO catalysts are partially limited by their wide band gap energy (Eg > 3 eV), fast recombination and low charge transfer rates of their photoinduced electron–hole pairs [14]. The heterojunction between semiconductor catalysts via properly matching band energy levels can enhance light absorption to the UV–visible region and enable the efficient separation of photo-induced charge carriers [15]. Several metals (Cu, Au, Pt and Ag) have been used to improve the photocatalytic activity of TiO_2 and ZnO [16]. Among them, Ag has proven its ability to reduce the recombination of photogenerated pairs, thereby extending their lifetime [17]. The reason for this is that (i) the Fermi level of Ag is lower than that of semiconductors; (ii) when Ag is on the surface of a catalyst, it acts as an electron trap and (iii) these e^- can enhance the rate of oxygen photoreduction, increasing the amount of photogenerated hydroxyl radicals [18]. All this leads to an improved photocatalytic efficiency of the semiconductors. ZnO- and TiO_2-nanostructured films can be prepared by different methods, and the most important are metal–organic chemical vapor deposition [19], sol–gel [20], magnetron sputtering [21] and atomic layer deposition [22]. Among all these methods, sol–gel is the simplest and most advanced process [23].

Here, we report a new, simple and inexpensive sol–gel method for the preparation of co-catalyst layers of TiO_2 and ZnO films with silver-modified surfaces using, for the first time, a combination of a dip-coating technique and the photo-fixation of Ag (I) ions with different concentrations in water under ultraviolet irradiation. The optical and structural properties of the synthesized materials were characterized by scanning electron microscopy (SEM), X-ray diffraction (XRD), X-ray photoelectron spectroscopy (XPS), electron spin resonance (EPR) and ultraviolet–visible spectroscopy (UV–Vis). The relationship between all these properties and the photocatalytic efficiency of the pure and modified films was clearly established. The influence of the presence of Ag^+ on the photocatalytic properties of the semiconductors for the degradation of malachite green and methylene blue was also investigated using UV–vis and TOC analysis.

2. Results and Discussion

2.1. Structural Characterization

The crystal structure of the prepared pure semiconductor (TiO_2, ZnO) and co-catalytic-silver (10^{-2} M)-modified films were identified by XRD analysis (Figure 1).

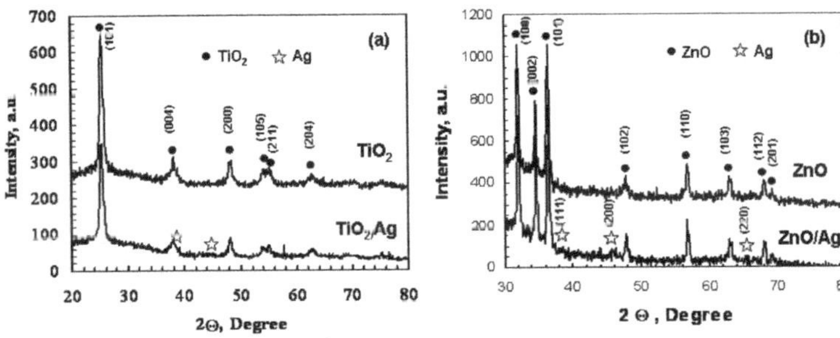

Figure 1. XRD patterns of semiconductor films: TiO_2 and TiO_2/Ag (**a**), ZnO and ZnO/Ag (**b**).

The recorded peaks indicated that the main phase of TiO_2 and TiO_2/Ag films was anatase, as indicated in the standard JCPDS card (96-500-0224). The intense diffraction peaks at 25.45°, 38.35°, 48.35°, 53.91°, 55.07° and 63.2° corresponded to the crystal planes (101), (004), (200), (105), (211) and (204), respectively. The modifying with silver did not alter the anatase structure of the TiO_2/Ag films, as shown in Figure 1a. For this reason, the modified TiO_2 film did not present additional peaks, which suggested that the Ag was incorporated into the TiO_2 structure. Similar patterns have been reported for co-catalytic TiO_2 modified with silver ions [24].

The X-ray patterns for the pure and silver-modified ZnO films (Figure 1b) were in concordance with the characteristic peaks of ZnO with a hexagonal structure, and the peaks were indexed with a JCPDS card (96-230-0117). The peaks were observed at 31.94°, 34.67°, 36.51°, 56.84° and 68.18°, which corresponded to the (100), (002), (101), (110) and (201) planes, respectively. Well-defined peaks in the pure ZnO showed a crystalline film with no impurities or other phases indicated in its structure. For the co-catalytic ZnO films modified with Ag, it was observed that the hexagonal crystalline structure remained present. However, when Ag was presented with a concentration 10^{-2} M, additional peaks appeared located at 38.46, 45.86 and 64.72°, respectively. These peaks corresponded to the planes (111, 200 and 220) of the metallic form of Ag (JCPDS 96-901-3048). The diffraction patterns for the ZnO films were similar, and no pattern distortion or peaks were observed arising from the co-catalysts modified with Ag^+ [25].

The XRD results showed that the addition of silver ions did not significantly affect the crystal size. However, when the film was co-catalytic-modified, the intensity of the TiO_2 and ZnO peaks decreased. The average crystallite size decreased slightly with the silver concentration ($d(TiO_2)$ = 31.2 nm, $d(TiO_2/Ag)$ = 26.9 nm, d(ZnO) = 48.4 nm and d(ZnO/Ag) = 42.9 nm), indicating that the crystalline lattice remained almost constant. Table 1 summarizes the structural parameters of the pure and silver-modified (10^{-2} M) thin films. The unit cell parameters of the silver-modified samples were found to be very close to the unit cell parameters of the pure TiO_2 and ZnO films. This study also used the c-axis lattice parameter to estimate the film strain along the c-axis.

The calculations showed positive values, which represented the tensile strain. The estimations presented in Table 1 showed that the strain in the sol–gel-derived films was tensile, and that the modified co-catalytic films exhibited a lower-magnitude tensile strain than the pure samples.

Table 1. The structural parameters calculated from XRD patterns of the films.

Sample	Unit Cell Parameters, Å	Crystallite Size, nm	Microstrains, $\times 10^{-3}$ a.u.
TiO_2	a = b = 3.7873; c = 9.5075	31.2	7.6
TiO_2/Ag	a = b = 3.7795; c = 9.4986	26.9	6.7
ZnO	a = b = 3.2524; c = 5.2124	48.4	0.9
ZnO/Ag	a = b = 3.2518; c = 5.2105	42.9	0.8

In order to investigate the elements presented on the surface of the samples and their oxidation states, we applied X-ray photoelectron spectroscopy. In particular, we were interested in the oxidation state of the Ag nanoclusters. In Figure 2 the high-resolution XP spectra of the Ag3d photoelectron region (Figure 2a) and the relevant Ag peak, AgMNN (Figure 2b), are shown, which were measured for Ag/TiO_2 film.

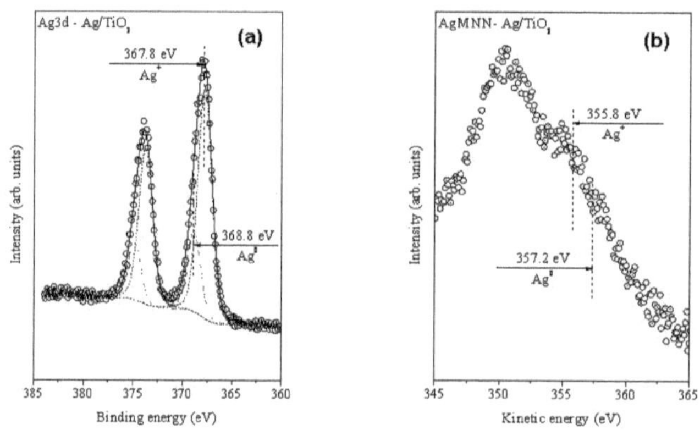

Figure 2. High-resolution XPS spectra of Ag3d (**a**) and AgMNN (**b**) lines measured for Ag/TiO_2 film.

A closer look at the peak line shape of both the photoelectron and Ag line revealed that silver existed on the surface of the samples with two oxidation states, namely Ag^0 and Ag^+. The curve-fitting procedure of the Ag3d line showed that the Ag^0 peak had a binding energy (BE) of 368.8 eV, whereas the Ag^+ had a BE of 367.8 eV. Additionally, we could recognize two features of the Auger peak that showed a 357.2 kinetic energy (KE), characteristic of Ag^0, and a 355.8 KE, characteristic of Ag^+. A more-sensitive parameter regarding the oxidation state of silver is the modified Auger parameter, which is defined as the sum of the BE of the 3d5/2 photoelectron peak and the KE of the Auger MNN peak ($\acute{\alpha}$ = EK(AgMNN) + EB(Ag3d5/2). The calculated numbers were 726.0 eV and 723.6 eV. These values were also calculated for metal silver and silver with a 1^+ oxidation state, respectively [26,27]. Using the results of the curve-fitting procedure, we could estimate that the concentration of the Ag^+ clusters was three times higher than that estimated for the Ag0 particles. The existence of oxidized silver could also be supposed from the relative concentrations of the presented elements on the surface of the Ag/TiO_2 film. We calculated the ratio between the atomic percentage of Ti:O:Ag = 68:24:8. Silver particles were photofixed on the TiO_2substrate. Theoretically, we expected a ratio between Ti:O = 1:2, which was not observed in our case. The surface of the sample was enriched with oxygen. The excess of oxygen was attributed to oxygen bonded to the silver and/or oxygen from OH-groups. The existence of Ag^+ clusters was also confirmed by EPR measurements.

Figure 3 shows the high resolution XP spectra of the Ag3d photoelectron region (Figure 3a) and the relevant Ager peak, AgMNN (Figure 3b), measured for the Ag/ZnO film. The line shapes of both the Ag3d photoelectron region and the Ager AgMNN peak revealed that photofixed metal silver particles were present on the surface of the ZnO. The BE of the 3d5/2 peak was defined at 368.4 eV, whereas the KE of the AgMNN peak was at 356.5 eV. The modified Auger parameter was calculated to be 724.9 eV. The value of all the above-mentioned parameters together with the loss structure in the BE ranging from 370.5 eV to 372.5 eV observed in the Ag3d region showed that the detected silver on the ZnO surface had a metallic character. The relative concentration of the elements presented on the surface showed a ratio of Zn:O:Ag = 47:48:5. The ratio between zinc and oxygen was close to the expected theoretical one. The negligible difference was attributed again to oxygen in the OH groups and eventually oxidized silver, which was with a concentration below the resolution of the spectrometer analyzer. Additional speculation was made according to the value of modified Auger parameter, because it was slightly lower than that measured for pure metallic silver, which could be the reason that one could expect the existence of a small quantity of Ag^+ clusters or metallic particles covered by thin oxide layer. This speculation was supported by the very low EPR signal of the Ag^+ clusters.

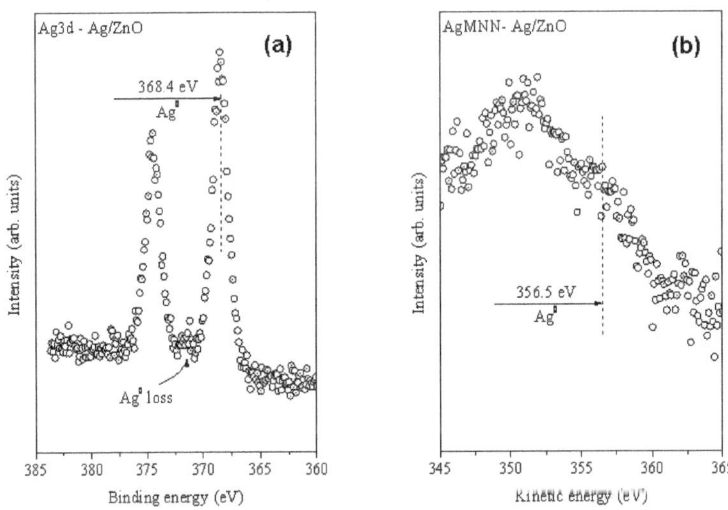

Figure 3. High-resolution XPS spectra of Ag3d (**a**) and AgMNN lines measured for Ag/ZnO (**b**) film.

In the EPR spectrum of the ZnO films (Figure 4), two signals are recorded at room temperature. The more intensive line recorded at g = 1.957 (S1) is associated with the so-called shallow effective mass donor (SD) center in the ZnO single crystals. Some authors have associated the SD center with Zn-related defects. The second less-intensive line (S2) is detected at a g value of 2.0014. Some authors have attributed this signal with a g factor close to the free electron value of 2.0023 in ZnO to singly ionized oxygen vacancies ($V_0{}^+$) defects. In other works, this has been detected with a similar signal with g = 2.003 and has been attributed to Zn vacancies.

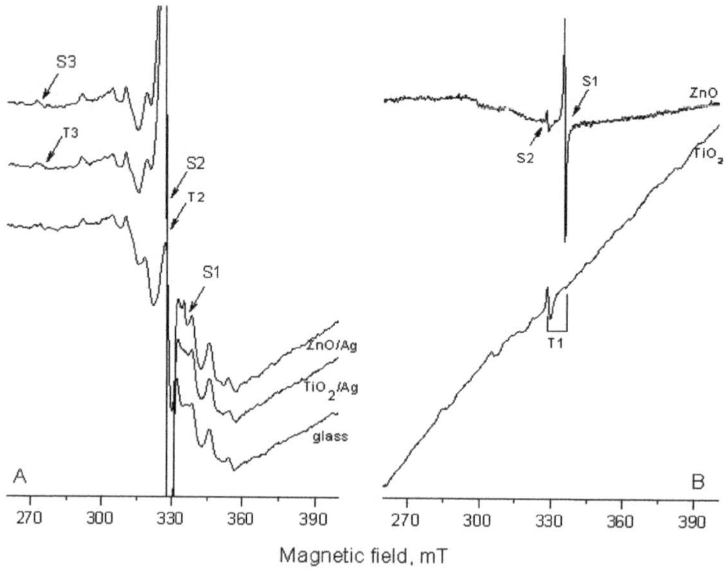

Figure 4. Low-temperature (123 K) EPR spectra of Ag-modified ZnO and TiO_2 (**A**); room-temperature (298 K) EPR spectra of pure ZnO and TiO_2 (**B**).

In the EPR spectrum of the Ag-modified ZnO films, the line S3 recorded at g = 2.004 was assigned to the Ag^{2+} center. Silver entered the TiO_2 lattice as an Ag^+ ion. Ag^+ ions are diamagnetic with a closed 4d10 electronic configuration and do not exhibit any ESR spectra. However, Ag^+ ions can lose an electron upon UV irradiation, resulting in the formation of paramagnetic Ag^{2+} [28]. The lower intensity of the SD center peak (S1) in the ZnO:Ag films indicated a decreased concentration of shallow donors, while the EPR line S2 showed a significant increase in intensity and a slight shift toward a lower magnetic field. The weak intensive signals from 285 mT to 360 mT are due to the glass used to prepare the films.

A narrow signal, T1, with $g_{\perp}$ = 1.99 and $g_{//}$ = 1.96 was recorded in the TiO_2 films. This signal has also been detected in colloidal TiO_2 anatase, and it was assigned to substitution Ti^{3+} (T3) in hydrated anatase [29]. After doping with silver, an intensive signal, T2, at g = 2.0038 was recorded. The EPR peak at g ≈ 2.001–2.004 is attributed to oxygen vacancies [29].

The surface micrograph analyses showed that the morphology of the TiO_2 films consisted of micro aggregates (sandwich structures), and the ZnO films had a ganglia-like structure, as shown in Figure 5. It was observed that all the samples possessed a homogeneous surface. It was evident from the SEM images that the co-catalytic modified films showed few nanoparticle aggregates on the top over the entire surface, which suggested a better photocatalytic performance of the TiO_2/Ag and ZnO/Ag samples.

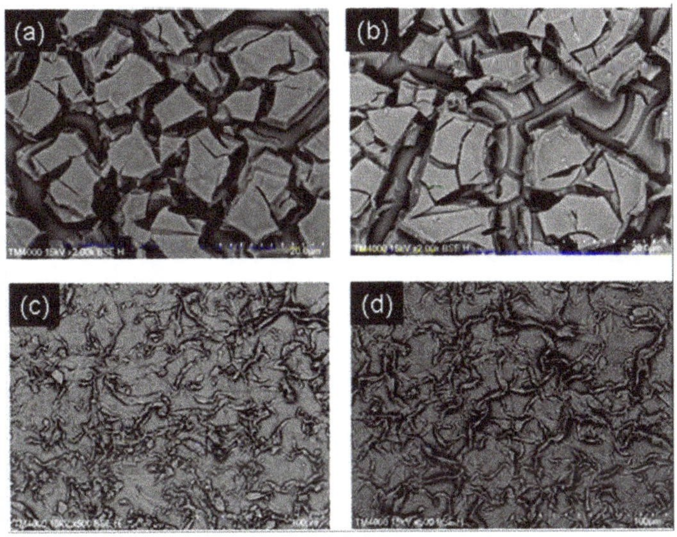

Figure 5. SEM images of TiO_2 (**a**), TiO_2/Ag (**b**), ZnO (**c**) and ZnO/Ag (**d**) films.

2.2. Optical Characterization

The optical properties of the semiconductors were investigated by UV–vis absorption spectroscopy. It can be seen in Figure 6 that the pure films presented an absorption band around 330–370 nm in the UV region (TiO_2 λ_{max} = 325 nm, ZnO λ_{max} = 361 nm), while the co-catalytic silver-modified samples showed a slight shift in the absorption band (TiO_2/Ag λ_{max} = 338 nm, ZnO λ_{max} = 364 nm), which was caused by the interaction between the Ag and the semiconductors [30]. This interaction was due to the strong interfacial electronic coupling between the TiO_2, ZnO and Ag. This helped to further reduce the loss of light energy [31].

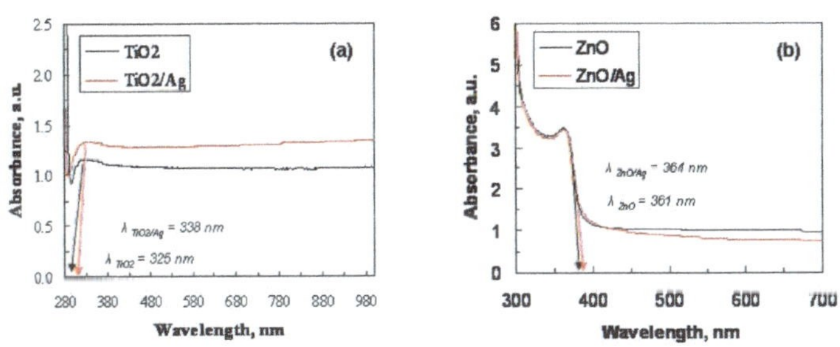

Figure 6. UV–visible patterns of semiconductor films: TiO_2 and TiO_2/Ag (**a**), ZnO and ZnO/Ag (**b**).

The band gap energy of the prepared semiconductors was calculated using the following equation:

$$(ahv)^2 = A(hv - E_g) \quad (1)$$

where the absorption coefficient, Planck's constant, photon frequency, energy band gap and a constant are α, h, v, E_g and A, respectively. The E_g values were obtained from plots of $(\alpha hv)^2$ verses hv, extrapolating a straight line to the x-axis. The values from the Kubelka–Munk extrapolation vs. the energy band gap of the films are presented in Figure 7 and showed the change in the absorption edge that was attributed to the incorporation of the silver ions into the TiO_2 and ZnO. This led to narrowing of the optical band gap with the co-catalytic modification of the semiconductors with Ag.

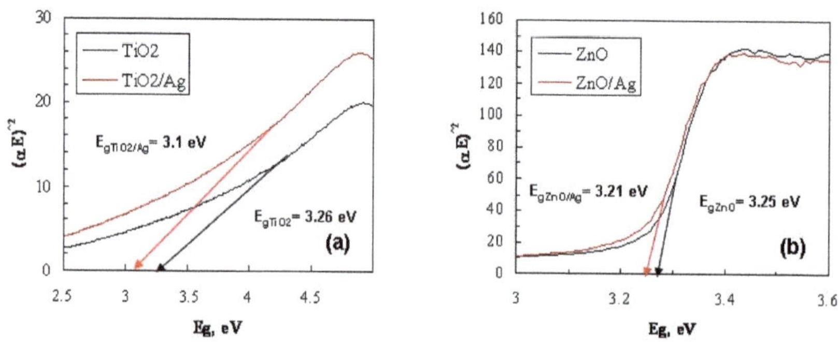

Figure 7. Linear plots of $(\alpha h\nu)^2$ versus E_g for semiconductors films: TiO_2 and TiO_2/Ag (**a**), ZnO and ZnO/Ag (**b**), calculated by Equation (1).

The decrease in the semiconductor band gap with silver may be due to the renormalization effect of the band gap [31]. The renormalization was due to the exchange interaction (sp-d) in the electrons of the band and the localized electrons (d) of the Ag^+. After the possible exchange interaction, there was a modification in the band gap of the TiO_2/Ag and ZnO/Ag with reference to pure semiconductors [31]. The TiO_2 and ZnO showed a band gap of 3.26 and 3.25 eV, while the band gap value of the co-catalytic silver-modified films decreased to 3.10 and 3.21 eV. The co-catalytic silver-modified TiO_2 and ZnO films were able to capture more light-generated electrons, which made them more efficient light-trapping materials in the oxidation and reduction reactions that took place on the surface. Thus, the silver ions reduced the rate of recombination of the photogenerated charge pairs (electron–hole). Therefore, a better photocatalytic activity was expected for the TiO_2/Ag and ZnO/Ag films.

2.3. Photocatalytic Activity

The photocatalytic activities of the pure and co-catalytic silver-modified films were investigated by MG and MB degradation under UV illumination. In order to examine the effect of the co-catalytic modification, we prepared two parallel sol–gel method procedures for the preparation of the TiO_2 and ZnO catalysts. For both samples, the efficiency of the catalyst was enhanced significantly upon the addition of silver ions (from 10^{-4} to 10^{-2} M), with the degradation rate increased from 0.1277 to 0.6723 h^{-1} for the titanium dioxide (Figure 8a) and from 0.3629 to 0.7977 h^{-1} for the zinc oxide samples (Figure 8b). The results obtained, as shown in Figure 8, showed that the photocatalytic reaction was of a pseudo-first-order kinetics. The apparent first-order reaction rate constants were evaluated from the experimental data using a linear regression. All of the correlation coefficients (R^2) were higher than 0.947. It is well-known that the photocatalytic efficiency of co-catalytic modified TiO_2 and ZnO depends on many factors, such as the synthesis method, light illumination, crystallite size, surface area, etc. Obviously, in our photocatalytic experiments, the presence of silver was beneficial to the photocatalytic behavior of the samples under UV irradiation. No bleaching of the malachite green solution was observed in the absence of the pure and co-catalytic silver-modified semiconductor films.

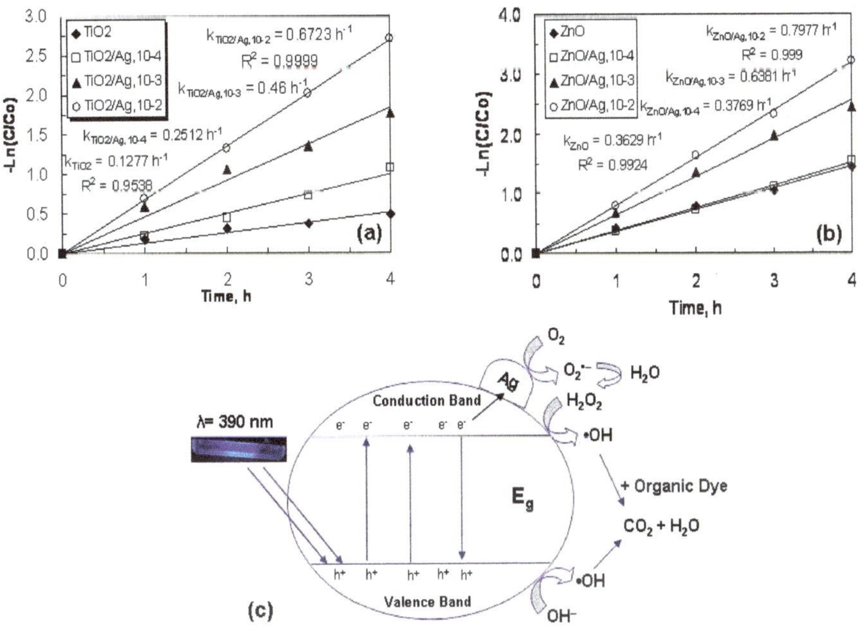

Figure 8. Logarithmic graphs of malachite green concentration vs. illumination time in the presence of TiO_2 and TiO_2/Ag (**a**). ZnO and ZnO/Ag (**b**) sol–gel films. (**c**) Probable mechanism of photocatalysis in the co-catalytic Ag-modified TiO_2 and ZnO under UV light illumination.

Figure 8 shows that the silver-modified samples were more efficient than the pure TiO_2 and ZnO films in decolorizing the MG and MB. The UV irradiation of the semiconductor generated e^-/h^+ pairs at the illuminated surface. The recombination of these charge carriers reduced the rate of photocatalysis. The positive effect from the co-catalytic modification with silver ions on the efficiency of the TiO_2 and ZnO for the photodegradation of the dyes may be explained by silver's ability to trap electrons. Thus, a more effective electron transfer occurred through adsorption onto the surface molecules than in the case of the pure semiconductors. Oxygen adsorbed onto the photocatalyst surface trapped the electrons and produced superoxide anions. On the other hand, the holes at the semiconductor (TiO_2, ZnO) surface could oxidize the adsorbed water or hydroxide ions to produce hydroxyl radicals. Szabo-Bardos et al. [32] showed that electron scavenging by oxygen at the surface of excited semiconductor particles cannot efficiently compete with electron transfer to silver ions. So, the electron transfer to silver ions is rather fast compared to electron transfer to oxygen molecules, and therefore the formation of $O_2^{\bullet-}$ is reduced. But in the co-catalytic silver-modified semiconductors with the sol–gel method (TiO_2/Ag, ZnO/Ag), the loading of Ag metal on the TiO_2 and ZnO surfaces could expedite the transport of photogenerated electrons to the outer systems. The transfer of electrons to the metal deposits resulted in partially negatively charged deposits. The deposits of silver ions on the surface enhanced the photoactivity by accelerating the transfer of electrons to the dissolved oxygen molecules. Therefore, superoxide anion radicals were formed as a result of oxygen reduction by the transfer of trapped electrons from the Ag metal to the oxygen, as can be seen in Figure 8c. The enhancement of the efficiency of the photocatalysts with Ag co-catalytic modification was marked by absorption spectra for these experiments, and these are shown in Figure 9. A significant shift in the absorption maximum wa observed in the silver-modified ZnO.

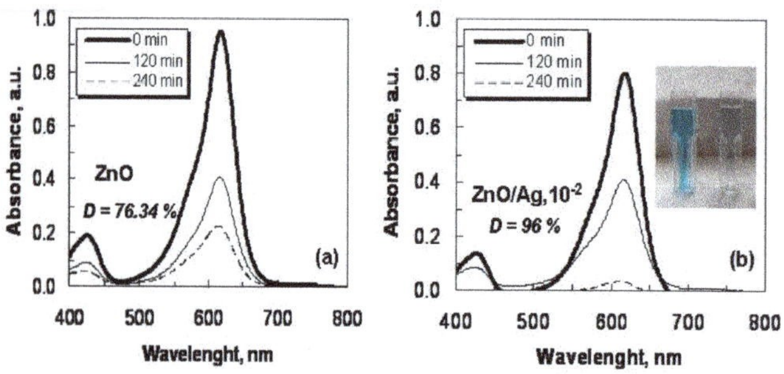

Figure 9. Absorbance spectra of the degradation of malachite green using ZnO (**a**) and ZnO/Ag (**b**) films.

Figure 10 compares the photocatalytic efficiency of the TiO_2 and TiO_2/Ag, and the ZnO and ZnO/Ag sol–gel films and shows the time course of the decrease in the concentration of MG under UV and visible light irradiation. The initial concentration of the dye was 5 ppm. No bleaching of the malachite green solution was observed in the absence of the pure and silver-modified films. The photocatalytic experiments showed that the decolorization of the dye by the sol–gel films under UV and visible irradiation followed pseudo-first-order kinetics. The slope of the logarithmic scale linear fits represents the rate constant of photocatalysis *k*.

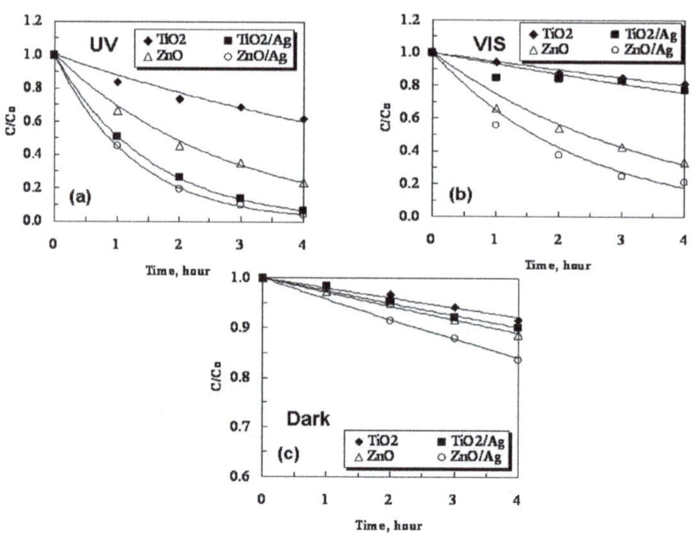

Figure 10. Photocatalytic activity of pure semiconductor and silver-modified (10^{-2} M) films for degradation of malachite green under (**a**) UV light, (**b**) visible illumination and (**c**) darkness. The initial dye concentration is 5 ppm.

As can be seen from Figure 10a,b, the same trend was observed in the photocatalytic processes carried out in the presence of UV and visible light, where the ZnO/Ag films had the highest photocatalytic activity (under UV and visible irradiation, $k_1 = 0.7977\ h^{-1}$ and $k_2 = 0.4319\ h^{-1}$). The titanium dioxide modified with silver had a higher efficiency (under UV and visible irradiation, $k_1 = 0.6723\ h^{-1}$ and $k_2 = 0.0695\ h^{-1}$) in comparison with the pure TiO_2 film (under UV and visible irradiation, $k_1 = 0.1277\ h^{-1}$ and $k_2 = 0.0545\ h^{-1}$).

We compared the photocatalytic results with the degradation of dye by the semiconductors in darkness (Figure 10c). The concentration of the organic pollutant decreased without irradiation but was slower in comparison with the respective photocatalytic reactions. The rate of dye degradation by the sol–gel films reached almost a constant value after 4 h (~18%). The decrease in the pollutant concentration was considered as the adsorption of the MG onto the sol–gel films or some kind of dye destruction process taking part on the semiconductors even in darkness (without UV illumination). All these assumptions need further evidence and will be the subject of further future research.

Finally, the tuning of the electronic structure of the TiO_2 and ZnO samples by co-catalytic modifying with Ag using the sol–gel technique was successfully achieved. Moreover, their unusual effectiveness encouraged us to apply this test to another organic pollutant in the presence of ultraviolet illumination. The photocatalytic activity of the sol–gel films was investigated for the decolorization of MB by the same procedure as MG and is shown in Figure 11. The photocatalytic experiments related to MB degradation showed the same trend, where the modified films possessed a higher activity (MG—$D_{TiO2/Ag}$ = 93.29%, $D_{ZnO/Ag}$ = 96%; MB—$D_{TiO2/Ag}$ = 61.77%, $D_{ZnO/Ag}$ = 62.51%) compared to the pure films. This result confirmed the positive effect of Ag on the photoinduced efficiency of TiO_2 and ZnO.

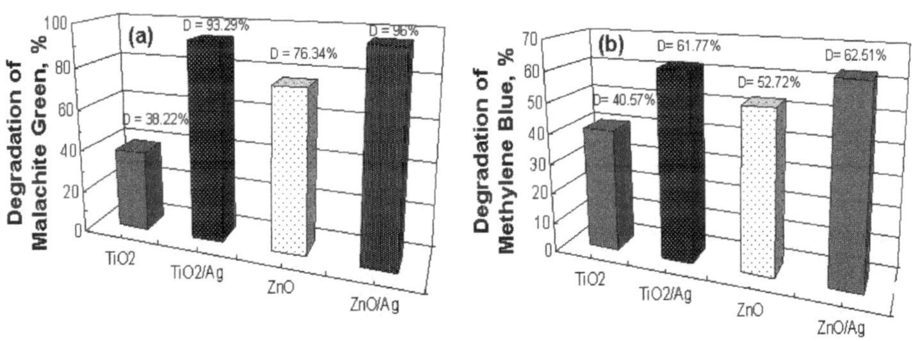

Figure 11. Photocatalytic degradation of the organic dyes: Malachite Green (**a**) and Methylene Blue (**b**), using pure and co-catalytic Ag^+ (10^{-2} M)-modified semiconductor films.

The silver co-catalytically modified sol–gel films showed a faster degradation of malachite green compared to methylene blue due to the formation of stable intermediates by the reaction of $OH^•$ with triarylmethane dye (C=C bond) during the photocatalytic process. The probable mechanisms of the degradation of methylene blue and malachite green, induced by a reaction with $OH^•$, have been reported in the literature [33,34]. Houas et al. [33] suggested that in the initial step the hydroxyl radical cleaves the S+=C bond, which in turn induces a ring-opening reaction of the central heterocyclic part. This leads to the degradation of the MB solution. Regarding MG, Ju et al. [34] showed that the initial step of dye decolorization includes a radical attack on the central C atom, which transforms it into a malachite green carbinol base and again leads to the degradation of the pollutant.

Taking into account the above reasoning, it can be concluded that the faster degradation of malachite green, observed in the present study, could be attributed to the facilitated attack of the hydroxyl radical toward the S atom. The reason for this was the increased electron density.

The photocatalytic results were confirmed by TOC analysis. TOC measurements were carried out in order to examine the degradation of both of the dyes (malachite green and methylene blue) after 240 min of UV light illumination, as shown in Figure 12. The TOC percentages for the mineralization of MG and MB were found to be 20.28% and 22.84%, respectively. In comparison with the TiO_2, the ZnO catalyst showed higher percentages of TOC removal of the dyes (51.37% and 40.73%), but the trend was the same.

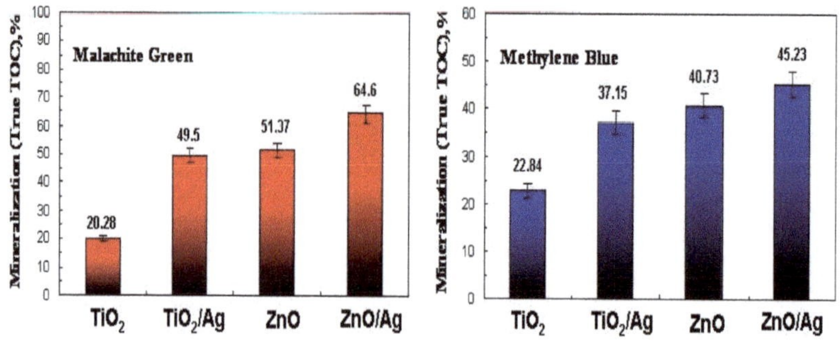

Figure 12. Histograms of TOC conversion after 4 hours of UV illumination.

When the semiconductor films were co-catalytically modified with silver ions, the TOC conversion rates of dye photodegradation were increased (MG—49.5% TiO_2/Ag, 64.6% ZnO/Ag; MB—37.15% TiO_2/Ag, 45.23% ZnO/Ag). On the other hand, the TOC removal percentages of the MG and MB were found to be correspondingly slightly lower than their degradation rate in the photocatalytic processes. The reason for this was the formation of intermediate byproducts.

The TOC measurements showed lower values for malachite green and methylene blue removal than the UV–vis spectroscopy. This was a clear indication that the mineralization went through a complex multistep process including several intermediate products. For example, S. Ray et al. and S. Nayak identified several intermediate products during the photocatalytic oxidation of malachite green and methylene blue [35,36]. All these byproducts have different oxidation potentials when reacting with OH^{-} radicals produced during the photocatalytic process. This leads to different speeds of mineralization for each of the intermediates and therefore their detection during the TOC analysis.

3. Materials and Methods

Nanostructure TiO_2 and ZnO thin films were deposited by two sol–gel method procedures using the dip-coating technique on glass substrates (ca. 76 mm Ч 26 mm, ISO-LAB, Wertheim, Germany). Titanium (IV) isopropoxide (Sigma Aldrich, ≥97.0%, Bulington, MA, USA) and zinc acetate dehydrate (Fluka, >99.5%, Buchs, Switzerland) were used as the initial materials for both syntheses. 2-methoxyethanol (Fluka, >99.5%, Buchs, Switzerland) and monoethanolamine (MEA, Fluka, >99.5%, Buchs, Switzerland) were used as a solvent and a stabilizer, respectively. Titanium isopropoxide and zinc acetate dihydrate were dissolved in 2-methoxyethanol in a round-bottomed flask and stirred at room temperature for 15 min. The substances were mixed together with monoethanolamine (Ti:MEA:ME and Zn:MEA:ME, molar ratio 1:4:1). The two resulting solutions were stirred at 60 °C for 1 h to obtain clear and homogenous solutions and then were aged for 1 day at room temperature before application as coating solutions. No visible changes were observed in the solutions upon storing them at room temperature for at least 3–4 months (Figure 13).

The deposition of TiO_2 and ZnO films consisted of dip-coating and drying of the material. The glass slides were immersed in the precursor sol and withdrawn at a speed of 0.9 cm/min at room temperature. The films were dried at 100 °C for 10 min after each successive coating to evaporate the solvent and remove organic compounds. After repeating the dip coating and drying procedures five times, the films were annealed at 500 °C for 1 h in the oven to enhance the crystallinity of the dried TiO_2 and ZnO sol–gel films.

The heterostructures of TiO_2/Ag and ZnO/Ag were prepared by a simple chemical photodeposition method for the first time. The pure semiconductor films were immersed in an aqueous silver nitrate solution with different molar concentrations (10^{-2}–10^{-4} M) for 20 min in the presence of UV light illumination. After the photo-fixation, the TiO_2/Ag and

ZnO/Ag films were washed with distilled water and dried at 100 °C for 10 min in order to obtain the final films for the photocatalytic tests.

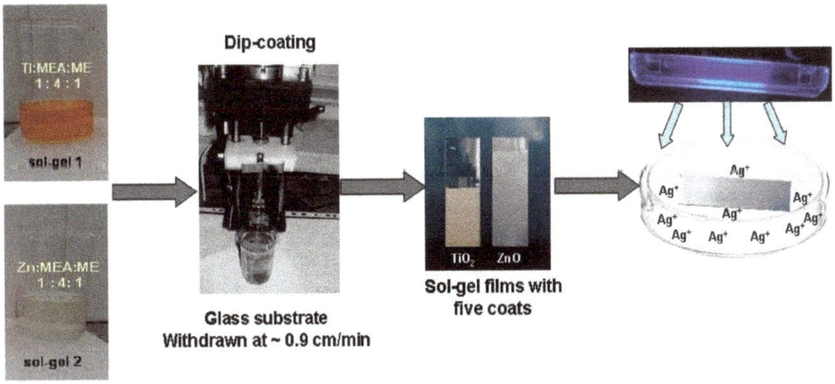

Figure 13. Schematic diagram of sol–gel method for the synthesis of pure and co-catalytic Ag-modified films.

Powder X-ray diffraction (XRD) was used to confirm the purity and the crystallinity of the as-prepared films using a Siemens D500 (Karlsruhe, Germany)with CuK_α radiation within a 2θ range of 20–80° at a step of 0.05° 2θ and a counting time of 2 s/step. The average crystallite sizes were estimated according to Scherrer's equation. X-ray photoelectron measurements were carried out on an ESCALAB MkII (VG Scientific, now Thermo Scientific, (Manchester, UK) electron spectrometer with a base pressure in the analysis chamber of 5×10^{-10} mbar equipped with a twin-anode $MgK\alpha/AlK\alpha$ non-monochromated X-ray source using excitation energies of 1253.6 and 1486.6 eV, respectively. The measurements were only provided with a $AlK\alpha$ non-monochromated X-ray source. The instrumental resolution was about 1 eV. The data were analyzed using the SpecsLab2 CasaXPS software (2.3.25PR1.). Because of electrostatic sample charging, the energy scale was calibrated by normalizing the Ti2p line to 458.5 eV for the titania-containing sample and that of Zn2p to 1022.0 eV for the zinc-containing sample, respectively. The processing of the measured spectra included a subtraction of the X-ray satellites and a Shirley-type background [37]. The peak positions and areas were evaluated by a symmetrical Gaussian–Lorentzian curve-fitting. The relative concentrations of the different chemical species were determined based on the normalization of the peak areas to their photoionization cross-sections, as calculated by Scofield [38].

The EPR spectra of TiO_2 and ZnO were recorded at room temperature. TiO_2:Ag and ZnO:Ag films were recorded at 298 K and 123 K with a JEOL JES-FA 100 EPR (Tokyo, Japan) spectrometer operating in the X-band with a standard TE011 cylindrical resonator. A Varied Temperature Controller ES-DVT4 (Tokyo, Japan) was used to record the EPR spectra at low temperatures. The desired temperature was achieved by sending cold gas (liquid nitrogen) to the sample area. The EPR spectra were detected at the following conditions: modulation frequency: 100 kHz, microwave power: 3 mW, modulation amplitude: 0.2 mT, time constant: 0.3 s and sweep time: 2 min.

Each film's surface morphology was examined with a Hitachi TM4000 (Krefeld, Germany) with an accelerating voltage of 15 kV. The investigated samples were coated with gold by a JFC-1200 fine coater (JEOL, Krefeld, Germany) before observation. The optical transmittance spectra of these films were investigated with an Evolution 300 Thermo Scientific, UV–vis spectrophotometer in a range of 270–1000 nm.

The photocatalytic degradation of the malachite green (MG) and methylene blue (MB) was conducted at room temperature in a photoreactor (200 mL capacity) equipped with a magnetic stirrer (rotating speed controlled by stroboscope) and UV (Sylvania BLB18 W,

315–400 nm of emission range) and visible (150 W LED, V-Tac, emitting 5700 K white, 420–700 nm) lamps. The light power density at the film position was 0.66 mW/cm^2 (measured with a research radiometer from Ealing Electro-optics, Inc., (Ettlingen, Germany). The concentration of the dyes after the different times of irradiation was measured using an Evolution 300 Thermo Scientific (Madison, USA), UV-visible spectrophotometer. The concentrations of MG and MB in the aqueous solutions were determined as a function of the irradiation time from the absorbance change at 615 and 666 nm corresponding to the maximum absorption wavelength of the dyes. The kinetics curves were plotted by measuring the concentration in 2 mL aliquots taken periodically. The photocatalytic efficiency D (%) was calculated by the following Equation (2):

$$D(\%) = \frac{C_O - C}{C_O} * 100 \quad (2)$$

C_0 is the initial concentration of the dyes and C is the concentration of the dyes after photo-illumination.

The experiments were carried out with four types of samples: pure semiconductor films and films modified with different concentrations of Ag^+ (10^{-2}, 10^{-3}, 10^{-4} M) prepared with five coatings each. The effect of the concentration of Ag^+ was evaluated. All photocatalytic tests were performed at a constant stirring rate (500 rpm) at room temperature (25 °C).

To investigate the stability of the pure and modified thin films, some of the photocatalytic experiments were repeated three times using a new dye solution in each measurement with the same 5 ppm concentration. All the photocatalytic results were reproducible, which allowed us to reach a conclusion about the stability of the films. No changes were observed in the morphology of the films, which we found by SEM analysis.

The total organic carbon (TOC) measurements were made for the treated dye solutions using a high-temperature (850 °C) catalytic oxidation method and and Elementar Vario Select TOC analyzer (Langenselbold, Germany). The observed standard deviation was calculated based on three measurements of each sample. The percentage of TOC removal was calculated using the following formula:

$$TOC(\%) = \frac{TOC_0 - TOC_t}{TOC_0} * 100 \quad (3)$$

where TOC_0 is the initial TOC of the malachite green and methylene blue solution and TOC_t is the final TOC at time t.

4. Conclusions

In conclusion, we successfully prepared a new, cost-effective and simple fabrication method of co-catalytic silver-modified TiO_2 and ZnO sol–gel films using, for the first time, a combination of a dip-coating technique and the photo-fixation of silver ions onto the catalyst film surfaces. The modified nanostructure films had varying concentrations of Ag (10^{-2}, 10^{-3}, 10^{-4} M). The characterization showed that the semiconductors' crystal structures were not affected by the modifying process. Both anatase and hexagonal wurtzite phases were identified by X-Ray diffraction. The UV–vis analysis showed that the TiO_2/Ag and ZnO/Ag efficiently narrowed their band gaps to 3.1 and 3.21 eV, respectively. The prepared pure and co-catalytic modified catalysts were stable and recyclable and were active under UV light illumination. The trend of the degradation of malachite green in the TiO_2 and ZnO species was as follows: 10^{-2} M Ag-modified > 10^{-3} M Ag-modified > 10^{-4} M Ag-modified > pure semiconductor. The photocatalytic results showed that the silver-modified TiO_2 and ZnO thin films were promising materials for the degradation of the organic wastes present in polluted water.

Author Contributions: N.K.: methodology, investigation, data curation, writing—original draft, writing—reviewing scientific contents and editing. N.K., A.B., K.P.—All authors have read and agreed to the published version of the manuscript.

Funding: This research was funded by the Bulgarian NSF project KП-06-H59/11.

Data Availability Statement: Not applicable.

Acknowledgments: This work was financially supported by the Bulgarian NSF project KП-06-H59/11. The authors thank the their colleagues at the Faculty of Chemistry and Pharmacy (A. Apostolov, M. Tzvetkov), Institute of Catalysis, Bulgarian of Academic Science (H. Kolev, R. Mladenova) for the support in characterizing and analyses.

Conflicts of Interest: The authors declare no conflict of interest.

References

1. Robinson, T.; McMullan, G.; Marchant, R.; Nigam, P. Remediation of dyes in textile effluent: A critical review on current treatment technologies with a proposed alternative. *Bioresour. Technol.* **2001**, *77*, 247–255. [CrossRef] [PubMed]
2. Ajibade, F.O.; Adelodun, B.; Lasisi, K.H.; Fadare, O.O.; Ajibade, T.F.; Nwogwu, N.A.; Wang, A. Environmental pollution and their socioeconomic impacts. In *Microbe Mediated Remediation of Environmental Contaminants*; Woodhead Publishing: Cambridge, UK, 2021; pp. 321–354.
3. Wallace, J.S. Increasing agricultural water use efficiency to meet future food production. *Agric. Ecosyst. Environ.* **2000**, *82*, 105–119. [CrossRef]
4. Lahmar, H.; Benamira, M.; Douafer, S.; Messaadia, L.; Boudjerda, A.; Trari, M. Photocatalytic degradation of methyl orange on the novel hetero-system La2NiO4/ZnO under solar light. *Chem. Phys. Lett.* **2021**, *742*, 137132. [CrossRef]
5. Sidra, B.; Awais, A.; Mohsin, A.R.A.; Abdul, H.; Muhammad, S.; Syed, S.S.; Abdullah, A.K. Photocatalytic degradation of malachite green and methylene blue over reduced graphene oxide (rGO) based metal oxides (rGO-Fe_3O_4/TiO_2) nanocomposite under UV-visible light irradiation. *J. Environ. Chem. Eng.* **2021**, *9*, 105580.
6. Dong, X.; Li, Y.; Li, D.; Liao, D.; Qin, T.; Prakash, O.; Kumar, A.; Liu, J. A new 3D 8-connected Cd(ii) MOF as a potent photocatalyst for oxytetracycline antibiotic degradation. *CrystEngComm* **2022**, *24*, 6933–6943. [CrossRef]
7. Kaneva, N.; Bojinova, A.; Papazova, K.; Dimitrov, D.; Zaharieva, K.; Cherkezova-Zheleva, Z.; Eliyas, A. Effect of thermal and mechano-chemical activation on the photocatalytic efficiency of ZnO for drugs degradation. *Arch. Pharm. Res.* **2016**, *39*, 1418–1425. [CrossRef] [PubMed]
8. Zheng, M.; Chen, J.; Zhang, L.; Cheng, Y.; Lu, C.; Liu, Y.; Singh, A.; Trivedi, M.; Kumar, A.; Liu, J. Metal organic frameworks as efficient adsorbents for drugs from wastewater. *Mater. Today Commum.* **2022**, *31*, 103514. [CrossRef]
9. Rao, C.; Zhou, L.; Pan, Y.; Lu, C.; Qin, X.; Sakiyama, H.; Muddassir, H.; Liu, J. The extra-large calixarene-based MOFs-derived hierarchical composites for photocatalysis of dye: Facile syntheses and contribution of carbon species. *J. Alloys Compd.* **2022**, *897*, 163178. [CrossRef]
10. Singh, A.; Singh, A.K.; Liu, J.; Kumar, A. Syntheses, design strategies, and photocatalytic charge dynamics of metal–organic frameworks (MOFs): A catalyzed photo-degradation approach towards organic dyes. *Catal. Sci. Technol.* **2021**, *11*, 3946–3989. [CrossRef]
11. Kaneva, N.; Bojinova, A.; Papazova, K.; Dimitrov, D. Effect of the substrate on photocatalytic efficiency of ZnO for Malachite Green degradation. *J. Chem. Tech. Metall.* **2014**, *49*, 149–156.
12. Simsek, E.B. Solvothermal synthesized boron doped TiO_2 catalysts: Photocatalytic degradation of endocrine disrupting compounds and pharmaceuticals under visible light irradiation. *Appl. Catal. B Environ.* **2017**, *200*, 309–322. [CrossRef]
13. Kumar, S.G.; Rao, K.S.R.K. Comparison of modification strategies towards enhanced charge carrier separation and photocatalytic degradation activity of metal oxide semiconductors (TiO_2, WO_3 and ZnO). *Appl. Surf. Sci.* **2017**, *391*, 124–128. [CrossRef]
14. Jiang, G.; Li, X.; Wei, Z.; Jiang, T.; Du, X.; Chen, W. Growth of N-doped BiOBr nanosheets on carbon fibers for photocatalytic degradation of organic pollutants under visible light irradiation. *Powder Technol.* **2014**, *260*, 84–89. [CrossRef]
15. Wang, N.; Zhou, Y.; Chen, C.; Cheng, L.; Ding, H. A g-C_3N_4 supported graphene oxide/Ag_3PO_4 composite with remarkably enhanced photocatalytic activity under visible light. *Catal. Commun.* **2016**, *73*, 74–79. [CrossRef]
16. Pérez-González, M.; Tomás, S.A.; Santoyo-Salazar, J.; Gallardo-Hernández, S.; Tellez-Cruz, M.M.; Solorza-Feria, O. Sol-gel synthesis of Ag-loaded TiO_2-ZnO thin films with enhanced photocatalytic activity. *J. Alloys Compd.* **2019**, *779*, 908–917. [CrossRef]
17. Tao, J.; Gong, Z.; Yao, G.; Cheng, Y.; Zhang, M.; Lv, J.; Shi, S.; He, G.; Jiang, X.; Chen, X.; et al. Effects of hydrothermal time on the morphologies of rutile TiO_2 hierarchical nanoarrays and their optical and photocatalytic properties. *J. Alloys Compd.* **2016**, *688*, 605–612. [CrossRef]
18. Espino-Estévez, M.R.; Fernández-Rodríguez, C.; González-Díaz, O.M.; Araña, J.; Espinós, J.P.; Ortega-Méndez, J.A.; Doña-Rodríguez, J.M. Effect of TiO_2–Pd and TiO_2–Ag on the photocatalytic oxidation of diclofenac, isoproturon and phenol. *Chem. Eng. J.* **2016**, *298*, 82–95. [CrossRef]
19. Rozyyev, V.; Murphy, J.G.; Barry, E.; Mane, A.U.; Sibener, S.J.; Elam, J.W. Vapor-phase grafting of a model aminosilane compound to Al_2O_3, ZnO, and TiO_2 surfaces prepared by atomic layer deposition. *Appl. Surf. Sci.* **2021**, *562*, 149996. [CrossRef]

20. Hakki, H.K.; Allahyari, S.; Rahemi, N.; Tasbihi, M. Surface properties, adherence, and photocatalytic activity of sol–gel dip-coated TiO_2–ZnO films on glass plates. *CR Chim.* **2019**, *22*, 393–405. [CrossRef]
21. Pérez-González, M.; Tomás, S.A.; Santoyo-Salazar, J.; Morales-Luna, M. Enhanced photocatalytic activity of TiO_2-ZnO thin films deposited by dc reactive magnetron sputtering. *Ceram. Int.* **2017**, *43*, 8831–8838. [CrossRef]
22. Fouad, S.S.; Parditka, B.; Atyia, H.E.; Baradács, E.; Bekheet, A.E.; Erdélyi, Z. AC conductivity and dielectric parameters studies in multilayer TiO_2/ZnO thin films produced via ALD technique. *Chin. J. Phys.* **2022**, *77*, 73–80. [CrossRef]
23. Brinker, C.J.; Scherer, G.W. *Sol-Gel Science—The Physics and Chemistry of Sol-Gel Processing*, 3rd ed.; Academic Press: San Diego, CA, USA; London, UK, 1991; pp. 1–881.
24. Perumal, S.; Gnana, C. Synthesis and characterization studies of solvothermally synthesized undoped and Ag-doped TiO_2 nanoparticles using toluene as a solvent. *J. Eng. Res.* **2014**, *4*, 184–187.
25. Yıldırım, O.A.; Unalan, H.E.; Durucan, C. Highly efficient room temperature synthesis of silver-doped zinc oxide (ZnO: Ag) nanoparticles: Structural, optical, and photocatalytic properties. *J. Am. Ceram. Soc.* **2013**, *96*, 766–773. [CrossRef]
26. Pargar, F.; Kolev, H.; Koleva, D.; Breugel, K. Potentiometric Response of Ag/AgCl Chloride Sensors in Model Alkaline Medium. *Adv. Mat. Sci. Eng.* **2018**, *3*, 8135492. [CrossRef]
27. Todorova, S.; Kolev, H.; Shopska, M.; Kadinov, G.; Holgado, G.; Caballero, A. Silver-based catalysts for preferential CO oxidation in hydrogen-rich gases (PROX). *Bulg. Chem. Commun.* **2018**, *50*, 17–23.
28. Bhatt, B.C.; Dhabekar, B.; Kumar, R.; Gundu Rao, T.K.; Lakshmanan, A.R. Defect centres and thermoluminescence in $CaSO_4$:Dy, Ag phosphor. *Radiat. Prot. Dosim.* **2006**, *119*, 53–56. [CrossRef]
29. Coronado, J.M.; Maira, A.J.; Conesa, J.C.; Yeung, K.L.; Augugliaro, V.; Soria, J. EPR Study of the Surface Characteristics of Nanostructured TiO_2 under UV Irradiation. *Langmuir* **2001**, *17*, 5368–5374. [CrossRef]
30. Singh, R.; Barman, P.B.; Sharma, D. Synthesis, structural and optical properties of Ag doped ZnO nanoparticles with enhanced photocatalytic properties by photo degradation of organic dyes. *J. Mater. Sci. Mater. Electron.* **2017**, *28*, 5705–5717. [CrossRef]
31. Kotlhao, K.; Mtunzi, F.M.; Pakade, V.; Laloo, N.; Ejidike, I.P.; Modise, S.J.; Moutloali, R.M.; Klink, M.J. Enhancing the photocatalytic degradation of selected chlorophenols using Ag/ZnO nanocomposites. *MRS Adv.* **2018**, *3*, 2129–2136. [CrossRef]
32. Szabó-Bárdos, E.; Czili, H.; Horváth, A. Photocatalytic oxidation of oxalic acid enhanced by silver deposition on a TiO_2 surface. *J. Photochem. Photobiol. A* **2003**, *154*, 195–201. [CrossRef]
33. Houas, A.; Lachheb, H.; Ksibi, M.; Elaloui, E.; Guillard, C.; Herrmann, M. Photocatalytic Degradation Pathway of Methylene Blue in Water. *Appl. Cat. B Environ.* **2001**, *31*, 145–157. [CrossRef]
34. Ju, Y.; Qiao, J.; Peng, X.; Xu, Z.; Fang, J.; Yang, S.; Sun, C. Photodegradation of malachite green using UV–vis light from two microwave-powered electrodeless discharge lamps (MPEDL-2): Further investigation on products, dominant routes and mechanism. *Chem. Eng. J.* **2013**, *221*, 353–362. [CrossRef]
35. Ray, S.K.; Dhakal, D.; Lee, S. Insight into Malachite Green Degradation, Mechanism and Pathways by Morphology-Tuned α-NiMoO4 Photocatalyst. *Photochem. Photobiol.* **2018**, *94*, 552–563. [CrossRef] [PubMed]
36. Nayak, S.; Das, K.; Parida, K. Indulgent of the physiochemical features of MgCr-LDH nanosheets towards photodegradation process of methylene blue. *J. Colloid Interface Sci.* **2023**, *634*, 121–137. [CrossRef] [PubMed]
37. Shirley, D. High-Resolution X-Ray Photoemission Spectrum of the Valence Bands of Gold. *Phys. Rev. B* **1972**, *5*, 4709–4714. [CrossRef]
38. Scofield, J.H. Hartree-Slater subshell photoionization cross-sections at 1254 and 1487 eV. *J. Electron. Spectrosc. Relat. Phenom.* **1976**, *8*, 129–137. [CrossRef]

catalysts

MDPI

Article

Sol-Gel Obtaining of TiO_2/TeO_2 Nanopowders with Biocidal and Environmental Applications

Albena Bachvarova-Nedelcheva [1,*], Reni Iordanova [1], Anton Naydenov [1], Angelina Stoyanova [2], Nelly Georgieva [3], Veronica Nemska [3] and Tsvetelina Foteva [3]

[1] Institute of General and Inorganic Chemistry, Bulgarian Academy of Sciences, Acad. G. Bonchev Str., bld. 11, 1113 Sofia, Bulgaria
[2] Department Chemistry and Biochemistry, Faculty of Pharmacy, Medical University-Pleven, Kl. Ohridski Str., 1, 5800 Pleven, Bulgaria
[3] Department Biotechnology, Faculty of Chemical and Systems Engineering, University of Chemical Technology and Metallurgy, Kl. Ohridski Blvd, 8, 1756 Sofia, Bulgaria
* Correspondence: albenadb@svr.igic.bas.bg

Abstract: TiO_2/TeO_2 powders were obtained by an aqueous sol-gel method. Telluric acid (H_6TeO_6) and titanium butoxide were used as precursors. The as-prepared gel was step-wisely heated in the temperature range 200–700 °C and subsequently characterized by XRD, IR, and UV-Vis analysis and SEM. Mixtures containing TiO_2 (anatase), α-TeO_2 (paratellurite), and $TiTe_3O_8$ were established by XRD as final products, depending on heating temperature. The thermal stability of the obtained gels in the temperature range 100–400 °C was investigated. It was found by IR spectroscopy that the samples heated up to 300–400 °C consist mainly of an organic–inorganic amorphous phase which is transformed into an inorganic one above these temperatures. The microstructure of the gels was verified by scanning electron microscopy (SEM). The photocatalytic degradation of the synthesized nanopowders toward Malachite green organic dye (MG) was examined in order to evaluate the potential applications for environmental remediation. The prepared TiO_2/TeO_2 samples showed up to 60% decoloration efficiency after 120 min exposure to UV-light. The composition exhibited good antimicrobial activity against *E. coli K12*. The properties of the obtained material were investigated by the reactions of complete catalytic oxidation of different alkanes and toluene, and it could be suggested that TiO_2/TeO_2 powders are promising material for use as an active phase in environmental catalysts.

Keywords: sol-gel; powders; antibacterial properties; photocatalytic properties

1. Introduction

Titania (TiO_2) is one of the most popular materials for (photo)catalytic applications and solar energy conversion. Its wide use for environmental remediation is due to the excellent properties such as non-toxicity, stability, and low cost. A number of excellent review papers devoted to various aspects of TiO_2 have been published [1–7]. The low temperature sol-gel method is widely used for production of TiO_2 nanoparticles with large surface area [8]. It is also known that due to nontoxicity, long-term photostability, and high effectiveness, TiO_2 has been widely utilized in mineralizing toxic and nonbiodegradable environmental contaminants [9]. TiO_2 also possesses good mechanical resistance as well as stability in acidic and oxidative environments. These properties make TiO_2 a desirable candidate for heterogeneous catalyst support [10].

The unceasing attention of scientists to the photocatalysis with metaloxide-based semiconductors is due to their ability for reducing the toxic substances and removal of heavy organic contaminants from wastewater [1,9,10]. The literature data revealed that the photocatalytic degradation to purify water is considered an effective and economical technique [1,9,11]. Among the various available metal oxides, TiO_2 is the preferable one as a prominent candidate for antibacterial and photocatalysis activity. Nowadays, new active photocatalysts have been obtained for Methylene blue and Oxytetracycline degradation under dark and visible light conditions [12,13]. Photocatalytic reactions at the surface of titanium dioxide have been attracting much attention in view of their practical applications to environmental cleaning such as self-cleaning of tiles, glasses, and windows [14]. As it is known TiO_2 absorbs only near-ultraviolet (UV) light and the doping with metal ions has been investigated to extend its absorption to the visible light region for better photocatalytic performance. The photocatalytic activity of metal-doped TiO_2 is determined by the type of dopant. It was reported that metal ion dopants strongly modify the anatase-rutile phase transition temperature, change the photoreactivity of TiO_2 nanosized particles, and enhance the catalytic properties of supported metals or oxides. Various methods for synthesis such as a hydrothermal route, sol-gel route, and wet impregnation have been studied for doped TiO_2 powders. Among them, the sol-gel process is the most widely used approach consisting of two steps: hydrolysis of a titanium salt and a further condensation reaction. It was found that this method leads to obtaining the greatest homogeneous distribution of the dopant in the host matrix and a high surface area of as-prepared TiO_2 particles [15]. TiO_2 shows relatively high reactivity and chemical stability under ultraviolet light (λ < 387 nm), and its energy exceeds the band gap of 3.3 eV in the anatase crystalline phase [14].

Malachite Green (MG), a triphenylmethane dye, is a compound that is mainly used in textile industries and partly used in aquaculture in fungicides and ectoparasiticides [16–18]. Several review papers concerning the wide range of toxicological effects of malachite green (MG) and its effect on the ecosystem have been published [19]. It has been generalized that the toxicity of this dye increases with exposure time, temperature, and concentration.

Additionally, TiO_2 NPs can be used as an antibacterial agent because of strong oxidation activity and superhydrophilicity. Specifically, their antimicrobial activity is greatly dependent on the photocatalytic performance of TiO_2, which depends strongly on its morphological, structural, and textural properties and these peculiarities have been described in details in several review papers and book chapters [20–23]. In recent years there has been an increase in the resistance of several bacterial strains to these substances, and therefore there is currently a great interest in the search for new antimicrobial substances. With this in mind, titanium dioxide nanoparticles are one of the antimicrobial NPs whose study still continues to gain interest. It is important to highlight that some research works have evidenced increased antimicrobial activity of TiO_2 NPs after irradiation with UV-A light due to the photocatalytic nature of this oxide. The mechanism referring to the antimicrobial action of TiO_2 is commonly associated to reactive oxygen species (ROS) which affect and destroy the bacterial cells by different mechanisms [24].

As a conditional network former, TeO_2 does not form glass alone, but tellurite glasses are attractive because of their excellent optical properties (high refractive index, high third order nonlinear properties, wide optical window from the ultraviolet to the infrared region, etc.) [25]. The sol-gel method is an alternative way to extend the application of new tellurite compositions. Powders and films in the TiO_2–TeO_2 and TeO_2–PbO–TiO_2 systems have been reported in several papers, irrespective of the anomalously high hydrolysis rate of Te(VI).

Nowadays, the most popular research field is designing of new nanomaterials using metal oxides with dual applications (photocatalyst and antibacterial agent). The nanocomposite potential's is due to their ease of synthesis, size, shape, and doping with countless metal and non-metal materials with different concentration which made them suitable to enhance the properties of existing material [10,26].

Our team has extensive experience in the sol-gel obtaining of composite powders containing TiO_2. Until now, we have obtained different binary, ternary, and multicomponent composites with good photocatalytic and antibacterial properties applying this technique [27–31]. This research is a continuation of our previous investigations on the sol-gel obtaining of TiO_2 nanocomposite powders by applying Te(VI) acid and Ti butoxide as a new combination of precursors. We continue searching for new combinations of compositions which have not previously been investigated with the focus on those with improved photocatalytic and antibacterial properties. To our knowledge, there are no reported data on TiO_2/TeO_2 sol-gel derived powders and study of their biocidal and environmental applications which emphasizes the novelty in the current work.

Progressing our studies, the aim of the present work is to synthesize TiO_2/TeO_2 composite by sol-gel technique and to investigate its thermal, structural, and optical properties. The investigation has been also focused on the antibacterial and photocatalytic properties of the as-prepared composition. Additionally, its catalytic properties for neutralization of waste gases containing VOCs were evaluated applying Pd.

2. Results and Discussion

2.1. *Phase Transformations and Thermal Stability of the Gel*

The XRD patterns of as-prepared TiO_2/TeO_2 sample heat treated from 200 to 700 °C are shown in Figure 1. The metallic tellurium (JCPDS 78-2312) is found predominantly up to 400 °C. At 400 °C it was registered that TiO_2 (anatase, JCPDS 78-2486) coexisted with metallic tellurium. With increasing the temperature (500 and 600 °C), the anatase became a dominant phase, while at 700 °C formation of crystalline $TiTe_3O_8$ phase (JCPDS 50-0250) along with TiO_2 (rutile) were observed. The average crystallite size of the particles, calculated from the broadening of the diffraction line using Sherrer's equation is about 60 nm. Irrespective of using a different precursor (Te(VI) acid) in the performed experiment, the obtained XRD data are very similar to those obtained by other teams [32–34].

The specific surface area (SSA) results utilizing Brunauer–Emmett–Teller (BET) nitrogen adsorption experiments showed that SSA for the TiO_2/TeO_2 sample was higher and had a value of 86 m^2/g, while it dropped to 71 m^2/g when the Pd was added.

DTA/TG curves of the as-prepared gel are presented in Figure 2. An endothermic effect appeared about 105 °C, which is attributed to the evaporation of organic solvent and desorption of physically adsorbed water. The observed weight loss is about 7%. Two broad exothermic peaks were registered at about 235 and 285 °C. The first one is related to the beginning of the decomposition of the organic groups (weight loss is ~8%) as in the pure TBT [35,36]. The second exothermic effect appeared at slightly higher temperature (285 °C) with around 15% weight loss, which suggests strong combustion of the organic components [37]. The noticed increase in the curve intensity about 400 °C suggests the presence of an exothermic effect due to the oxidation of Te to TeO_2. For comparison, other authors established that this oxidation proceeded at similar temperatures [32,38,39].

2.2. *IR and UV-Vis DRS Characterization*

The phase transformations during the heat treatment were also studied by means of IR spectroscopy which is a more sensitive method, and the IR spectra are shown in Figure 3. The assignment of the bands is made on the basis of the literature data as well as our previous investigations [40–42]. At low temperature (200 °C), intensive characteristic bands in the range 1460–880 cm^{-1} were observed. Generally, bands located between 1500–1300 cm^{-1} assigned to the bending vibrations of CH_3 and CH_2 groups [43,44]. The band at 1120 cm^{-1} is characteristic for the stretching vibrations of Ti-O-C, while those at 1080 and 1040 cm^{-1} are assigned to the vibrations of terminal and bridging C-O bonds in butoxy ligands. For the ethylene glycol the asymmetric and symmetric stretching vibrations of C-O bonds in CH_2-OH group are also at 1080 and 1040 cm^{-1}. In the absorption range below 1000 cm^{-1} there is an overlapping between the vibrations of different structural units TiO_6 and TeO_n and their assignments are difficult to be made because one and the same band could be related to the vibrations of different structural units. We can assume that the band centered about 790 cm^{-1} could be connected to the deformation vibrations of Ti-O-C bonds from butoxy groups [43,44]. On the other hand, bands in the range of 600–400 cm^{-1} could be also related to the vibrations of TiO_6 building units [45]. Additionally, the typical vibrations of TeO_4 groups building up the crystalline TeO_2 are situated about 770, 660, and 640–630 cm^{-1} [46–49].

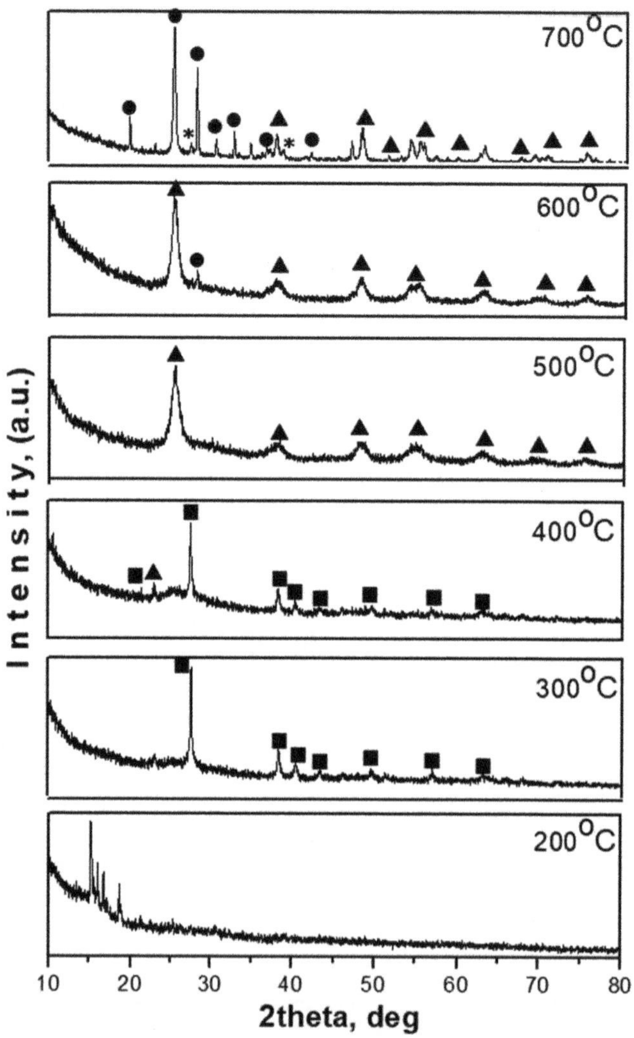

Figure 1. XRD patterns of sample with nominal composition $90TiO_2.10TeO_2$ heat treated at different temperatures: (■) Te, (▲) TiO_2-anatase, (✱) TiO_2-rutile, (•) $TiTe_3O_8$.

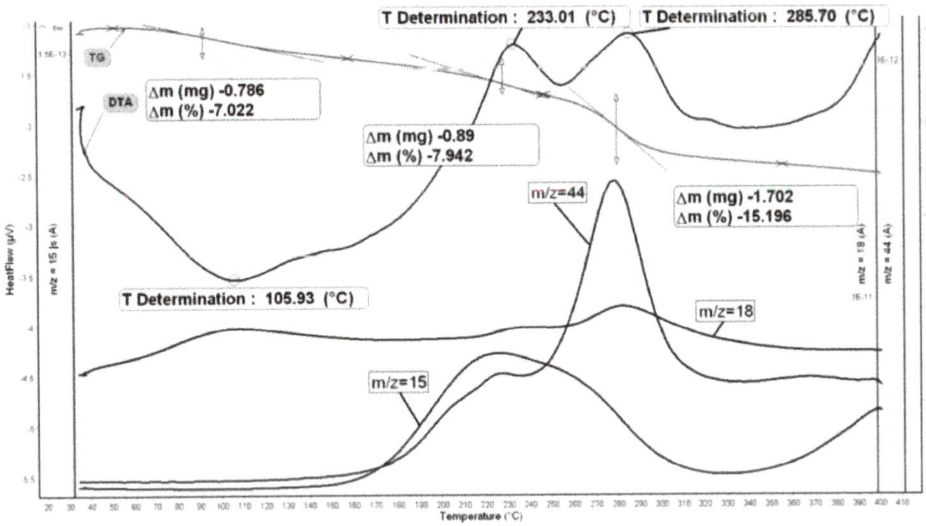

Figure 2. DTA/TG curves $90TiO_2.10TeO_2$ gel.

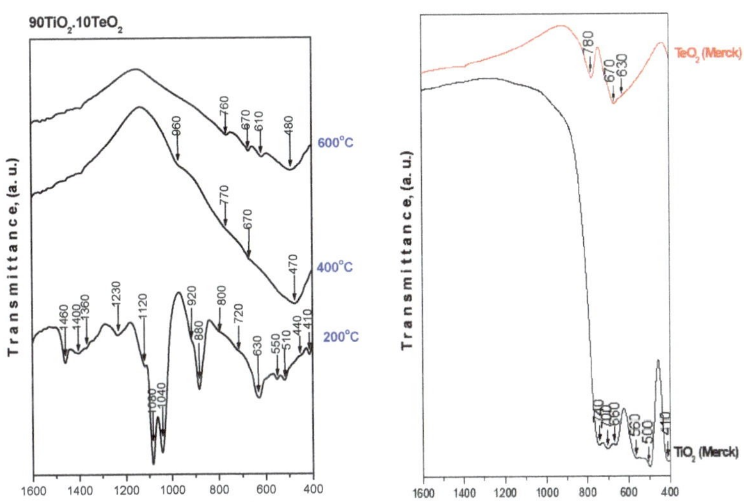

(a) (b)

Figure 3. IR spectra of: (a) investigated sample heat treated at different temperatures and (b) TeO_2 and TiO_2 commercial products.

The bands above 1000 cm^{-1}, characteristic of the organic groups, are not visible above 400 °C. The spectra at 400 and 600 °C are characterized mainly by bands below 900 cm^{-1}, typical for the inorganic units (TiO_6 and TeO_4 units). These bands are broadened, with the low intensity that is a peculiarity of disordered systems. At these temperatures the overlapping between these vibrations makes difficulties for the IR assignment. The structural data are in good accordance with the XRD data.

The UV–visible analysis is considered a non-destructive and it is a powerful tool for exploring the optical features of nanocomposite powders [10]. Numerous factors influence the absorbance, such as surface, impurity centers, oxygen deficiency, and bandgap [10]. The optical absorption spectra of both gels $90TiO_2.10TeO_2$ and Ti(IV) butoxide were shown in Figure 4. The absorption edge and optical band gap values of the samples are summarized in Table 1. The UV-Vis spectra exhibit two maxima 240–260 nm and 320–340 nm which could be assigned to the isolated TiO_4 and TiO_6 units, respectively [50]. As it seen from the figure, for TBT gel, the intensity of UV peaks in the spectra at 240 nm and that near 320 nm are compatible which is an indication of the compatible amount of TiO_4 and TiO_6 polyhedra in the gel network. In the $90TiO_2.10TeO_2$ gel the peak at 340 nm is dominant which is a result of additional activation of hydrolysis-condensation processes [50]. The other peculiarity is the red shift of the absorption edges in the spectra. These observations are consistent with the reports in the literature for Te— and C—modified TiO_2 compositions [51,52]. It was reported that [53] the doping of elementary Te into TiO_2 induced an increase in absorption due to smaller band gap of Te than that of TiO_2 (0.3 vs 3.2 eV). The UV-Vis spectra were also used to determine the optical band gap (E_{opt}) of investigated sample (Table 1). The obtained data are corroborated by the literature data [54].

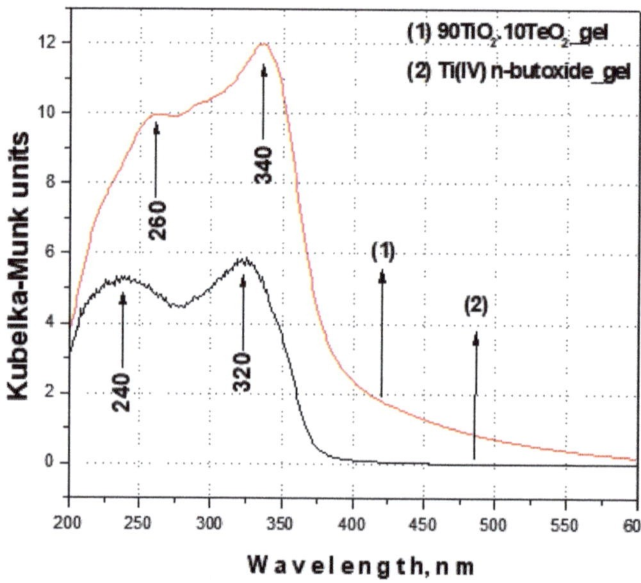

Figure 4. UV-Vis spectra of sample $90TiO_2.10TeO_2$ compared with Ti(IV) butoxide.

Table 1. Investigated TiO_2/TeO_2 gel, observed cut-off and calculated optical band gap values (E_g).

-	Sample	Cut-Off, nm	E_g, eV
1.	Ti(IV) butoxide	375	3.31
2.	$90TiO_2.10TeO_2$	395	3.14

2.3. *Transmission Electron Microscopy*

The HRTEM images performed in different parts of investigated samples $90TiO_2/10TeO_2$ and $90TiO_2/10TeO_2/Pd$ are shown in Figures 5–7. Predominantly spherical and well-shaped particles are observed with size between 50 nm to 100 nm, which corresponds to the results obtained by XRD. Bright field TEM micrographs from selected areas of the samples are shown in Figure 6. The composition map via STEM—XEDS analysis was carried out to examine the distribution of elements in the composite sample. The results of mapping studies (Figures 5–7) portray the coexistence of Ti, O, Te, and Pd elements and their homogeneous distribution. These results demonstrated also that Pd was uniformly doped in the TiO_2 nanoparticles and did not contain segregated particles. The obtained data from the qualitative elemental analysis performed in different points of the sample surface are summarized in Table 2.

Table 2. Elemental analysis according to TEM.

Sample	Map Sum Spectrum	wt %
$90TiO_2/10TeO_2$	Ti	51.09
	O	33.95
	Te	14.96
$90TiO_2/10TeO_2/Pd$	Ti	57.78
	O	22.08
	Te	18.84
	Pd	1.5

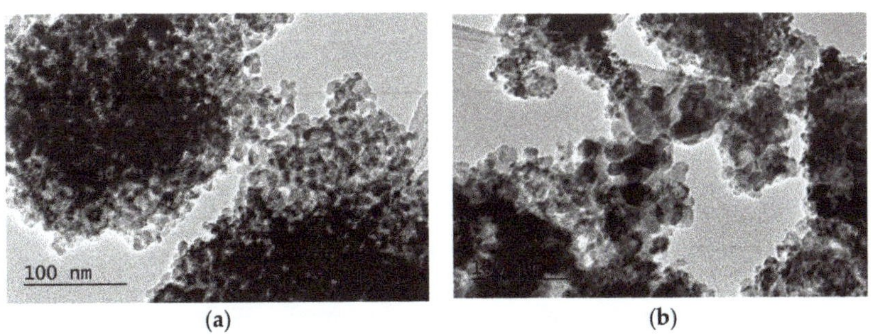

Figure 5. Bright field TEM micrographs from selected areas of samples $90TiO_2/10TeO_2$ heat treated at 600 °C (**a**) and $90TiO_2/10TeO_2/Pd$ (**b**).

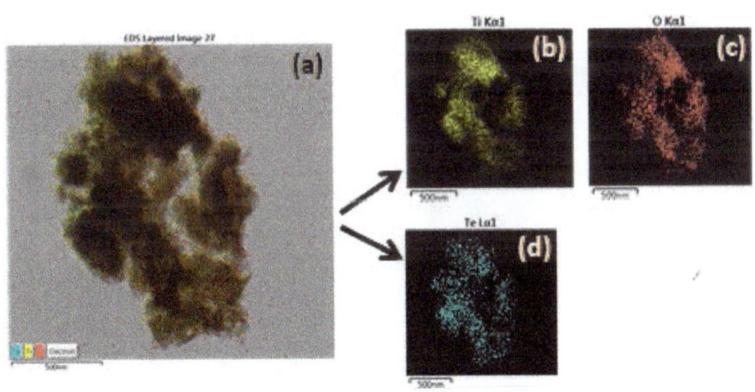

Figure 6. (**a**) Integral XEDS-STEM composition map from selected area of sample $90TiO_2/10TeO_2$ heat treated at 600 °C; (**b**) composition map of Ti; (**c**) composition map of O; (**d**) composition map of Te.

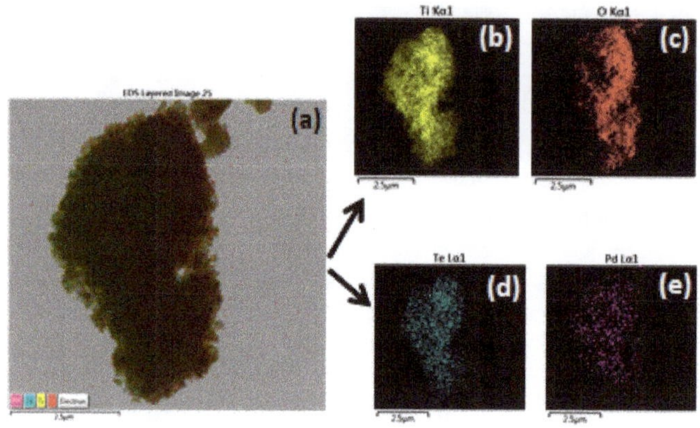

Figure 7. (**a**) Integral XEDS-STEM composition map from selected area of sample $90TiO_2/10TeO_2/Pd$ heat treated at 600 °C; (**b**) composition map of Ti; (**c**) composition map of O; (**d**) composition map of Te; (**e**) composition map of Pd.

2.4. Antibacterial Assessment

The investigated $90TiO_2.10TeO_2$ sample heat treated at 200 and 400 °C has been tested for antibacterial activity against *E. coli* K12 NBIMCC 407 (Figure 8). The testing has been performed by monitoring the cell reduction in their presence in a liquid medium and measuring the inhibition zones formed around the materials (Figure 8a,b).

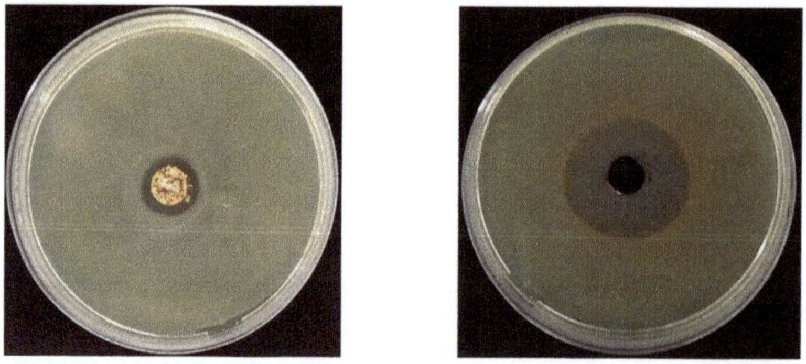

(**a**) (**b**)

Figure 8. Antibacterial test result of the sample $90TiO_2.10TeO_2$: (**a**) heat treated at 200 °C and (**b**) 400 °C against *E. coli* K12 407.

Both tested powdered materials showed good antibacterial activity against *E. coli* K12 NBIMCC 407. However, the highest level of bacterial cell reduction was achieved by the sample heated at 400 °C—68% reduction (Figure 8b). The other sample, heat treated at 200 °C showed weaker antibacterial activity—50%. Obviously, the temperature of the heat treatment does not greatly affect the antibacterial behavior of materials. Additionally, the antibacterial properties of the tested materials were estimated by second microbiological method refers to the measurement of the zones, free of bacterial growth. The obtained results revealed well-formed inhibition zones around both materials. Lower antibacterial activity was observed again by sample heated at the lower temperature (200 °C) (inhibition zone = 16 mm) (Figure 8a).

In interpretation of the obtained data, it must be underlined that there are many factors influencing the antimicrobial activity of the synthesized TiO_2 nanoparticles: morphology, crystal nature, size, and specific surface area [20,55,56]. Some authors [20] summarize that the enhancement of the specific surface area stimulates the antimicrobial reduction in TiO_2 nanoparticles. This could be explained by the nature of titanium dioxide, and one of the main mechanisms of its action is through the generation of reactive oxygen species (ROS) on its surface during the process of photocatalysis when it is exposed to light at an appropriate wavelength. During this process free radicals were produced that contribute to the biocidal activity of the samples by destruction of cellular organic compounds [57]. On the other hand, the antimicrobial activity of TiO_2 even in the absence of photo activation has been also reported [57]. It has been found that an electromagnetic attraction between microorganisms and the TiO_2 NPs occurred due to the positive charge carried by the TiO_2 and the negative charges carried by the microorganisms' surface which leads to oxidation reactions. In this way TiO_2 deactivates cellular enzymes and DNA by coordinating to electron-donating groups, such as: thiols, amides, carbohydrates, indoles, and hydroxyls. The resulting pits formed in bacterial cell walls lead to increased permeability and cell death [57].

Analyzing our results, it could be generalized that the samples heated at higher temperature (400 °C) showed better antibacterial activity (68%) compared to those heated at 200 °C (50%). Our suggestion is that these results could be explained bearing in mind the XRD results (Figure 1), where the presence of elemental tellurium at a higher temperature and metal-organic complexes at 200 °C were established. It has been reported that when metal nanoparticles and organics are present in the antibacterial agent, the release of their ions retards the Gram-negative bacteria [20,58]. On the other hand, the relatively good antibacterial activity at both temperatures could be related to the lower bandgap (3.14 eV) as well as the higher surface area (86 m^2/g) which contribute to the formation of highly reactive oxygen species (ROS) responsible for bacterial cell damage.

2.5. Photocatalytic Activity

The photocatalytic degradation of MG under UV irradiation has been tested in the presence of investigated $90TiO_2.10TeO_2$ sample. The photocatalytic properties were compared to those of TiO_2 obtained from Ti(IV) butoxide and to commercial TiO_2 (Fluka), used as a reference (Figure 9a). As can be seen from the figure, the best photocatalytic activity showed commercial pure TiO_2 (Fluka). The results of the photocatalytic test reveal that the addition of tellurium oxide did not improve the degradation of the MG dye at our experimental conditions and its effectiveness is comparable to those of TiO_2 obtained from Ti(IV) butoxide. These findings were confirmed by the application of pseudo-first-order kinetic model [59,60] for determination of the photodegradation rate constants in the degradation process of MG, commonly expressed by the equation:

$$\ln C_o/C_t = kt,$$

where k is the photodegradation rate constant (min^{-1}), C_o is the initial concentration and C_t is the concentration at any time, t.

Semi-logarithmic plots of the concentration data gave straight lines, from which slopes the rate constants are derived (Figure 9b). The rate constants of synthesized TiO_2/TeO_2 sample, pure TiO_2 and the commercial TiO_2 are 0.0072, 0.0065, and 0.0674 min^{-1}, respectively. All correlation coefficients (R^2) were higher than 0.947. As it is also seen from the figure, the best photocatalytic activity showed commercial pure TiO_2 (Fluka) with 100% decoloration efficiency after 60 min irradiation. The photodegradation efficiency of the TiO_2/TeO_2 sample was similar to that of synthesized from Ti(IV) butoxide pure TiO_2—both achieved only 35% bleaching after 60 min irradiation, and less than 60% after 120 min of illumination. It is well known that the photocatalytic activity of doped TiO_2 depends on many factors, such as synthesis method, dopant concentration, light source, particle size, and surface area. Obviously, at our experimental conditions, the presence of tellurium was not beneficial to the photoactivity under UV irradiation.

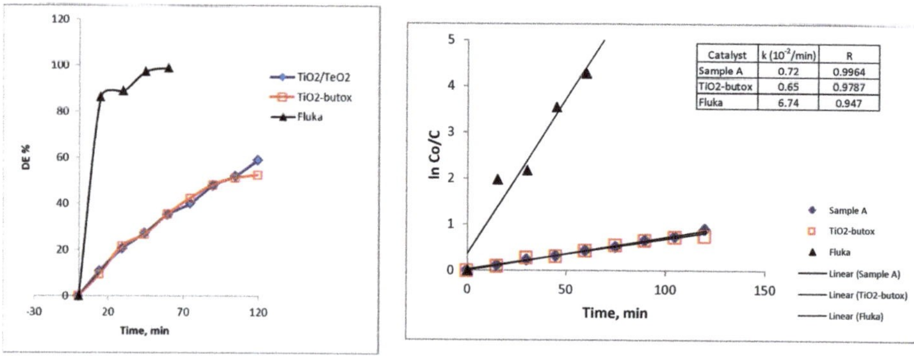

(**a**) (**b**)

Figure 9. (**a**) Photodegradation efficiency of TiO_2/TeO_2, TiO_2-butox and TiO_2-Fluka samples in decoloration of MG under UV irradiation; (**b**) Photocatalytic reaction kinetics of MG decoloration by TiO_2/TeO_2, TiO_2-butox and TiO_2-Fluka samples.

Girish et al. [61] generalized that it is difficult to compare the photocatalytic activities of doped with metal ion titania. They also stated that there are conflicting results on the effects of doping on the photoactivity of TiO_2. The wide variability of reports on the photoactivity of metal ion doped titania is due to the specific preparation methods, experimental conditions used, and the broad array of chemical reactions used to verify photoactivity over a range of wavelengths $\lambda > 400$ nm. Fundamentally, these issues can be understood in terms of: (i) lack of surface structure facilitating surface transfer and reactivity of carriers and (ii) the distribution of dopant-induced localized states not being optimal to facilitate the photo-oxidation reactions.

However, there is no study comparing the biocidal activity and photocatalytic properties of TiO_2/TeO_2 sample. In our investigations we observed different trends for photocatalytic and antibacterial activities. Although photocatalytic degradation and antibacterial activity are frequently attributed to the same mechanisms (reactive oxygen species (ROS) generation), it has been recently shown that they did not follow the same trends [60,62]. This phenomenon has been stated already in our previous paper concerning other systems containing TiO_2 [62]. The observed different trends could probably be explained by the direct oxidation/reduction processes occurring at the catalyst surface for the dye degradation while antibacterial activity may occur due to the involvement of ROS produced.

2.6. Catalytic Activity Test

The above-described properties of sol-gel derived $90TiO_2.10TeO_2$ offer the possibility to develop a thermally stable material with potential application in the environmental catalysis. In order to investigate this possibility, catalytic activity tests in the oxidation of CO and methane were performed. The modification of TiO_2 by Pd leads to remarkably high activity in the reaction of CO oxidation (Figure 10). At high gas hourly space velocity (GHSV = 46,150 h^{-1}) and in the presence of 1 vol. % water vapor, the reaction begins at almost ambient temperatures (50 °C). For practical application it is of interest to predict the activity at gas hourly space velocities ranged at 15,000 h^{-1}. The results, obtained by the mathematic modeling using the data at GHSV = 46,150 h^{-1} and recalculated further to GHSV = 15,000 h^{-1} are presented in Figure 11. Obviously, in the presence of water vapor one may achieve T_{90} (temperature for conversion of 90%) conversion of CO at 120 °C, T_{50} being 89 °C.

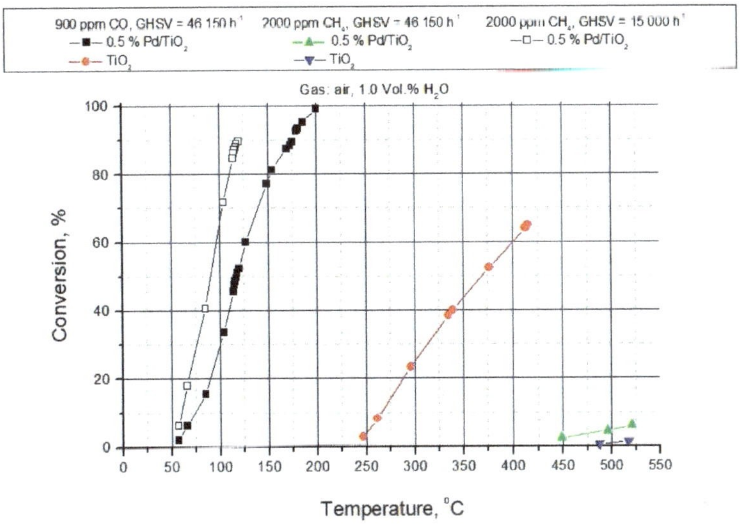

Figure 10. Temperature dependencies of the conversion during the catalytic oxidation of CO, methane over TiO_2 and TiO_2/0.5 wt% Pd catalysts.

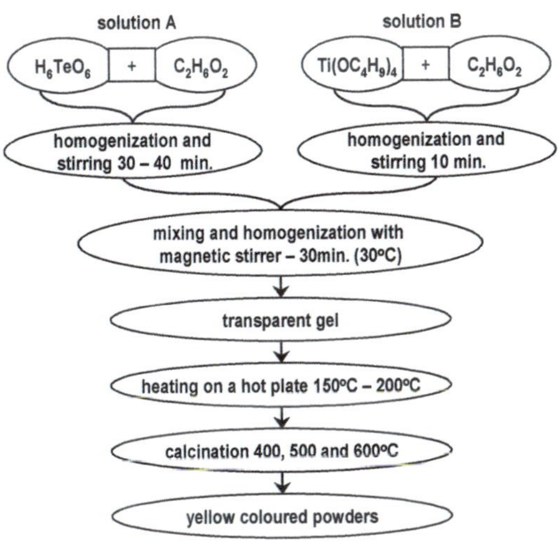

Figure 11. Scheme for the sol-gel synthesis of samples.

The activity towards methane combustion is relatively low, when compared with similar Pd-based catalysts; however, the use of TiO_2 as catalytic support increases the SO_2—tolerance of the exhaust gas catalysts [63,64], due to the preferred sulfite formation on the catalytic surface, rather than the formation of the stable and difficult to remove sulfates.

3. Materials and Methods

3.1. Samples Preparation

Based on our previous study on the gel formation in the binary TiO_2-TeO_2 system, a composition containing 90 mol% TiO_2 was selected ($90TiO_2.10TeO_2$) and subjected to detailed investigations. Knowing that tellurium alkoxides possess a high hydrolysis rate [37,65–67] we used Te(VI) acid (H_6TeO_6, Sigma-Aldrich, Darmstadt, Germany) in a combination of Ti(IV) butoxide (Fluka AG, Seelze, Germany) and ethylene glycol ($C_2H_6O_2$) (99% Sigma-Aldrich, Darmstadt, Germany). The scheme for synthesis is presented in Figure 11. The initial solutions were prepared by dissolving the precursors in ethylene glycol by vigorous magnetic stirring while keeping the molar ratio T(IV) butoxide/$C_2H_6O_2$ = 1:1 and $H_6TeO_6/C_2H_6O_2$ = 1:2. Finally, both solutions were mixed with vigorous stirring. During the experimental procedure, no additional water was added. The sol-gel hydrolysis reaction was accomplished only in the presence of air moisture. By doing this, a transparent gel was obtained and followed by heat treatment at 200 °C in air in order to hydrolyze unreacted -OR groups as well as to decrease their content. Aiming to verify the phase and structural transformations, the gels were subjected to further stepwise heating from 300 to 700 °C. The calcination of the samples was performed for 2 h exposure time in air, until obtaining powder. The calcination temperature was selected on the basis of our previous investigations [68]. The pH during the experiments was measured at 7.5–8.0. The photocatalytic properties of the synthesized powders were compared with those of a commercial TiO_2 (Fluka AG, Seelze, Germany).

Pd Impregnation over $90TiO_2/10TeO_2$

For verifying the catalytic properties, the so prepared $90TiO_2/10TeO_2$ powdered samples were modified by Pd (0.5 wt%) using the incipient wetness impregnation method. The impregnation was carried out at room temperature for 6 h using a solution of $Pd(NO_3)_2$ (99.9%, Alfa Aesar, Kandel, Germany), drying at room temperature overnight and calcination at 500 °C for 2 h in argon atmosphere [69].

3.2. *Samples Characterization*

Powder XRD patterns were registered at room temperature with a Bruker D8 Advance (Berlin, Germany) X-ray powder diffractometer with a Cu Ka radiation (k = 1.54056 Å) with a LynxEye solid position sensitive detector and X-ray tube operated at 40 kV and 40 mA. X-ray diffraction patterns were recorded in the range of 5.3–80° 2 h with a step of 0.02° 2 h. The decomposition process of the as prepared and aged in air gels was determined by differential thermal analysis (LABSYS^{TM} EVO apparatus, Setaram, Lyon, France) with Pt-Pt/Rh thermocouple at a heating rate of 10 K/min in air flow, using Al_2O_3 as a reference material. The accuracy of the temperature was ±5 °C. Heating of the samples was limited up to 600 °C. Gases evolved (EGA) during the thermal treatments were analyzed by mass spectrometry (MS) with a Pfeiffer OmniStarTM mass spectrometer (Pfeiffer Vacuum Technology AG, Wetzlar, Germany). Mass spectra recorded for samples A, C and D (Figure 3) show the m/z= 15, 18 and 44 signals, being ascribed to CH_3, H_2O and CO_2, respectively. The infrared spectra were registered in the range 1600–400 cm^{-1} using the KBr pellet technique on a Nicolet-320 FTIR spectrometer (Madison, USA) with 64 scans and a resolution of ±1 cm^{-1}. The optical absorption spectra of the powdered samples in the wavelength range 200–1000 nm were recorded by a UV–VIS diffused reflectance Spectrophotometer Evolution 300 (Thermo Electron Corporation, Madison, WI, USA) using a magnesium oxide reflectance standard as the baseline. The band gap energies (E_g) of the samples were calculated by the Planck's equation: $E_g = \frac{h.c}{\lambda} = \frac{1240}{\lambda}$, where E_g is the band gap energy (eV), h is the Planck's constant, c is the light velocity (m/s), and λ is the wavelength (nm). The morphology of the samples was examined by Transmission Electron Microscopy (TEM) investigations were performed on a JEOL JEM 2100 instrument (Tokyo, Japan) at an accelerating voltage of 200 kV. The specimens were prepared by grinding and dispersing them in ethanol by ultrasonic treatment for 6 min. The suspensions were dripped on standard carbon/Cu grids. Additional support for the existence of all participated elements in the investigated samples was performed by X-ray energy dispersive spectrometry (XEDS) elemental mapping studies. The specific surface areas (BETs) were determined by low-temperature (77.4 K) nitrogen adsorption in NOVA 1200e (Quantachrome Instruments NOVA 1200e apparatus, Anton Paar GmbH,

Graz, Austria) surface area and pore analyzer at relative pressures p/p0 = 0.1–0.3 using BET equation.

3.2.1. Photocatalytic Activity Experiments

The photocatalytic performance of the synthesized samples was evaluated by the extent of photodegradation of the organic dye Malachite green (MG) as a model pollutant. Reaction suspensions were prepared by adding 100 mg of synthesized photocatalysts to 150 mL MG water solution. For all experiments, the concentration of MG dye was 5 ppm.

Before irradiation, the suspensions were magnetically stirred and maintained for 30 min in the dark to attain the absorption–desorption equilibrium between the photocatalyst surface and dye molecules. Irradiation with UV-light was provided by a black light blue lamp (Sylvania BLB 50 Hz 8W T5) with the major fraction of irradiation occurring at 365 nm. The lamp was fixed 10 cm above the solution surface. Each MG/catalyst suspension has been exposed to UV-light with vigorous stirring. The zero timing was the moment of the UV lamp switch on. At specific time intervals, an aliquot (3 mL) of the suspension was extracted from the photoreacted mixture and centrifuged to remove the catalyst particles. At our experimental conditions the pH of the dye solution during the adsorption process varied in the range 5–6 and the pH values of supernatants after illumination were in the range 6–7. The progress of the gradual decomposition of dye was monitored by a Jenway 6505 UV-Vis spectrophotometer (designed and manufactured in UK) at the maximum absorption wavelength of MG—618 nm.

It is a fact that the absorption intensity is directly proportional to dye concentration. Thus, the percentage of degradation of dye was estimated using the following equation:

$$D\% = \frac{C_o - C_t}{C_o} \times 100$$

where D denotes the percentage of photodegradation efficiency, C_o is the dye concentration at zero time and C_t is the concentration after the illumination time t.

3.2.2. Antimicrobial Assay

The antibacterial activity of the tested materials was investigated using Escherichia coli K12 NBIMCC 407 as a bacterial reference strain. It was obtained from the National Bank for Industrial Microorganisms and Cell Cultures (NBIMCC, Bulgaria) and cultivated in Luria-Bertani (LB) broth on a shaker-incubator ES-20/60 (Biosan, Riga, Latvia, 120 rpm) at 37 °C for 24 h. The first method of antimicrobial assay consists of the determination of antibacterial potential of the materials by studying the cell reduction after exposure to them. For this purpose, 100 µL of the strain suspension (OD_{610} nm = 1.9) and 10 mg of each material were added to flasks, containing 100 mL LB broth. The control was a flask containing only bacterial cells (without materials added). After 24 h of incubation, 50 µL of the culture broth of each flask was seeded on sterile LB agar plates. The number of colonies grown after 24-h incubation in an incubator (Binder, Germany) was used to calculate the percentage of cell reduction, according to Bachvarova-Nedelcheva et al. [56]. In addition, the antibacterial potential of the materials against the test strain was also determined by the agar-well diffusion method [70]. Sterile LB agar plates were inoculated with 100 µL of an exponential culture of *E. coli* K12 NBIMCC 407, using the spread-plate method. A quantity of 100 mg of the materials was loaded onto marked wells, prepared in advance in the agar plates and were then cultured in an incubator at 37 °C for 24 h. The antibacterial activity was assessed by measuring the diameter of the obtained inhibition zones. All tests were performed in triplicate and the results obtained showed the mean values.

3.2.3. Catalytic Activity Tests

The following testing conditions were applied: catalyst bed volume of 0.5 cm^3, irregular shaped particles having an average diameter of 0.7 mm, reactor diameter of 8.0 mm, and quartz glass (Dreactor/Dparticles > 10). The gas hourly space velocity (GHSV) was fixed to 46,150 h^{-1}. For compensation of the adiabatic effect of the reaction, the catalyst bed temperature was kept almost constant (the deviation did not exceed ±1 °C). The pressure drop was measured to be below 0.21 m water column. The axial dispersion effect was not subtracted since the catalyst bed is corresponding to a chain of more than 10 ideal-mixing cells along the reactor length. Therefore, the geometrical characteristics and the flow conditions in the catalytic reactor allow the consideration that the reactor is operating as isothermal plug flow reactor (PFR). The inlet concentrations of reactants were fixed as follows: methane concentrations at 2000 ppm; CO 900 ppm; and oxygen: 20.9 vol%. All gas mixtures were balanced to 100% with nitrogen (4.0 grade). The ability to reproduce the results and the confidence intervals for the measured conversion degree were the subject of preliminary tests on repeating the experimental runs under conditions similar (but not identical) for each experimental run represented in the study. The calculated value for the standard deviation (±1.5%) was calculated on the basis of six measurements in each experimental point. The reported results are based on averaged values for the conversion degree within two parallel measurements. The gas analysis was performed by the using of mass spectrometer of the CATLAB system (Hiden Analytical), additionally supplemented by on-line gas-analyzers of $CO/CO_2/O_2$ (Maihak) and THC-FID (total hydrocarbon content with a flame ionization detector, Horiba). The possible formation of oxidized organics by products was monitored by a GC-MS-analyzer, model Agilent AL7890A (Santa Clara, CA, USA).

4. Conclusions

Aqueous sol-gel method was used for preparation of $90TiO_2/10TeO_2$ gel which was subjected to detailed investigations. It was established by XRD and IR analysis that the sample preserved the organic constituents up to 300 °C and the organic–inorganic amorphous phase transformed into inorganic one above 400 °C. The simultaneous existence of several crystalline phases TiO_2 (anatase), α-TeO_2 and $TiTe_3O_8$ were registered up to 700 °C. The UV—Vis spectroscopy showed two maxima about 240–260 nm and 320–340 nm related to the isolated TiO_4 units and condensed TiO_6 groups. The photocatalytic test reveals that the addition of tellurium oxide did not lead to better activity. The photodegradation efficiency of the TiO_2/TeO_2 sample was similar to that of synthesized from Ti(IV) butoxide pure TiO_2—both achieved only 35% bleaching after 60 min irradiation, and less than 60% after 120 min of illumination. The composition exhibited satisfactory antimicrobial activity against *E. coli* K12 as the samples heated at higher temperature (400 °C) showed better antibacterial activity (68%) compared to those heated at 200 °C (50%). The properties of the obtained material were investigated by the reactions of complete catalytic oxidation of different alkanes and toluene and it could be suggested that TiO_2/TeO_2 powders can be considered as prospective material for use as an active phase in the preparation of environmental catalysts. The mathematical model simulations show that the industrial adiabatic reactor for 99% abatement of CO at gas flow of 10,000 Nm^3/h should have the following dimensions: catalyst bed length 0.03 m; catalytic bed diameter 2.22 m (inlet temperature 118 °C; outlet temperature 126.7 °C). Therefore, further investigation is required to develop more efficiency of such powders aiming to improve their applications.

Author Contributions: Conceptualization, A.B.-N. and R.I.; methodology, A.B.-N. and R.I.; photocatalytic tests, A.S.; antibacterial properties N.G., V.N. and T.F.; catalytic tests, A.N.; writing—A.B.-N., A.S. and A.N.; supervision, R.I. and A.S. All authors have read and agreed to the published version of the manuscript.

Funding: This research received no external funding.

Data Availability Statement: The data presented in this study are available on request from the corresponding author.

Acknowledgments: This work was supported by the European Regional Development Fund within the Operational Programme "Science and Education for Smart Growth 2014–2020" under the Project CoE "National center of mechatronics and clean technologies" BG05M2OP001-1.001-0008. Research equipment of distributed research infrastructure INFRAMAT supported by Bulgarian Ministry of Education and Science under contract D01-284/17.12.2019 was used.

Conflicts of Interest: The authors declare no conflict of interest.

References

1. Carp, O.; Huisman, C.L.; Reller, A. Photoinduced reactivity of titanium dioxide. *Prog. Solid State Chem.* **2004**, *32*, 33–117. [CrossRef]
2. Fujishima, A.; Rao, T.N.; Tryk, D.A. Titanium dioxide photocatalysis. *J. Photochem. Photobiol. C Photochem. Rev.* **2000**, *1*, 1–21. [CrossRef]
3. Mor, G.K.; Varghese, O.K.; Paulose, M.; Shankar, K.; Grimes, C.A. A review on highly ordered, vertically oriented TiO_2 nanotube arrays: Fabrication, material properties, and solar energy applications. *Sol. Energy Mater. Sol. Cell* **2006**, *90*, 2011–2075. [CrossRef]
4. Chen, X.; Mao, S.S. Titanium dioxide nanomaterials: Synthesis, properties, modifications, and applications. *Chem. Rev.* **2007**, *107*, 2891–2959. [CrossRef] [PubMed]
5. Thompson, T.L.; Yates, J.T., Jr. Surface science studies of the photoactivation of TiO_2-New photochemical processes. *Chem. Rev.* **2006**, *106*, 4428–4453. [CrossRef]
6. Diebold, U. The surface science of titanium dioxide. *Surf. Sci. Rep.* **2003**, *48*, 53–229. [CrossRef]
7. Linsebigler, A.L.; Lu, G.; Yates, J.T. Photocatalysis on TiO_2 surfaces: Principles, mechanisms, and selected results. *Chem. Rev.* **1995**, *95*, 735–758. [CrossRef]
8. Gupta, M.; Tripathi, M. A review of TiO_2 nanoparticles. *Chin. Sci. Bull.* **2011**, *56*, 1639–1657. [CrossRef]
9. Nakata, K.; Fujishima, A. TiO_2 Photocatalysis: Design and Applications. *J. Photochem. Photobiol. C Photochem. Rev.* **2012**, *13*, 169–189. [CrossRef]
10. Mukhtar, F.; Munawar, T.; Nadeem, M.S.; Batool, S.; Hasan, M.; Riaz, M.; Iqbal, F. Highly efficient tri-phase TiO_2-Y_2O_3-V_2O_5 nanocomposite: Structural, optical, photocatalyst, and antibacterial studies. *J. Nanostruct. Chem.* **2022**, *12*, 547–564. [CrossRef]
11. Bagheri, S.; Julkapli, N.M.; Hamid, S.B.A. Titanium Dioxide as a Catalyst Support in Heterogeneous Catalysis. *Sci. World J.* **2014**, *2014*, 727496. [CrossRef] [PubMed]
12. Kumar, O.P.; Shahzad, K.; Nazir, M.A.; Farooq, N.; Malik, M.; Ahmad Shah, S.S.; ur Rehman, A. Photo-Fenton activated $C_3N_{4x}/AgO_y@Co_{1-x}Bi_{0\ 1-y}O_7$ dual s-scheme heterojunction towards degradation of organic pollutants. *Opt. Mater.* **2022**, *126*, 112199. [CrossRef]
13. Tóth, Z.-R.; Debreczeni, D.; Gyulavani, T.; Szekely, I.; Todea, M.; Kovacs, G.; Focsa, M.; Magyari, K.; Baia, L.; Pap, Z.; et al. Rapid Synthesis Method of Ag_3PO_4 as Reusable Photocatalytically Active Semiconductor. *Nanomaterials* **2023**, *13*, 89. [CrossRef] [PubMed]
14. Zaleska, A. Doped-TiO_2: A Review. *Recent Pat. Eng.* **2008**, *2*, 157–164. [CrossRef]
15. Chen, L.; Rahme, K.; Holmes, J.D.; Morris, M.A.; Slater, N.K.H. Non-solvolytic synthesis of aqueous soluble TiO_2 nanoparticles and real-time dynamic measurements of the nanoparticle formation. *Nanoscale Res. Lett.* **2012**, *7*, 297–307. [CrossRef]
16. Srivastava, S.; Sinha, R.; Roy, D. Toxicological effects of malachite green—Review. *Aquat. Toxicol.* **2004**, *66*, 319–329. [CrossRef] [PubMed]
17. Fu, X.-Y.; Zhao, W.; Xiong, A.-S.; Tian, Y.-S.; Zhu, B.; Peng, R.-H.; Yao, Q.-H. Phytoremediation of triphenylmethane dyes by overexpressing a *Citrobacter* sp. triphenylmethane reductase in transgenic Arabidopsis. *Appl. Microbiol. Biotechnol.* **2013**, *97*, 1799–1806. [CrossRef]
18. Nazir, M.A.; Najam, T.; Bashir, M.S.; Javed, M.S.; Bashir, M.A.; Imran, M.; Azhar, U.; Shah, S.S.A.; Rehman, A.U. Kinetics, isothermal and mechanistic insight into the adsorption of eosin yellow and malachite green from water via tri-metallic layered double hydroxide nanosheets. *Korean J. Chem. Eng.* **2022**, *39*, 216–226. [CrossRef]
19. Raval, N.P.; Shah, P.U.; Shah, N.K. Malachite green "a cationic dye" and its removal from aqueous solution by adsorption. *Appl. Water Sci.* **2017**, *7*, 3407–3445. [CrossRef]
20. Franco, D.; Calabrese, G.; Guglielmino, S.P.P.; Conoci, S. Metal-Based Nanoparticles: Antibacterial Mechanisms and Biomedical Application. *Microorganisms* **2022**, *10*, 1778. [CrossRef]
21. Hamimed, S.; Chatti, A. Chapter 7: Photocatalytic metal bionanocomposites for biomedical applications. In *Bionanotechnology: Emerging Applications of Bionanomaterials*; Elsevier Inc.: Amsterdam, The Netherlands, 2022; pp. 227–258. [CrossRef]
22. Ishiguro, H.; Nakano, R.; Yao, Y.; Kajioka, J.; Fujishima, A.; Sunada, K.; Minoshima, M.; Hashimoto, K.; Kubota, Y. Photocatalytic inactivation of bacteriophages by TiO2-coated glass plates under low-intensity, long-wavelength UV irradiation. *Photochem. Photobiol. Sci.* **2011**, *10*, 1825–1829. [CrossRef] [PubMed]
23. De Dicastillo, C.L.; Correa, M.G.; Martínez, F.B.; Streitt, C.; Galotto, M.J. Antimicrobial Effect of Titanium Dioxide Nanoparticles. In *Antimicrobial Resistance-A One Health Perspective*; IntechOpen: London, UK, 2020. [CrossRef]
24. Verdier, T.; Coutand, M.; Bertron, A.; Roques, C. Antibacterial activity of TiO_2 photocatalyst alone or in coatings on *E. coli*: The influence of methodological aspects. *Coatings* **2014**, *4*, 670–686. [CrossRef]
25. Vogel, W. *Chemistry of Glass*; Wiley-American Ceramic Society: Hoboken, NJ, USA, 1985.

26. Mukhtar, F.; Munawar, T.; Shahid Nadeem, M.; Naveedur Rehman, M.; Mahmood, K.; Batool, S.; Hasan, M.; Rehman, K.; Iqbal, F. Enhancement in carrier separation of ZnO-Ho_2O_3-Sm_2O_3 hetrostuctured nanocomposite with rGO and PANI supported direct dual Z-scheme for antimicrobial inactivation and sunlight driven photocatalysis. *Adv. Powder Technol.* **2021**, *32*, 3770–3787. [CrossRef]
27. Stoyanova, A.; Bachvarova-Nedelcheva, A.; Iordanova, R.; Ivanova, N.; Hitkova, H.; Sredkova, M. Synthesis, photocatalytic and antibacterial properties of nanosized $ZnTiO_3$ powders obtained by different sol–gel methods. *Dig. J. Nanomater. Biostruct.* **2012**, *7*, 777–784.
28. Shalaby, A.; Bachvarova-Nedelcheva, A.; Iordanova, R.; Dimitriev, Y.; Stoyanova, A.; Hitkova, H.; Ivanova, N. Sol-gel synthesis and properties of nanocomposites in the Ag/TiO_2/ZnO system. *J. Optoelectron. Adv. Mater.* **2015**, *17*, 248–256.
29. Bachvarova-Nedelcheva, A.; Iordanova, R.; Stoyanova, A.; Georgieva, N.; Angeova, T. Sol-gel synthesis of Se and Te containing TiO_2 nanocomposites with photocatalytic and antibacterial properties. *J. Optoel. Adv. Mater.* **2016**, *18*, 5–9.
30. Gegova, R.; Bachvarova-Nedelcheva, A.; Iordanova, R.; Dimitriev, Y. Synthesis and crystallization of gels in the TiO_2-TeO_2 -ZnO system. *Bulg. Chem. Commun.* **2015**, *47*, 378–386.
31. Stoyanova, A.; Hitkova, H.; BachvarovaNedelcheva, A.; Iordanova, R.; Ivanova, N.; Sredkova, M. Synthesis and antibacterial activity of TiO_2/ZnO nanocomposites prepared via nonhydrolytic route. *J. Chem. Technol. Metall.* **2013**, *48*, 154–161.
32. Weng, L.; Hodgson, S. Sol-gel processing of tellurite powder. *J. Mater. Sci.* **2001**, *36*, 4955–4959. [CrossRef]
33. Hodgson, S.N.B.; Weng, L. Processing and characterization of sol-gel glasses and powders in the system TeO_2-TiO_2. *J. Mater. Sci.* **2002**, *37*, 3059–3066. [CrossRef]
34. Hayakawa, T.; Koyama, H.; Nogami, M.; Thomas, P. Optical properties of TeO_2-TiO_2 thin films doped with Eu^{3+} ions fabricated by sol-gel processing. *J. Univ. Chem. Technol. Metall.* **2012**, *47*, 381–386.
35. Iordanova, R.; Bachvarova-Nedelcheva, A.; Gegova, R.; Dimitriev, Y. Sol-gel synthesis of composite powders in the TiO_2-TeO_2-SeO_2 system. *J. Sol-Gel Sci. Technol.* **2016**, *79*, 12–28. [CrossRef]
36. St I, Y.; Bachvarova-Nedelcheva, A.D.; Iordanova, R.S. Influence of ethylene glycol on the hydrolysis-condensation behavior of Ti(IV) butoxide. *Bulg. Chem. Commun. Spec. Issue A* **2017**, *49*, 265–270.
37. Lecomte, A.; Bamiere, F.; Coste, S.; Thomas Ph Champarnaud-Mesjard, J.C. Sol-gel processing of TeO_2 thin films from citric acid stabilized tellurium isopropoxide precursor. *J. Eur. Ceram. Soc.* **2007**, *27*, 1151–1158. [CrossRef]
38. Saylikan, F.; Asilturk, M.; Saylikan, H.; Onal, Y.; Akarsu, M.; Aprac, E. Characterization of TiO_2 synthesized in alcohol by a sol-gel process: The effects of annealing temperature and acid catalyst. *Turk. J. Chem.* **2005**, *29*, 697–706.
39. Madarász, J.; Brăileanu, A.; Crişan, M.; Pokol, G. Comprehensive evolved gas analysis (EGA) of amorphous precursors for s-doped titania by in situ TG–FTIR and TG/DTA–MS in air: Part 2. Precursor from thiourea and titanium(IV)-n-butoxide. *J. Anal. Appl. Pyrolysis* **2009**, *85*, 549–556. [CrossRef]
40. Doeuff, S.; Henry, M.; Sanchez, C.; Livage, J. Hydrolysis of titanium alkoxides: Modification of the molecular precursor by acetic acid. *J. Non-Cryst. Solids* **1987**, *89*, 206–216. [CrossRef]
41. Wei, H.Y.; Lin, J.; Huang, W.H.; Feng, Z.B.; Li, D.W. Preparation of TeO_2 based thin films by nonhydrolytic sol–gel process. *Mater. Sci. Eng. B* **2009**, *164*, 51–59. [CrossRef]
42. Kato, K.; Tsuge, A.; Niihara, K. Microstructure and Crystallographic Orientation of Anatase Coatings Produced from Chemically Modified Titanium Tetraisopropoxide. *J. Am. Ceram. Soc.* **1996**, *79*, 1483–1488. [CrossRef]
43. Barboux-Doeuff, S.; Sanchez, C. Synthesis and characterization of titanium oxide-based gels synthesized from acetate modified titanium butoxide precursors. *Mater. Res. Bull.* **1994**, *29*, 1–13. [CrossRef]
44. Uzunova-Bujnova, M.; Dimitrov, D.; Radev, D.; Bojinova, A.; Todorovsky, D. Effect of the mechanoactivation on the structure, sorption and photocatalytic properties of titanium dioxide. *Mater. Chem. Phys.* **2008**, *110*, 291–298. [CrossRef]
45. Murashkevich, N.; Lavitskaya, A.S.; Barannikova, T.I.; Zharskii, I.M. Infrared absorption spectra and structure of TiO_2. *J. Appl. Spectrosc.* **2008**, *75*, 730–734. [CrossRef]
46. Bachvarova-Nedelcheva, A.; Iordanova, R.; Kostov, K.L.; Yordanov St Ganev, V. Structure and properties of a Non-Traditional Glass Containing TeO_2, SeO_2 and MoO_3. *Opt. Mater.* **2012**, *34*, 1781–1787. [CrossRef]
47. Dimitriev, Y.; Dimitrov, V.; Arnaudov, M. IR spectra and structures of tellurite glasses. *J. Mater. Sci.* **1983**, *18*, 1353–1358. [CrossRef]
48. Arnaudov, M.; Dimitrov, V.; Dimitriev, Y. Infrared-spectral investigation of tellurites. *Mater. Res. Bull.* **1982**, *17*, 1121–1129. [CrossRef]
49. Iordanova, R.; Gegova, R.; Bachvarova-Nedelcheva, A.; Dimitriev, Y. Sol-gel synthesis of composites in the ternary TiO_2-TeO_2-B_2O_3 system. *Phys. Chem. Glasses Eur. J. Glass Sci. Technol. B* **2015**, *56*, 128–138. [CrossRef]
50. Barlier, V.; Bounor-Legare, V.; Boiteux, G.; Davenas, J. Hydrolysis–condensation reactions of titanium alkoxides in thin films: A study of the steric hindrance effect by X-ray photoelectron spectroscopy. *Appl. Surf. Sci.* **2008**, *254*, 5408–5412. [CrossRef]
51. Nagaveni, K.; Hegde, M.S.; Ravishankar, N.; Subbanna, G.N.; Madras, G. Synthesis and Structure of Nanocrystalline TiO_2 with Lower Band Gap Showing High Photocatalytic Activity. *Langmuir* **2004**, *20*, 2900–2907. [CrossRef]
52. Ngyen Vi Nu Hoai Beydoun, D.; Amal, R. Photocatalytic reduction of selenite and selenate using TiO_2 photocatalyst. *J. Photochem. Photobiol. A Chem.* **2005**, *171*, 113–120.
53. Meng, Z.D.; Zhu, L.; Ullah, K.; Ye, S.; Oh, W.-C. Detection of oxygen species generated by WO_3 modification fullerene/TiO_2 in the degradation of 1,5-diphenyl carbazide. *Mater. Res. Bull.* **2014**, *56*, 45–53. [CrossRef]

54. Weng, L.; Hodgson, N.B. Multicomponent tellurite thin film materials with high refractive index. *Opt. Mater.* **2002**, *19*, 313–317. [CrossRef]
55. Saidin, S.; Jumat, M.A.; Amin, N.A.A.M.; Al-Hammadi, A.S.S. Organic and inorganic antibacterial approaches in combating bacterial infection for biomedical application. *Mater. Sci. Eng. C* **2021**, *118*, 111382. [CrossRef] [PubMed]
56. Bachvarova-Nedelcheva, A.; Yordanov St Iordanova, R.; Stambolova, I.; Stoyanova, A.; Georgieva, N.; Nemska, V. The influence of Nd and Sm on the structure and properties of sol-gel derived TiO_2 powders. *Molecules* **2021**, *26*, 3824. [CrossRef] [PubMed]
57. Nakano, R.; Hara, M.; Ishiguro, H.; Yao, Y.; Ochiai, T.; Nakata, K.; Murakami, T.; Kajioka, J.; Sunada, K.; Hashimoto, K.; et al. Broad spectrum microbicidal activity of photocatalysis by TiO_2. *Catalysts* **2013**, *3*, 310–323. [CrossRef]
58. Godoy-Gallardo, M.; Eckhard, U.; Delgado, L.M.; de Roo Puente, Y.J.; Hoyos-Nogués, M.; Gil, F.J.; Perez, R.A. Antibacterial Approaches in Tissue Engineering Using Metal Ions and Nanoparticles: From Mechanisms to Applications. *Bioact. Mater.* **2021**, *6*, 4470–4490. [CrossRef]
59. Kumar, K.; Porkodi, K.; Rocha, F. Langmuir-Hinshelwood kinetics—A theoretical study. *Catal. Comm.* **2008**, *9*, 82–84. [CrossRef]
60. Chan, C.M.N.; Ng, A.M.C.; Fung, M.K.; Cheng, H.S.; Guo, M.Y.; Djurišić, A.B.; Leung, F.C.C.; Chan, W.K. Antibacterial and photocatalytic activities of TiO_2 nanotubes. *J. Exp. Nanosci.* **2013**, *8*, 859–867. [CrossRef]
61. Girish, S.; Gomathi Devi, L. Review on Modified TiO_2 Photocatalysis under UV/Visible Light: Selected Results and Related Mechanisms on Interfacial Charge Carrier Transfer Dynamics. *J. Phys. Chem. A* **2011**, *115*, 13211–13241.
62. Bachvarova-Nedelcheva, A.; Iordanova, R.; Georgieva, N.; Nemska, V. Preparation, characterization and antibacterial assessment of sol-gel derived $TiO_2/TeO_2/ZnO$ powders. *J. Chem. Technol. Metall.* **2022**, *57*, 589–597.
63. Kanno, Y.; Hihara, T.; Watanabe, T.; Katoh, K. *Low Sulfate Generation Diesel Oxidation Catalyst*; SAE Technical Paper, 2004-01-1427; SAE: Warrendale, PA, USA, 2004.
64. Hirata, H.; Hachisuka, I.; Ikeda, Y.; Tsuji, S.; Matsumoto, S. NOx storage-reduction three-way catalyst with improved sulfur tolerance. *Top. Catal.* **2001**, *16*, 145–149. [CrossRef]
65. Weng, L.; Hodgson, S.; Bao, X.; Sagoe-Crentsil, K. Achieving controllable sol–gel processing of tellurite glasses through the use of Te(VI) precursors. *Mater. Sci. Eng.* **2004**, *B107*, 89–93. [CrossRef]
66. Hodgson, S.; Weng, L. Chemical and sol–gel processing of tellurite glasses for optoelectronics. *J. Mater. Sci. Mater. Electron.* **2006**, *17*, 723–733. [CrossRef]
67. Weng, L.; Hodgson, S. Sol-gel processing of tellurite materials from tellurium ethoxide precursor. *Mater. Sci. Eng. B* **2001**, *87*, 77–82. [CrossRef]
68. Shalaby, A.; Dimitriev, Y.; Iordanova, R.; Bachvarova-Nedelcheva, A.; Iliev, T. Modified sol-gel synthesis of submicron powders in the system $ZnO\text{-}TiO_2$. *J. Univ. Chem. Technol. Metall.* **2011**, *46*, 137–142.
69. Hagen, J. *Industrial Catalysis a Practical Approach*; WileyVCH GmbH: Weinheim, Germany, 2006; pp. 227–228.
70. Valgas, C.; de Souza, S.M.; Smania, E.F.A., Jr. Screening methods to determine antibacterial activity of natural products. *Braz. J. Microbiol.* **2007**, *38*, 369–380. [CrossRef]

catalysts

Article

Application of $BiVO_4/TiO_2$/CNT Composite Photocatalysts for Membrane Fouling Control and Photocatalytic Membrane Regeneration during Dairy Wastewater Treatment

Elias Jigar Sisay [1,2], Szabolcs Kertész [2], Ákos Fazekas [2], Zoltán Jákói [2], Endre Zsolt Kedves [2], Tamás Gyulavári [3], Áron Ágoston [4], Gábor Veréb [2] and Zsuzsanna László [2,*]

1 Doctoral School of Environmental Sciences, University of Szeged, Rerrich Béla Sqr. 1, 6720 Szeged, Hungary
2 Department of Biosystems Engineering, Faculty of Engineering, University of Szeged, Moszkvai Blvd. 9, 6725 Szeged, Hungary
3 Department of Applied and Environmental Chemistry, Institute of Chemistry, University of Szeged, Rerrich Béla Sqr. 1, 6720 Szeged, Hungary
4 Department of Physical Chemistry and Materials Science, University of Szeged, Rerrich Béla Sqr. 1, 6720 Szeged, Hungary
* Correspondence: zsizsu@mk.u-szeged.hu; Tel.: +36-62-546-525

Abstract: This study aimed to investigate the performance of composite photocatalytic membranes fabricated by incorporating multiple nanoparticles (TiO_2, carbon nanotubes, $BiVO_4$) into polyvinylidene fluoride membrane material for real dairy wastewater treatment. The composite photocatalytic membranes exhibited superior antifouling behavior, lower filtration resistance, better flux, and higher flux recovery ratio than the pristine membrane. Salinity, pH, and lactose concentration are determinant factors that affect filtration resistance and rejection performance during the ultrafiltration of dairy wastewater. Generally, higher irreversible and total resistances and slightly lower chemical oxygen demand (COD) rejections were found at higher salinity (expressed by electric conductivity values of >4 mS/cm) than lower salinity (<4 mS/cm) levels. The presence of lactose in dairy wastewater increased irreversible resistance and severely reduced COD rejection during ultrafiltration due to the ability of lactose to pass through the membranes. It was ascertained that membranes require further treatment after filtrating such wastewater. Lower resistances and slightly better COD rejections were observed at pH 7.5 and pH 9.5 compared to those observed at pH 4. Photocatalytic membranes fouled during the ultrafiltration of real dairy wastewater were regenerated by visible light irradiation. The membrane containing all constituents (i.e., TiO_2, carbon nanotubes, and $BiVO_4$) showed the best regeneration performance, exceeding that of the pristine membrane by 30%.

Keywords: photocatalytic membranes; antifouling; PVDF; bismuth vanadate; carbon nanotubes; visible light

1. Introduction

The dairy industry generates huge amounts of highly polluted wastewater globally. Generally, it produces 1–10 L of effluent for every liter of milk processed. The effluent contains numerous organic components (such as lactose, casein, fat, and whey protein), inorganic salts, and nitrogenous compounds. In addition, dairy wastewater also contains large amounts of detergents and sanitizers used for washing and a considerable amount of nutrients. These wastewaters can be characterized by high biochemical oxygen demand (BOD) and chemical oxygen demand (COD) [1,2].

The discharge of untreated or partially treated dairy wastewater results in serious environmental problems. For example, eutrophication, occurring as a result of high nutrient content, promotes the growth of algae and bacteria, which depletes oxygen in the aquatic environment. This can result in the gradual loss of aquatic life. Hence, it is essential to treat such effluents before they are discharged into water bodies. Dairy wastewater is generally treated by physico-chemical and biological methods [1–3]. However, these methods have their drawbacks, including high costs, the need to use chemicals, inefficient removal of pollutants, high energy requirement, and various operational difficulties [1,3].

Recently, membrane technologies have become increasingly prominent in wastewater remediation, offering numerous benefits such as the efficient removal of pollutants, environmental friendliness, cost efficiency, clean and easy operation, flexibility during system design, and compact equipments [4]. Among membrane processes, ultrafiltration (UF) is widely used for the dairy industry due to its good price and performance [1]. Polyvinylidene fluoride (PVDF) is widely used as a membrane material because of its high mechanical strength, thermal stability, and chemical resistance [5,6]. It is soluble in organic solvents and can be used to fabricate porous membranes [7].

Membrane fouling restricts the application of membranes by increasing operating costs and decreasing their lifetime [1]. Many strategies have been developed to prevent fouling, such as optimizing operational conditions [8], functionalization with polymers [9,10], and incorporating nanoparticles [11] and catalysts [8,12–16]. Recently, several catalytic polymeric membranes have been reported as possessing antifouling [12,13] and self-cleaning properties [12,17] and the ability to degrade pollutants [17,18]. Such benefits can be realized by combining filtration and advanced oxidation processes (AOP). During the latter, reactive oxygen species are generated, such as hydroxyl radicals (•OH), to degrade organic pollutants into inorganic molecules. These species can be generated through Fenton-type processes or photocatalysis. Fenton processes include the catalytic decomposition of an oxidant such as hydrogen peroxide (H_2O_2) by a ferrous (Fe^{2+}) or ferric (Fe^{3+}) salt in an acidic medium (pH~3) to form •OH radicals without the application of external energy [19]. Photocatalysis is a process of generating •OH radicals and other reactive oxygen species utilizing light energy and semiconductors such as ZnO, TiO_2, WO_3, V_2O_5, ZrO_2, or Fe_2O_3 [19]. Photocatalysis has been widely studied for membrane cleaning to avoid the excessive use of chemicals and harness the environmentally friendly solar energy [13].

Nowadays, TiO_2-based photocatalytic membranes are widely developed and investigated [13]. However, the relatively high e^-/h^+ recombination rate and low light-utilization rate of TiO_2 restricts their practical application. Numerous attempts have been made to dope TiO_2-based semiconductors to address the problems described above. These attempts include doping with various metals such as Ag [20,21], Au, Pt [20], and zirconia (ZrO_2) [12], as well as non-metals such as graphene [22] and GO [1]. Carbon nanotubes (CNTs) have also received considerable attention due to their outstanding properties, such as high stiffness, flexibility, thermal and electrical conductivities, and large specific surface area. They are broadly used as electron acceptors to reduce the recombination rate of photogenerated charge carriers [23,24]. As TiO_2 has a relatively wide band gap of ~3.15 eV, it can only be efficiently activated under UV light ($\lambda < 390$ nm), while visible light is not utilized [25,26]. To overcome this problem, TiO_2 can be combined with semiconductors having narrower band gaps (such as $BiVO_4$) [27,28], resulting in efficient visible light absorption. Another plausible strategy is to modify TiO_2-containing photocatalysts with good electron acceptors such as CNTs, which may result in more effective visible light activation [29].

In this work, PVDF-based UF membranes were modified with TiO_2, CNT, and $BiVO_4$ nanoparticles according to our previous work [29] via the phase inversion method and applied for synthetic and real dairy wastewater treatment.

2. Results

2.1. Zeta Potential of Prepared Membranes

Zeta potential analysis was performed based on the linear dependence between the streaming potential and pressure gradient. Zeta potentials were calculated using Equation (1). Figure 1a–c illustrates the zeta potentials of unused membranes as a function of pH. The zeta potentials of all membranes are negative at neutral pH. Both modified membranes (Figure 1b,c) had more negative charges than the pristine membrane. This result is significant considering the pH of dairy wastewater, which is within the range of 7.2–8.8 [30,31]. At these operating pHs, membranes repel negatively charged feed constituents more strongly than the pristine membranes; thus, milk proteins (with an isoelectric point around pH 4.7) are negatively charged in this pH range. This has a considerable contribution to reducing the fouling of membranes.

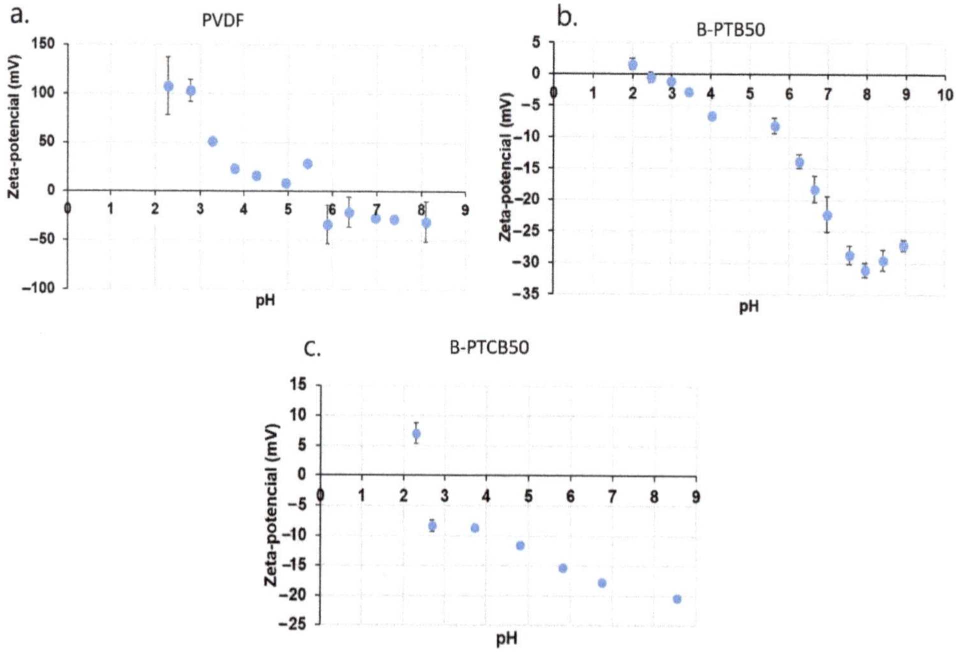

Figure 1. Zeta potentials of unused PVDF, blended PVDF-TiO_2/$BiVO_4$ (50–50%) composite (PTB50), and blended PVDF-TiO_2/CNT/$BiVO_4$ (48–2–50%) composite (PTCB50) membranes.

2.2. *Application of PTCB Membranes for Synthetic Dairy Wastewater Treatment*

The experiment aimed to investigate the effect of pH, salinity, and lactose content on the filtration performance of PVDF-, B-PTB50-, and B-PTCB50-blended membranes during the filtration of Bovine serum albumin (BSA) containing synthetic dairy wastewater (SW-BSA).

2.2.1. Effects of Salinity on Fouling and Retention

The effect of salinity on membrane fouling and rejection is shown in Figure 2. The irreversible and total resistances of the modified membranes (B-PTB50 and B-PTCB50) at all salinity levels were lower than those of the pristine PVDF membrane (Figure 2a). This is important from a fouling mitigation perspective. The COD and turbidity rejections of pristine PVDF, B-PTB50, and B-PTCB50 membranes were above 91% at all salinity levels (Figure 2b); however, these membranes show slightly lower COD rejection at the highest salinity level (EC > 4) than at the medium and lowest salinity levels. This could be due to the shielding effect of saline ions. The rejection for turbidity was above 98% for all membranes.

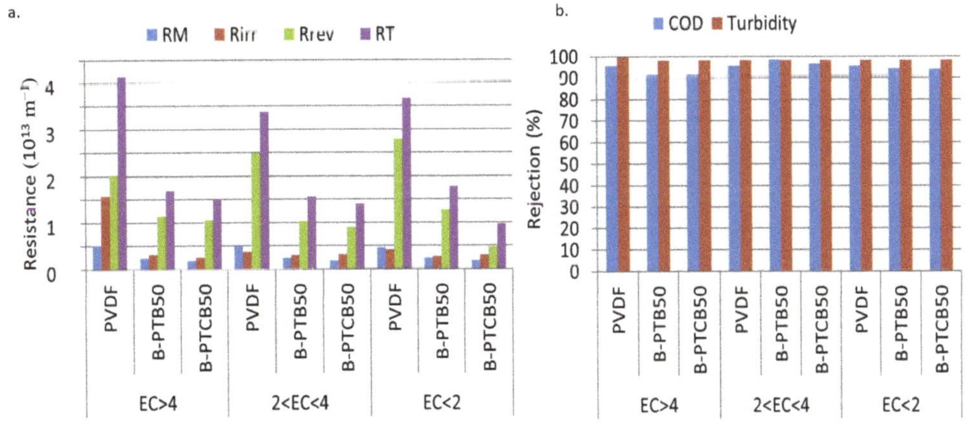

Figure 2. Effect of salinity on membrane fouling during synthetic dairy wastewater membrane filtration expressed by filtration resistances (where *RM* is membrane resistance, *Rirr* is irreversible resistance, *Rrev* is reversible resistance, and *RT* is total resistance) obtained from Equations (Section 3.6.2) (**a**) or COD and turbidity rejections (**b**).

2.2.2. Effects of Lactose on Fouling and Retention

The effects of lactose on membrane fouling and rejection were investigated in the next series of experiments. A lactose level of 0.5 g/L increased the irreversible and total resistances of all membranes, whereas at 1 g/L, the total resistances decreased (Figure 3a). As expected, when the synthetic dairy wastewater contained 1 g/L lactose, the COD rejection of all membranes was reduced to about 50% (Figure 3b). This is because lactose could easily pass through the membranes during filtration. Moreover, in all cases, modifying the membranes resulted in a better performance; the fouling resistances decreased. This means that the blended membranes had considerable antifouling properties even in the presence of lactose.

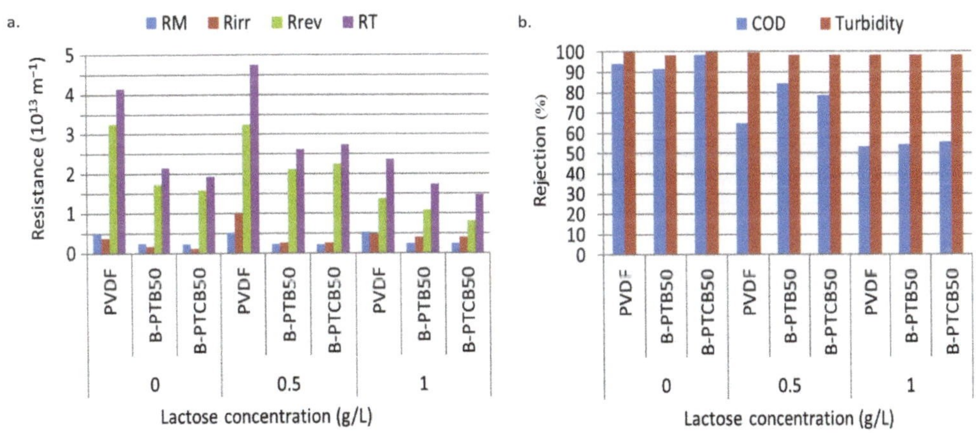

Figure 3. Effect of lactose on membrane fouling during the membrane filtration of BSA and lactose-containing synthetic dairy wastewater (SW-BSA-L) (**a**) and COD and turbidity rejections (**b**).

2.2.3. Effects of pH on Fouling and Rejection

Three different pH values (4, 7.5, and 9.5) were selected to investigate the effect of pH on fouling and rejection (Figure 4). Lower resistances were observed for B-PTB50 and B-PTCB50 at pH 7.5 and pH 9.5, respectively (Figure 4a), compared to the unmodified

PVDF membrane. This is due to the strong repulsion between the negatively charged surface of the membranes and the negatively charged protein (BSA, with an isoelectric point around pH 4.7). At the lower pH (pH 4), higher resistances were observed due to the attraction (van der Waals) forces between the nearly isoelectric membranes (pristine PVDF) and BSA. It is worth noting that irreversible fouling is the parameter that was most efficiently reduced by the membrane modification, which resulted in more negative membrane surfaces. Figure 4b shows the effect of pH on COD and turbidity rejections. It was found that the pH only slightly affected COD rejection, while it did not affect turbidity rejection. Turbidity rejection of all membranes was above 98%, while the COD rejection was above 90%.

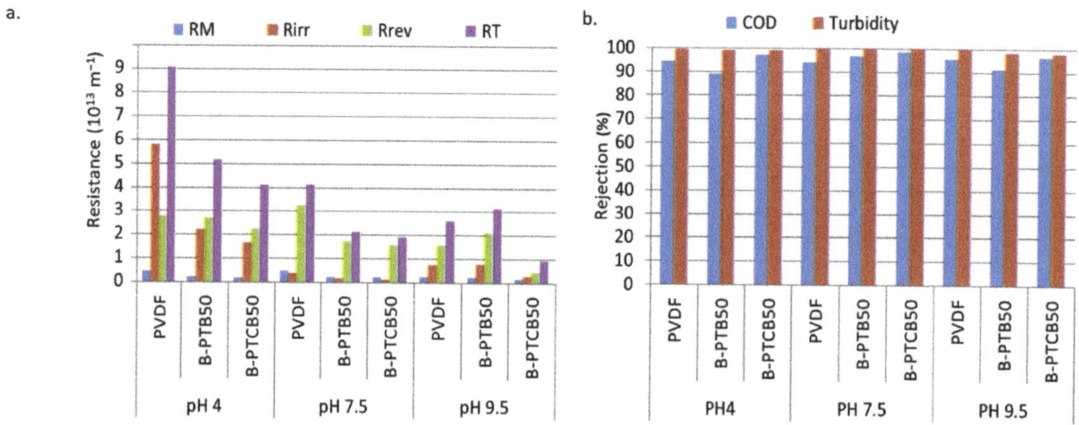

Figure 4. Effect of pH on membrane fouling during the membrane filtration of synthetic dairy wastewater (SW-BSA) (**a**) or COD and turbidity rejections (**b**).

2.3. *Application of PTCB Membranes for Real Dairy Wastewater Treatment*

This experiment aimed to evaluate the applicability of PVDF-, PTB50-, and PTCB50-blended membranes for real dairy wastewater treatment. The filtration and regeneration performance of PVDF-, B-PTB50-, and B-PTCB50-blended membranes were investigated at pH 7.09.

2.3.1. Filtration Resistances

Filtration resistances of PVDF, B-PTB50, and B-PTCB50 membranes during real dairy wastewater filtration are shown in Figure 5. Higher resistances were observed for unfiltered wastewater (Figure 5a) than for pre-filtered wastewater (Figure 5b). These results indicate the need for pre-filtration during real dairy wastewater treatment by membranes. However, in both cases, the total and irreversible resistances of PVDF membranes were higher than those of the improved (B-PTB50 and B-PTCB50) membranes.

2.3.2. Rejection

The rejection performance of PVDF, B-PTB50, and B-PTCB50 membranes during real dairy wastewater filtration is shown in Figure 6. During the UF of unfiltered and pre-filtered real dairy wastewater, the highest and the lowest COD rejections were 60% and 30% for B-PTCB50 (Figure 6a) and pristine PVDF (Figure 6b), respectively. Almost 100% turbidity rejection was observed in all membranes for unfiltered real dairy wastewater (Figure 6a). The lower rejection performances of the membranes were due to the ability of lactose to pass through the membranes (Figure 6b) which require further treatment.

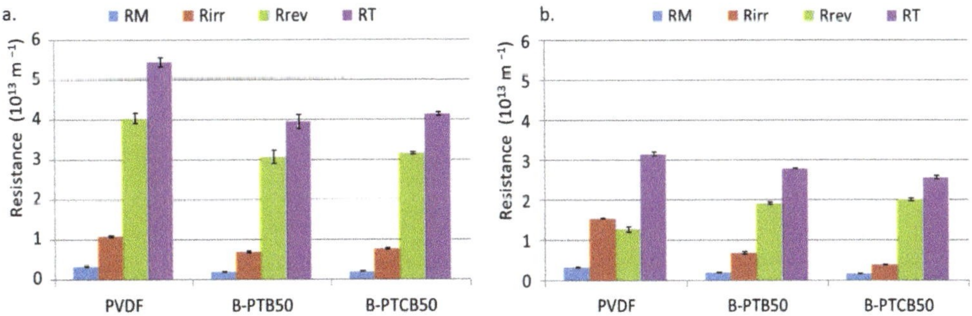

Figure 5. Filtration resistances of PVDF, PTB50, and PTCB50 membranes for real dairy wastewater, using unfiltered (**a**) or pre-filtered (0.2 μm filter paper) wastewater (**b**).

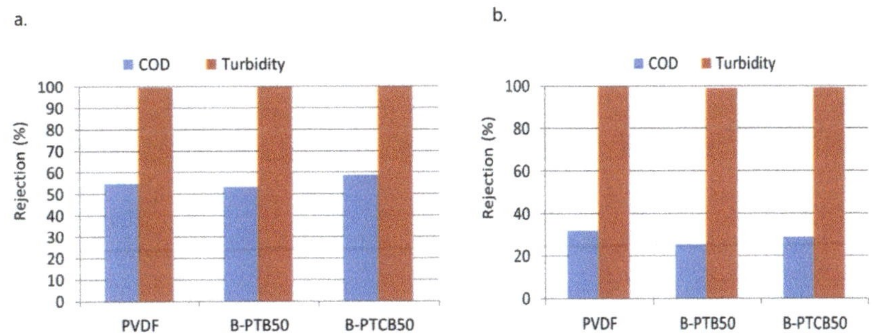

Figure 6. Rejection performance of pristine PVDF, PTB50, and PTCB50 membranes during real dairy wastewater filtration, using unfiltered (**a**) or pre-filtered (0.2 μm filter paper) wastewater (**b**).

2.3.3. Membrane Regeneration

The results of the regeneration of fouled PVDF, B-PTB50, and B-PTCB50 membranes during real dairy wastewater filtration are shown in Figure 7. The regeneration that could be obtained for the modified membranes under 3 h of visible light exposure during the filtration of pre-filtered real dairy wastewater (Figure 7b) was twice/half of what could be achieved during the filtration of unfiltered real dairy wastewater (Figure 7a). Fouled CNT/$BiVO_4$-containing membranes exhibited a better flux restoration after flushing with water than $BiVO_4$-containing membranes (Figure 7b). Moreover, the incorporation of 2% CNTs resulted in a slightly better regeneration performance, probably due to their ability to suppress electron recombination (Figure 7b).

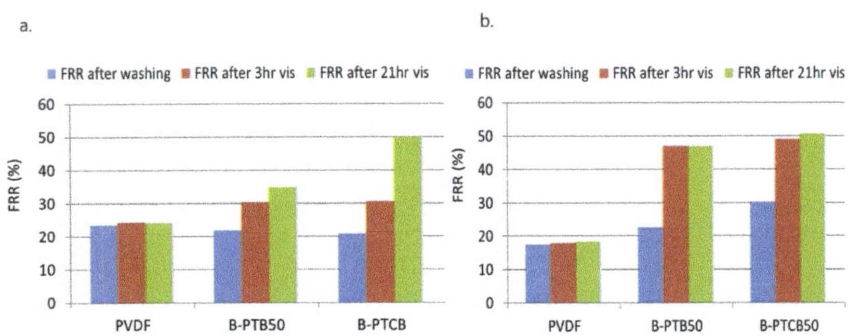

Figure 7. Regeneration performance of fouled PVDF, PTB50, and PTCB50 membranes during real dairy wastewater filtration, using unfiltered (**a**) or pre-filtered (0.2 μm filter paper) wastewater (**b**).

3. Materials and Methods

3.1. Wastewater Collection and Preparation

Membrane filtration experiments were carried out using synthetic and real dairy wastewater. The synthetic dairy wastewater was prepared in our laboratory, while the real dairy wastewater was collected from a nearby milk-processing business.

3.2. Synthetic Dairy Wastewater

Synthetic dairy wastewater was prepared according to the publication of Muniz et al. [3]. It contained bovine serum albumin (BSA) (VWR International KFt, Debrecen, Hungary) and other chemical compounds (synthetic waste, SW) and was labeled as SW-BSA (Table 1) (purchased from VWR International, Debrecen, Hungary).Various concentrations of lactose were added (0, 0.5, and 1 g/L) to the synthetic dairy wastewater to investigate their effect. The SW-BSA samples containing lactose were labeled as SW-BSA-L. The effect of pH on the filtration performance of membranes was investigated by adjusting the pH of the synthetic dairy wastewater with sulfuric acid (1 M) and Na_2HPO_4 (0.1 M).

Table 1. Composition of synthetic dairy wastewater (SW-BSA).

Number	Chemicals	Concentration in g/L
1	BSA	1
2	Ammonium chloride (NH_4Cl)	0.5833
3	Sodium dihydrogen phosphate (NaH_2PO_4)	0.9
4	Sodium bicarbonate ($NaHCO_3$)	1.560
5	Magnesium sulfate heptahydrate ($MgSO_4 \times H_2O$)	0.6
6	Ferrous sulfate heptahydrate ($Fe(SO_4) \times 7H_2O$)	0.024
7	Manganese sulfate monohydrate ($MnSO_4 \times H_2O$)	0.024
8	Calcium chloride ($CaCl_2$ without water)	0.036

The characteristics of synthetic dairy wastewater without lactose and with lactose are presented in Tables 2 and 3, respectively. The absorbance of the model protein solution was measured at a wavelength of 280 nm with a UV–visible spectrophotometer (Hitachi Co., U-2000, Chiyoda City, Japan). EC (electrical conductivity), salinity, and total dissolved solids (TDS) were analyzed with a multi-parameter analyzer (Consort BVBA, Turnhout, Belgium). Turbidity and pH values were measured with a nephelometer (Hach 2100N) and a pH meter (Consort), respectively. Chemical oxygen demand (COD) was analyzed by the potassium dichromate oxidation method. For this purpose, 2 mL of samples was added to test tubes (0–1500 mg/L; (Merck KGaA, Darmstadt, Germany)) and digested at 150 °C for 2 h in a digester (Lovibond ET108; Tintometer, Dortmund, Germany). Last, the values were obtained using a COD photometer (Lovibond PC-CheckIt; Tintometer, Germany).

Table 2. Characteristics of synthetic dairy wastewater (SW-BSA) with various salinity levels.

Level of Salinity	COD	Turbidity	EC (mS)	SAL	TDS (g/L)	pH
High salinity (EC > 4)	1154	46.67	4.14	2.2	2.165	7.5
Medium salinity (2 < EC < 4)	1148	21.83	2.33	1.1	1.25	7.5
Low salinity (EC < 2)	1155	8.16	1.59	0.8	0.85	7.5

Table 3. Characteristics of SW-BSA and SW-BSA-L.

Level of Lactose (g/L)	COD	Turbidity	EC (mS)	SAL	TDS (g/L)	pH
0	1154	46.67	4.14	2.2	2.165	7.5
0.5	1653	158.67	3.83	2.1	2.07	7.88
1	2316	188.33	3.80	2.1	2.23	7.81

3.3. Real Dairy Wastewater

Real dairy wastewater was collected from a milk-processing business (Sole-Mizo, Szeged, Hungary) and kept at 0 °C. The Ca, $CaCO_3$, NH_4^+, NH_3, total N, total P, and PO_4^{3-} contents were analyzed by spectrophotometry (Spectroquant Nova 60; Merck KGaA, Darmstadt, Germany). Biological oxygen demand (BOD) was analyzed with a Lovibond BOD device (Lovibond Oxidirect; Tintometer, Dortmund, Germany). All parameters were expressed as the average of three measurements. The characteristics of the dairy wastewater are shown in Table 4.

Table 4. Physico-chemical characteristics of real industrial dairy wastewater.

Parameter	Average	SD
pH	7.09	0.02
Color	milky white	
EC (mS)	2.1	0.01
TDS (g/L)	1.11	0.01
BOD (mg/L)	2181	70.71
COD (mg/L)	3770	20.00
Ca (mg/L)	159.67	4.04
CaO (mg/L)	231	1.00
$CaCO_3$ (mg/L)	412.33	2.52
NH_4 (mg/L)	56.87	0.31
NH_4N (mg/L)	44.15	0.39
NH_3 (mg/L)	53.31	0.38
NO_3 (mg/L)	10.05	0.83
NO_2N (mg/L)	2.2	0.20
TOTAL N (mg/L)	74.33	1.15
PO_4^{3-} (mg/L)	1178.33	3.51
PO_4-P (mg/L)	40.43	2.60
P_2P_5 (mg/L)	89.61	1.65
TOTAL P (mg/L)	39.10	1.68

3.4. Membrane Preparation

In this study, the phase-inversion method was used to prepare both pristine and improved UF PVDF membranes [29,32]. The polymer-to-solvent ratio during the preparation of UF PVDF membranes was 17.5% to 82.5%. To prepare photocatalytic membranes, we used a photocatalyst: polymer ratio of 1:99. In the TiO_2-$BiVO_4$-based PVDF membrane (B-PTB50), each photocatalyst accounts for 50% of the total photocatalyst amount in the polymer. When this membrane was further modified with CNTs (B-PTCB50), the amounts of TiO_2, CNT, and $BiVO_4$ used were 48%, 2%, and 50% of the total photocatalyst amount in the polymer, respectively. Before membrane preparation, the nanoparticles and PVDF powder were dried in an oven at 80 °C for 4 h. The dried nanoparticles were dissolved in an N-Methyl-2 pyrrolidone (NMP) solution and ultrasonicated for 1 min. Then, the powder was added to the solution under continuous magnetic stirring at 50 °C for 12 h. Afterward, the solution was kept in the dark without stirring for another 12 h to remove air bubbles. For the same purpose, the casting dope solutions were ultrasonicated for 30 min. Figure 8 shows the scheme of membrane preparation using the phase-inversion method. In this method, the solution poured on the glass plate was cast by a casting blade (400 µm thickness) and kept at rest for 30 s for skin layer formation. Then, the glass plates containing the casted solution were put into a bath containing a 3 g/L surfactant solution (sodium dodecyl sulfate) at 15 °C for 3 h. Last, the system was stored in distilled water overnight.

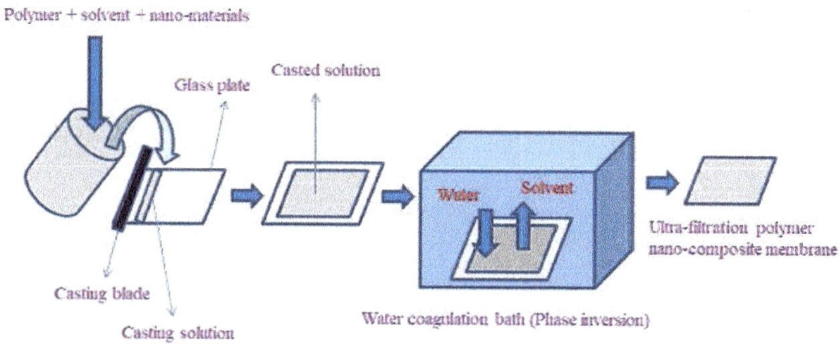

Figure 8. Schematic representation of the phase inversion method.

3.5. Membrane Characterization

Zeta Potential Analysis

Membranes were already characterized in our earlier work (Sisay et al., 2022). Zeta potentials of membranes were calculated using the Helmholtz-Smoluchowski equation (Equation (1)) [33]:

$$\zeta = \frac{\Delta E}{\Delta P} \cdot \frac{\eta}{\varepsilon_{rel}\,\varepsilon_0} \cdot K_B \tag{1}$$

where ζ is the apparent zeta potential; $\Delta E/\Delta P$ is the streaming potential developed as a result of an applied pressure gradient; η and ε_{rel} are the dynamic viscosity and dielectric coefficient of water, respectively; ε_0 is the permittivity of vacuum; K_B is the electric conductivity of the aqueous solution.

3.6. Membrane Filtration Experiments

3.6.1. Water Flux and Contaminant Rejection Performance

The filtration performance of the membranes was evaluated using synthetic or real dairy wastewater. Figure 9 presents a dead-end filtration setup involving a Millipore dead-end cell (Millipore, XFUF04701,Merck KGaA, Darmstadt, Germany). Prior to the filtration experiments (compaction), distilled water was allowed to pass through a 0.0035 m^2 membrane for 30 min. The volume reduction ratio (VRR) of each filtration experiment was fixed to be five, and each experiment was performed at 0.1 MPa and 350 rpm with a built-in magnetic stirrer. The flux and rejection performance of the prepared membranes was calculated based on Equations (2) and (4):

$$J = \frac{W}{A \cdot t} \tag{2}$$

where J refers to the flux (kg/m^2 h); W refers to the weight of permeate (kg); A means the area of effective membrane (m^2); and τ means the time (s).

The VRR was obtained by Equation (3):

$$VRR = \frac{V_0}{V_0 - V_f} \tag{3}$$

where V_0 stands for the initial volume, while V_f stands for the final volume.

The rejection of contaminants was calculated by Equation (4):

$$\textbf{Rejection } (\%) = \frac{c_1 - c_2}{c_1} \times \mathbf{100\%} \tag{4}$$

where c_1 and c_2 are the concentrations of contaminants in the feed and permeate, respectively.

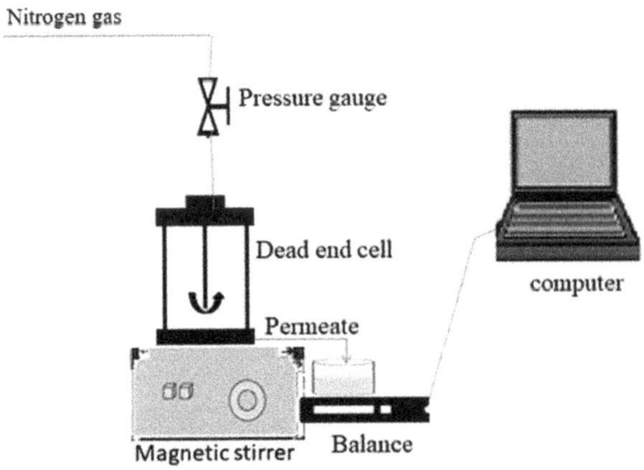

Figure 9. Graphical illustration of the dead-end filtration setup.

3.6.2. Fouling Models

Information about membrane fouling was obtained from filtration resistances which were calculated using the resistances-in-series model [34]. This model estimates individual and overall resistances (Table 5).

Table 5. Formulas for filtration resistances.

Resistances (m^{-1})	Formula
Membrane resistance	$RM = \frac{\Delta P}{J_0 \times \eta w}$
	$Rirrev = \frac{\Delta P}{JW \times \eta w} - RM$
Reversible resistance	$Rrev = \frac{\Delta P}{Jw \times \eta ww} - RF - RM$
Overall resistance	$RT = RM + Rirr + Rrev$

In Equations (Table 5), Δp is the change of pressure (Pa), J_0 is initial water flux ($Lm^{-2}\ h^{-1}$), Jw is the flux of the fouled membrane, JW is the water flux after the fouled membrane is rinsed (L/m^2 h), ηW is the viscosity of water (Pas), and ηww is the viscosity of wastewater.

Antifouling performances of the prepared membranes were examined based on the same fouling experiment using the flux recovery ratio (FRR) and obtained by the following equation (Equation (5)):

$$FRR = \frac{J_c}{J_0} \cdot 100 \quad (5)$$

where J_0 is initial water flux ($Lm^{-2}\ h^{-1}$) and Jc is the water flux of the used membrane after cleaning ($Lm^{-2}\ h^{-1}$).

Membrane regeneration efficiency was investigated by performing flux recovery experiments in the photocatalytic membrane reactors. For this purpose, the water fluxes of the fouled and flushed (with distilled water) membranes were measured after the filtration of dairy wastewater. The regeneration experiments were performed using the filtration cell as a photoreactor. This cylindric reactor had a diameter of 7.45 cm and was equipped with a 1 m long LED strip (5050 SMD, "cool white," 600 lm/m light intensity). The membrane was placed at the bottom of the cylinder. The fluxes were measured after 3 h and 21 h of visible light exposure.

4. Conclusions

Industrial dairy wastewater treatment using composite photocatalytic PVDF membranes offers promising solutions for many environmental concerns, specifically water scarcity and pollution. These membranes also provide a golden opportunity to use visible light for membrane cleaning, reducing the use of environmentally unfriendly chemicals and costly UV light.

This study aimed to fabricate composite photocatalytic membranes by incorporating multiple nanoparticles (TiO_2, CNT, $BiVO_4$) into PVDF membranes and investigate their performance for real dairy wastewater treatment. The membranes were prepared by the phase-inversion method.

The composite photocatalytic membranes exhibited superior antifouling behavior, lower filtration resistance, better flux, and higher FRR than the pristine membrane. Salinity, pH, and lactose concentration are determinant factors that affect filtration resistance and rejection performance. Generally, higher irreversible and total resistances and slightly lower COD rejection were observed at a higher EC level (>4). Irreversible and total resistances of the modified (B-PTB50 and B-PTCB50) membranes at various salinity levels were lower than those of the pristine PVDF membrane, proving that the modification resulted in better fouling-mitigation properties.

The presence of lactose increased the irreversible resistance and severely reduced COD rejection.

Negatively charged membranes showed lower resistances and slightly better COD rejection at pH 7.5 and pH 9.5 as compared to pH 4. The lower resistance at higher pH was due to the strong repulsion between the negatively charged surface of the membranes and the negative charge of the protein.

Both B-PTB50 and B-PTCB50 showed lower total resistance and better membrane cleaning (FRR) properties during the UF of real dairy wastewater than the pristine membrane. However, all the membranes exhibited lower COD rejection due to the ability of lactose to pass through the membranes, which consequently require further treatment.

Fouled TiO_2-CNT-$BiVO_4$-PVDF membrane showed the best regeneration performance, improving that of the pristine membrane by 30%.

Author Contributions: E.J.S.: Conceptualization, Methodology, Formal analysis, Investigation, Resources, Writing—Original Draft, Writing—Review and Editing, Visualization, Project administration; E.Z.K.: Software, Data Curation, Validation. T.G.: Writing—Review and Editing; Á.Á.: Investigation; S.K.: Writing—Review and Editing, Resources; G.V.: Conceptualization, Methodology, Investigation, Resources, Writing—Review and Editing, Visualization; Á.F.: Investigation, Writing—Review and Editing, Visualization; Z.J.: Investigation, Writing—Review and Editing, Visualization; Z.L.: Supervision, Resources, Visualization, Funding acquisition, Methodology, Investigation, Project administration, Writing—Review and Editing. All authors have read and agreed to the published version of the manuscript.

Funding: The authors would like to thank the Science and Research Foundation of Hungary for financial support, 2017-2.3.7-TÉT-IN-2017-00016. Elias Jigar Sisay's grateful acknowledgement also goes to the Stipendium Hungaricum Scholarship for financial provision to the research. The research was co-financed by the Hungarian National Research, Development and Innovation Office—NKFIH (FK_20_135202).

Data Availability Statement: Not applicable.

Conflicts of Interest: The authors declare no conflict of interest.

References

1. Zinadini, S.; Vatanpour, V.; Zinatizadeh, A.A.; Rahimi, M.; Rahimi, Z.; Kian, M. Preparation and characterization of antifouling graphene oxide/polyethersulfone ultrafiltration membrane: Application in MBR for dairy wastewater treatment. *J. Water Process Eng.* **2015**, *7*, 280–294. [CrossRef]
2. Ahmad, T.; Aadil, R.M.; Ahmed, H.; ur Rahman, U.; Soares, B.C.V.; Souza, S.L.Q.; Cruz, A.G. Treatment and utilization of dairy industrial waste: A review. *Trends Food Sci. Technol.* **2019**, *88*, 361–372. [CrossRef]

3. Muniz, G.L.; Borges, A.C.; da Silva, T.C.F. Performance of natural coagulants obtained from agro-industrial wastes in dairy wastewater treatment using dissolved air flotation. *J. Water Process Eng.* **2020**, *37*, 101453. [CrossRef]
4. Catenacci, A.; Bellucci, M.; Yuan, T.; Malpei, F. Dairy wastewater treatment using composite membranes. In *Current Trends and Future Developments on Bio Membranes*; Elsevier: Amsterdam, The Netherlands, 2020; pp. 261–288. [CrossRef]
5. Sobola, D.; Kaspar, P.; Částková, K.; Dallaev, R.; Papež, N.; Sedlák, P.; Trčka, T.; Orudzhev, F.; Kaštyl, J.; Weiser, A.; et al. PVDF Fibers Modification by Nitrate Salts Doping. *Polymers* **2021**, *13*, 2439. [CrossRef]
6. Kaspar, P.; Sobola, D.; Částková, K.; Dallaev, R.; Šťastná, E.; Sedlák, P.; Knápek, A.; Trčka, T.; Holcman, V. Case Study of Polyvinylidene Fluoride Doping by Carbon Nanotubes. *Materials* **2021**, *14*, 1428. [CrossRef]
7. Ji, J.; Liu, F.; Hashim, N.A.; Abed, M.R.M.; Li, K. Poly(vinylidene fluoride) (PVDF) membranes for fluid separation. *React. Funct. Polym.* **2015**, *86*, 134–153. [CrossRef]
8. Fan, G.; Chen, C.; Chen, X.; Li, Z.; Bao, S.; Luo, J.; Yan, Z. Enhancing the antifouling and rejection properties of PVDF membrane by Ag3PO4-GO modification. *Sci. Total Environ.* **2021**, *801*, 149611. [CrossRef]
9. Wei, Q.; Wu, C.; Zhang, J.; Cui, Z.; Jiang, T.; Li, J. Fabrication of surface microstructure for the ultrafiltration membrane based on "active–passive" synergistic antifouling and its antifouling mechanism of protein. *React. Funct. Polym.* **2021**, *169*, 105068. [CrossRef]
10. Gholami, S.; Llacuna, J.L.; Vatanpour, V.; Dehqan, A.; Paziresh, S.; Cortina, J.L. Impact of a new functionalization of multi-walled carbon nanotubes on antifouling and permeability of PVDF nanocomposite membranes for dye wastewater treatment. *Chemosphere* **2022**, *294*, 133699. [CrossRef]
11. Farahani, M.H.D.A.; Vatanpour, V. A comprehensive study on the performance and antifouling enhancement of the PVDF mixed matrix membranes by embedding different nanoparticles: Clay, functionalized carbon nanotube, SiO_2 and TiO_2. *Sep. Purif. Technol.* **2018**, *197*, 372–381. [CrossRef]
12. Coelho, F.E.B.; Deemter, D.; Candelario, V.M.; Boffa, V.; Malato, S.; Magnacca, G. Development of a photocatalytic zirconia-titania ultrafiltration membrane with antifouling and self-cleaning properties. *J. Environ. Chem. Eng.* **2021**, *9*, 106671. [CrossRef]
13. Wang, X.; Li, S.; Chen, P.; Li, F.; Hu, X.; Hua, T. Photocatalytic and antifouling properties of TiO_2-based photocatalytic membranes. *Mater. Today Chem.* **2022**, *23*, 100650. [CrossRef]
14. Yang, Y.; Yang, L.; Yang, F.; Bai, W.; Zhang, X.; Li, H.; Duan, G.; Xu, Y.; Li, Y. A bioinspired antibacterial and photothermal membrane for stable and durable clean water remediation. *Mater. Horiz.* **2023**, *10*, 268–276. [CrossRef] [PubMed]
15. Xu, Y.; Hu, J.; Zhang, X.; Yuan, D.; Duan, G.; Li, Y. Robust and multifunctional natural polyphenolic composites for water remediation. *Mater. Horiz.* **2022**, *9*, 2496–2517. [CrossRef]
16. Bai, H.; He, P.; Hao, L.; Fan, Z.; Niu, R.; Tang, T.; Gong, J. Waste-treating-waste: Upcycling discarded polyester into metal–organic framework nanorod for synergistic interfacial solar evaporation and sulfate-based advanced oxidation process. *Chem. Eng. J.* **2023**, *456*, 140994. [CrossRef]
17. Chen, L.; Yang, B.; Zhou, P.; Xu, T.; He, C.; Xu, Y.; Zhao, W.; Zhao, C. A polyethersulfone composite ultrafiltration membrane with the in-situ generation of CdS nanoparticles for the effective removal of organic pollutants and photocatalytic self-cleaning. *J. Membr. Sci.* **2021**, *638*, 119715. [CrossRef]
18. Zhang, H.; Zhang, J.; Luo, J.; Wan, Y. A novel paradigm of photocatalytic cleaning for membrane fouling removal. *J. Membr. Sci.* **2022**, *641*, 119859. [CrossRef]
19. Titchou, F.E.; Zazou, H.; Afanga, H.; El Gaayda, J.; Ait Akbour, R.; Nidheesh, P.V.; Hamdani, M. Removal of organic pollutants from wastewater by advanced oxidation processes and its combination with membrane processes. *Chem. Eng. Process. Process. Intensif.* **2021**, *169*, 108631. [CrossRef]
20. Nasr, O.; Mohamed, O.; Al-Shirbini, A.-S.; Abdel-Wahab, A.-M. Photocatalytic degradation of acetaminophen over Ag, Au and Pt loaded TiO_2 using solar light. *J. Photochem. Photobiol. A Chem.* **2019**, *374*, 185–193. [CrossRef]
21. Nair, A.K.; Jagadeesh Babu, P.E. Ag-TiO_2 nanosheet embedded photocatalytic membrane for solar water treatment. *J. Environ. Chem. Eng.* **2017**, *5*, 4128–4133. [CrossRef]
22. Jiang, M.; Zhang, M.; Wang, L.; Fei, Y.; Wang, S.; Núnez-Delgado, A.; Bokhari, A.; Race, M.; Khataee, A.; Klemes, J.; et al. Photocatalytic degradation of xanthate in flotation plant tailings by TiO_2/graphene nanocomposites. *Chem. Eng. J.* **2022**, *431*, 134104. [CrossRef]
23. Trapalis, A.; Todorova, N.; Giannakopoulou, T.; Boukos, N.; Speliotis, T.; Dimotikali, D.; Yu, J. TiO_2/graphene composite photocatalysts for NOx removal: A comparison of surfactant-stabilized graphene and reduced graphene oxide. *Appl. Catal. B Environ.* **2016**, *180*, 637–647. [CrossRef]
24. Selvaraj, M.; Hai, A.; Banat, F.; Haija, M.A. Application and prospects of carbon nanostructured materials in water treatment: A review. *J. Water Process Eng.* **2020**, *33*, 100996. [CrossRef]
25. Akhavan, O. Lasting antibacterial activities of Ag–TiO_2/Ag/a-TiO_2 nanocomposite thin film photocatalysts under solar light irradiation. *J. Colloid Interface Sci.* **2009**, *336*, 117–124. [CrossRef] [PubMed]
26. Zouzelka, R.; Kusumawati, Y.; Remzova, M.; Rathousky, J.; Pauporté, T. Photocatalytic activity of porous multiwalled carbon nanotube-TiO_2 composite layers for pollutant degradation. *J. Hazard. Mater.* **2016**, *317*, 52–59. [CrossRef] [PubMed]
27. Malathi, A.; Arunachalam, P.; Kirankumar, V.S.; Madhavan, J.; Al-Mayouf, A.M. An efficient visible light driven bismuth ferrite incorporated bismuth oxyiodide ($BiFeO_3$/BiOI) composite photocatalytic material for degradation of pollutants. *Opt. Mater.* **2018**, *84*, 227–235. [CrossRef]

28. Ratova, M.; Redfern, J.; Verran, J.; Kelly, P.J. Highly efficient photocatalytic bismuth oxide coatings and their antimicrobial properties under visible light irradiation. *Appl. Catal. B Environ.* **2018**, *239*, 223–232. [CrossRef]
29. Sisay, E.J.; Veréb, G.; Pap, Z.; Gyulavári, T.; Ágoston, Á.; Kopniczky, J.; Hodúr, C.; Arthanareeswaran, G.; Sivasundari, G.K.; László, Z. Visible-light driven photocatalytic PVDF-TiO_2/CNT/$BiVO_4$ hybrid nanocomposite ultrafiltration membrane for dairy wastewater treatment. *Chemosphere* **2022**, *307*, 135589. [CrossRef]
30. Qazi, J.I.; Nadeem, M.; Baig, S.S.; Baig, S.; Syed, Q. Anaerobic fixed-film biotreatment of dairy wastewater. *Middle East J. Sci. Res.* **2011**, *8*, 590–593.
31. Deshannavar, U.B.; Basavaraj, R.K.; Naik, N.M. High-rate digestion of dairy industry effluent by upflow anaerobic fixed-bed reactor. *J. Chem. Pharm. Res.* **2012**, *4*, 2895–2899.
32. Srivastava, H.P.; Arthanareeswaran, G.; Anantharaman, N.; Starov, V.M. Performance of modified poly(vinylidene fluoride) membrane for textile wastewater ultrafiltration. *Desalination* **2011**, *282*, 87–94. [CrossRef]
33. Lawrence, N.D.; Perera, J.M.; Iyer, M.; Hickey, M.W.; Stevens, G.W. The use of streaming potential measurements to study the fouling and cleaning of ultrafiltration membranes. *Sep. Purif. Technol.* **2006**, *48*, 106–112. [CrossRef]
34. Vatanpour, V.; Yekavalangi, M.E.; Safarpour, M. Preparation and characterization of nanocomposite PVDF ultrafiltration membrane embedded with nanoporous SAPO-34 to improve permeability and antifouling performance. *Sep. Purif. Technol.* **2016**, *163*, 300–309. [CrossRef]

catalysts

MDPI

Article

Photodynamic Light-Triggered Release of Curcumin from Hierarchical FAU Zeolite

Ewelina Musielak [1], Agnieszka Feliczak-Guzik [1,*], Mietek Jaroniec [2] and Izabela Nowak [1,*]

1 Faculty of Chemistry, Adam Mickiewicz University, Uniwersytetu Poznańskiego 8, 61-614 Poznań, Poland
2 Department of Chemistry and Biochemistry, Kent State University, Kent, OH 44242, USA
* Correspondence: agaguzik@amu.edu.pl (A.F.-G.); nowakiza@amu.edu.pl (I.N.); Tel.: +48-618291747 (A.F.-G.); +48-618291580 (I.N.)

Abstract: Photodynamic therapy (PDT) is a method used in the treatment of various diseases and represents a promising alternative to classical active substance release methods. According to the United States National Institute of Health, this therapy exploits the health-promoting potential of active substances, referred to as "photosensitizers", and radiation of specific energy. Curcumin is a natural "photosensitizer" that exhibits anti-inflammatory, antioxidant, and anti-cancer effects. Despite the novelty and innovation of PDT, the method is limited by the difficulty of using "photosensitizers" such as curcumin because of its low solubility in water, which hinders the clinical application of this technique. The use of a "photosensitizer" along with suitable carriers such as hierarchical zeolites is a potential strategy to offset this difficulty. Therefore, the objective of this study was to develop the photodynamic release of curcumin from hierarchical zeolites using the classical open-circuit release method and a novel closed-circuit release technique involving light of different wavelengths. Based on the obtained data, it was possible to unequivocally state that the speed and efficiency of curcumin release from the zeolite carrier is influenced by light with a wavelength in the range of 430–525 nm (blue-cyan-green). As much as 53.24% of the active substance was released upon light exposure.

Keywords: photodynamic therapy; nanotechnology; curcumin; hierarchical zeolites; drug delivery systems

1. Introduction

The intensive development of medicine in recent years has undeniably contributed to an increase in life expectancy, as well as to people's awareness of taking care of their health. The result of these changes is an increase in demand for new and improved drugs and an improvement in the way they are delivered [1]. An important part of each therapy is to deliver a given drug to the specific site in the body and to release it during preferred times. Classical methods of active substance release are based on the fact that the delivered active molecules rapidly increase their concentration in plasma, and next, their concentration is exponentially decreased as the drug is excreted or metabolized. The disadvantage of such a system is the limited control over the rate of drug release and the site of administration of the therapeutic substance. A way to maintain an adequate therapeutic level of the drug for a longer period and its release in a controlled manner is the use of active substance carriers [2,3]. Such carriers should have a well-defined relationship between the dose and the therapeutic effect of the active substance. Furthermore, the effect achieved by administration of active substance should be at least comparable to the effect obtained with the use of a conventional drug/supplement form [4].

Among others, the following materials are used as drug carriers: carbon materials, polymeric materials, and in general, porous materials [5,6]. Porous materials, according to the IUPAC (International Union for Pure and Applied Chemistry) classification, are solids having internal spaces that can be filled with a gas or liquid with chemical properties different from the solid [7]. These spaces are called pores, channels, or fissures, and their length is much greater than their width (ranges from 1–100 nm). Among porous materials, the most popular types of active substance carriers are structurally ordered materials such as zeolites, or hierarchical zeolites, which are recently popular [6,8].

In recent years, hierarchical zeolites have been the object of intensive studies because they can be very good carriers for various active substances [9–12]. Hierarchical zeolites are materials that possess so-called secondary porosity (mainly mesopores, i.e., pores between 2 and 50 nm) in addition to micropores, which have sizes below 2 nm according to the IUPAC classification [13]. Hierarchical zeolites can be successfully used, for example, in catalysis [14], adsorption, or medicine [15]. Hierarchical zeolites are also used as carriers for therapeutic substances that can be released in a sustained manner due to their unique properties [16]. These materials have high biocompatibility and nontoxicity to various types of cells and tissues [17,18]. It is important to find carriers and therapeutic substances with the lowest possible toxicity and the highest possible efficacy. One of the strategies with high promise is the use of active substances isolated from plants [19].

Plants have played a very important role in history. In addition to their use as food and spices, they have been used as medicinal substances for thousands of years [20,21]. The most popular plant-derived substance used as a therapeutic agent in recent years is curcumin, which was isolated from the root of *Curcuma longa* (Figure 1).

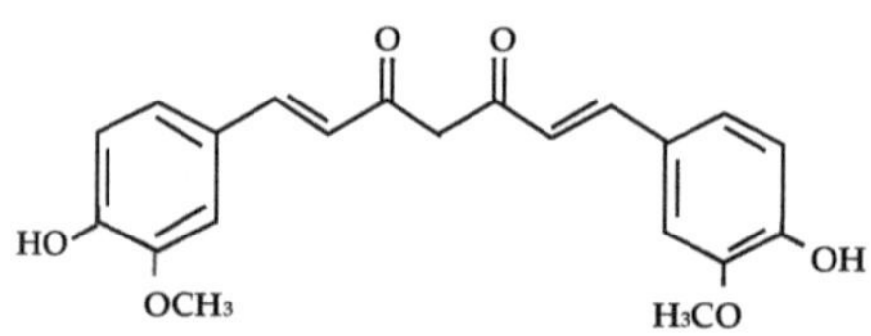

Figure 1. Structural formula of the keto form of curcumin (based on [22]).

It is a natural substance that belongs to the polyphenol group and has a long history of medicinal uses. The most used part of this plant is the rhizome [23], which consists of a wide range of compounds, including nonvolatile bioactive curcuminoids (curcumin, dimethoxycurcumin, and bisdemethoxycurcumin) and compounds present in the oil (mono- and sesquiterpenoids) [24,25]. It is currently used in the treatment of various diseases such as cardiovascular diseases, asthma, diabetes, bronchitis, rheumatoid arthritis, psoriasis, AIDS, and most importantly, it is used in the treatment of cancer [26–28]. Despite its numerous benefits, curcumin exhibits very poor bioavailability and very low blood and tissue concentrations; this is due to its poor absorption, rapid metabolism, and rapid systemic elimination [29,30]. To improve the properties of this active substance, special agents are used that can increase its bioavailability. A good example of a substance that improves curcumin properties is piperine, a known bioavailability enhancer, which is a major component of black pepper [31,32].

Plant-based controlled drug delivery systems have been developed to increase the therapeutic efficacy of drugs while reducing their side effects and patient discomfort. Parameters related to the material surface area, pore size, geometry, and possible reactions of the carrier with the surrounding environment are some of the many factors that are extremely important when designing drug delivery systems [2]. The classical methods for the release of active substances in the human body often do not produce the expected therapeutic effects; thus, new methods of delivery and release of the therapeutic substance are sought. Recent studies confirm that curcumin is a potential active substance used in photodynamic therapies [33].

Photodynamic therapy (PDT) is a minimally invasive therapy that can have significant advantages in the treatment of a variety of diseases compared to conventional methods, such as fewer side effects, high spatial–temporal precision, and rapid healing of healthy tissue [34]. This therapy uses a light-activated drug called a "photosensitizer" (PS) or "photosensitizing agent" to kill cancer cells. The light source can be a laser or, for example, a light-emitting diode (LED). The photodynamic reaction is based on the principle that a "photosensitizer" stored in target cells or tissues can be activated by absorbing sufficient light energy to produce radicals and reactive oxygen species (ROS). "Photosensitizers", when activated by light of the appropriate wavelength, probably undergo a chemical reaction by two types of mechanisms [35,36]. In the type I reaction, the radical ions and free radicals are formed by an electron/hydrogen transfer process between the "photosensitizer" and the substrate molecules [37]. In the type II reaction (the main process in PDT), highly cytotoxic singlet oxygen ($1O_2$) is formed by an energy transfer process from PS to molecular oxygen [35,38].

To better understand PDT, the mechanism of this process needs to be investigated. A detailed process of PDT involves a "photosensitizer" absorbing a sufficient dosage of light (hv), which causes an excitation from the ground state (S_0) to a high-energy singlet state (S_1). The latter state is extremely unstable, which results in the loss of energy and decay back to the ground state via fluorescence emission or heat production. Additionally, the excited singlet state can undergo the so-called "intersystem transition" to create a long-lived excited triplet state (T_1), which shows higher stability than S1. A PS molecule in the T_1 state can decay back to its ground state in two ways via: (1) transfer of energy to molecular oxygen ($3O_2$) to form singlet oxygen ($1O_2$) and PS in the ground state, and (2) formation of free radicals and "photosensitizer" anion radicals by hydrogen or electron transfer. Singlet oxygen and radical anions can react with oxygen to produce ROS, which can damage most types of biomolecules, including amino acids, lipids, and nucleic acids, thus inducing cell death and necrosis in proximal tissues (Figure 2) [39].

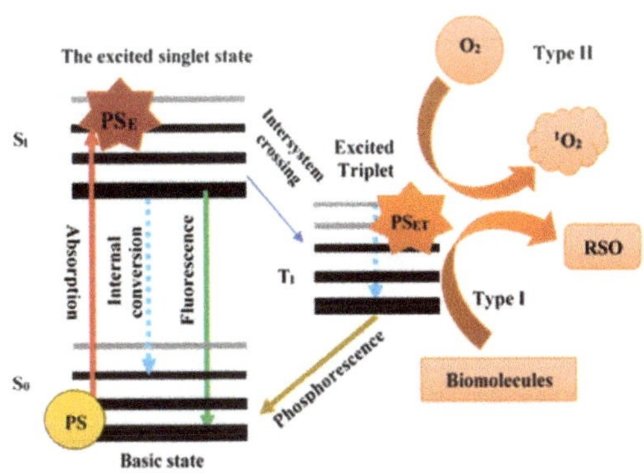

Figure 2. Schematic Jablonski-type diagram illustrating the mechanism of PDT (adapted with permission from ref. [39] 2022, Elsevier). Notation: **PS**—photosensitizer; $\mathbf{PS_{ES}}$—PS excited singlet state; $\mathbf{PS_{ET}}$—PS excited triplet state; **ROS**—reactive oxygen species; $\mathbf{^1O_2}$—singlet oxygen.

Currently, PDT is used in many localized precancerous conditions, solid tumors, oral precancerous lesions, and dermatologic squamous cell carcinoma of the head and neck [40–46]. In addition to cancer treatment, photodynamic therapy has become an intriguing new treatment modality in age-related macular degeneration (AMD) [47] and polypoidal choroidal vasculopathy (PCV) [41,48–50]. However, conventional photodynamic therapy also suffers from several problems, such as the depth of penetration of light into tissues and the efficiency of PS activation [51,52], oxygen dependence and oxygen consumption during PDT [53–55], PS biodistribution at the target site, and persistent skin and eye photosensitivity [56–59].

The aim of this study was to develop a method for the photodynamic release of curcumin from hierarchical zeolites, prepared based on faujasite-type zeolite and structure-directing agents, which enable the prolonged release of curcumin from the carrier under the influence of light in the range of 200 to 800 nm. Recently, we have published a paper regarding curcumin encapsulation [60]. In this paper, we present potential medical application of such a new curcumin-hierarchical zeolite hybrid.

2. Results

2.1. X-ray Diffraction Studies

XRD is a useful method for the preliminary evaluation of nanometer-sized materials, such as hierarchical zeolites.

The diffractograms presented in the low-angle range (panel A in Figure 3) juxtapose commercial FAU zeolite, pure curcumin, and hierarchical zeolites without curcumin and with curcumin applied. For the pure hierarchical zeolites, a broad and strong signal visible at 2θ~2.5° confirms the presence of additional mesopores in the hierarchical structure. This signal is not observed in the XRD pattern of commercial material and pure curcumin. A small reflection at an angle of 2θ at about 7.5~8.0° is observed on the diffractogram of the pure active substance; this is a characteristic reflection for curcumin. This signal is not visible for zeolite samples packed with the active substance. In the case of XRD patterns of curcumin-loaded samples, a small signal visible in the range of 2θ~2.5° confirms the presence of additional mesopores of the hierarchical materials [61]. Compared to pure hierarchical zeolite (FAU/CTABr), the diffractograms of curcumin-loaded materials show that the intensity of reflections weakens and shifts toward higher angles, which may be due to the pore filling or a decrease in the contrast of X-ray scattering between the organic part and the channel walls of the silicate skeleton [62].

By analyzing the high-angle diffractograms of the tested materials (panels B in Figure 3), one can see that the crystalline structure of the microporous commercial FAU-type zeolite is preserved. The pure curcumin diffractogram shows two distinct sharp reflections at angles of 2θ = 9.0° and 17.4°, which are not observed in the diffractograms of the curcumin-loaded materials. Analyzing the diffractograms of pure hierarchical zeolites and those loaded with curcumin, no significant changes are visible in the characteristic peaks of the zeolite after loading with the active substance, indicating that the structure did not change during the curcumin incorporation process [61].

2.2. Nitrogen Adsorption/Desorption Isotherms

Through sorption studies, it is possible to determine the textural properties of the obtained materials, namely, the specific surface area, the volumes of micropores and mesopores, the total pore volume, the pore size, and the pore size distribution.

A

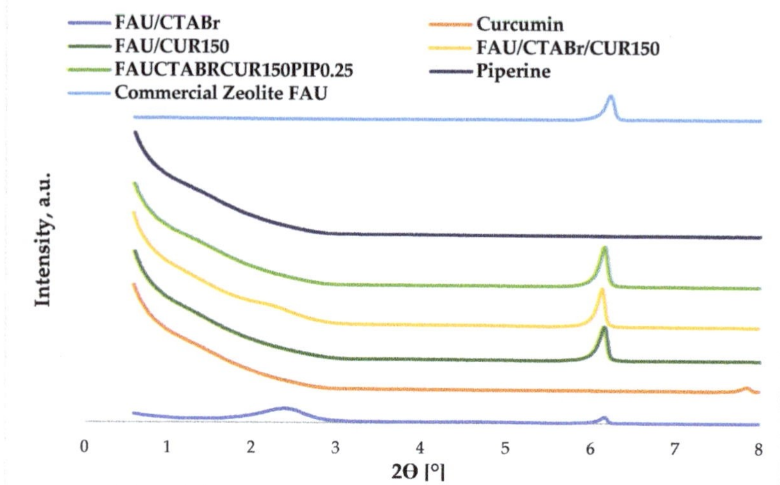

Figure 3. *Cont.*

B

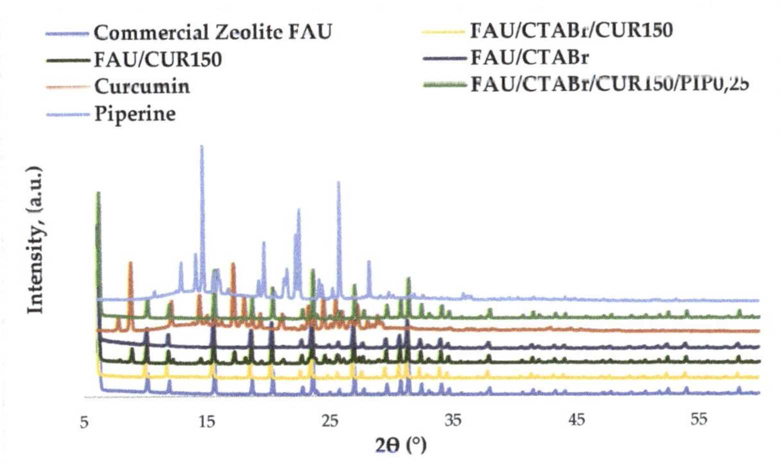

Figure 3. Low-angle diffractograms (panel (**A**)) and high-angle diffractograms (panel (**B**)) of pure curcumin and hierarchical materials based on commercial zeolite of FAU type, and high-angle diffractograms of pure hierarchical zeolite.

By analyzing individual nitrogen adsorption/desorption isotherms, one can estimate the type of isotherm. The isotherm for commercial FAU-type zeolite (panel A in Figure 4) is type I according to the IUPAC classification. Type I adsorption isotherms, which sharply rise at low pressures, are typical for microporous solids. For hierarchical zeolites synthesized based on CTABr and TEOS, the resulting isotherms are a mixture of type I and IVb. According to the IUPAC classification, the type IVb isotherm is characteristic for materials with conical and cylindrical mesopores (panel B in Figure 4). Zeolite materials with incorporated curcumin and those containing curcumin and piperine are also a mixture of type I and type IVb (panels C and D in Figure 4). Based on the isotherms shown in Figure 4, it is possible to confirm both the presence of micropores, which are directly related to the crystal structure of zeolites (primary porosity), and mesopores (secondary porosity) created as a result of the modification procedure [63].

The specific surface area (S_{BET}) for the zeolite samples was obtained by the Brunauer–Emmett–Teller (BET) method and varies from 203 to 892 m^2/g. The largest S_{BET} value is for the hierarchical zeolite based on CTABr. On the contrary, the zeolite materials containing curcumin and piperine showed the lowest specific surface area compared to the commercial zeolite. As can be seen from the adsorption measurements, the specific surface area and average pore size of the functionalized FAU-type hierarchical zeolite decrease with increasing amounts of active ingredient (Table 1). It is believed that during evaporation of acetone and incorporation of curcumin because of its hydrophobic nature, curcumin may agglomerate into nanoparticles of larger sizes. This is because hydrophilic silanol groups are present on the surface of the carrier material [64].

Table 1. Selected textural properties for materials based on commercial zeolite type FAU.

Materials	Surface Area (m^2/g)	Pore Volume (cm^3/g)		
	S_{BET}	Total Pore Volume	Volume of Micropores	Mesopore Volume
Commercial zeolite FAU	718	0.37	0.30	0.05
FAU/CTABr	892	0.49	0.19	0.30
FAU/CTABr/CUR150	688	0.35	0.31	0.04
FAU/CTABr/CUR150/PIP0,25	203	0.12	0.06	0.06

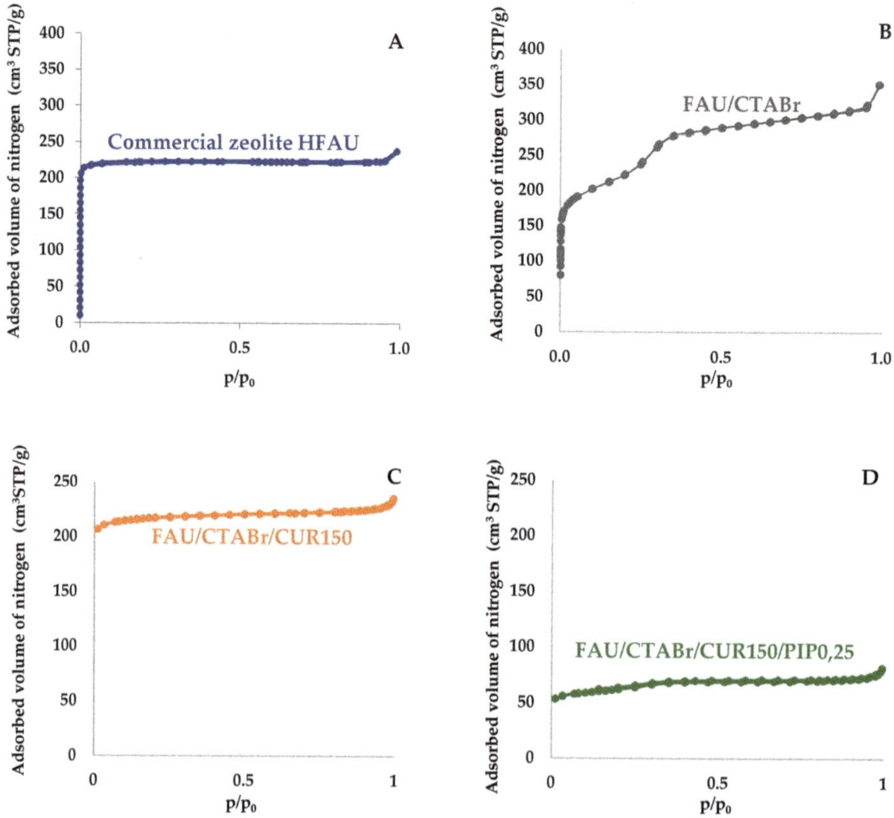

Figure 4. Nitrogen desorption/desorption isotherms for (**A**) commercial FAU-type zeolite, (**B**) hierarchical zeolite, (**C**) hierarchical zeolite with curcumin, and (**D**) hierarchical zeolite with curcumin and piperine.

Furthermore, partial amorphization of the layered phase of hierarchical zeolites during curcumin anchoring may also contribute to the observed decrease in the obtained parameters [65].

The data presented demonstrate the introduction of an additional modifying agent into zeolite materials. All the results confirm the successful application of curcumin to the zeolite carrier.

2.3. FT-IR Spectroscopy

The functional groups of the zeolite carriers were characterized by the FT-IR method (Figure 5). Infrared spectroscopy is one of the most versatile research methods that use infrared radiation absorption.

The absorption band at about 1230 cm^{-1} on the FT-IR spectrum of pure hierarchical zeolite (FAU/CTABr) can be attributed to the valence stretching vibrations for asymmetric Si-O-Si bonds. The band at ~796 cm^{-1} may come from valence-stretching vibrations of the symmetric Si-O-Si bonds. The absorption band at about 1650 cm^{-1} may refer to the bending vibrations in physically adsorbed water molecules. The band at about 1100 cm^{-1} is attributed to the asymmetric vibrations of the Si-O-Si of the silica matrix. Characteristic bands corresponding to lattice vibrations have also been observed between 1700 and 800 cm^{-1}. The broadband at 1013 cm^{-1} is related to the asymmetric stretching vibrations of the Si-O groups, while the band at 1640 cm^{-1} can be associated with O-H deformation [66].

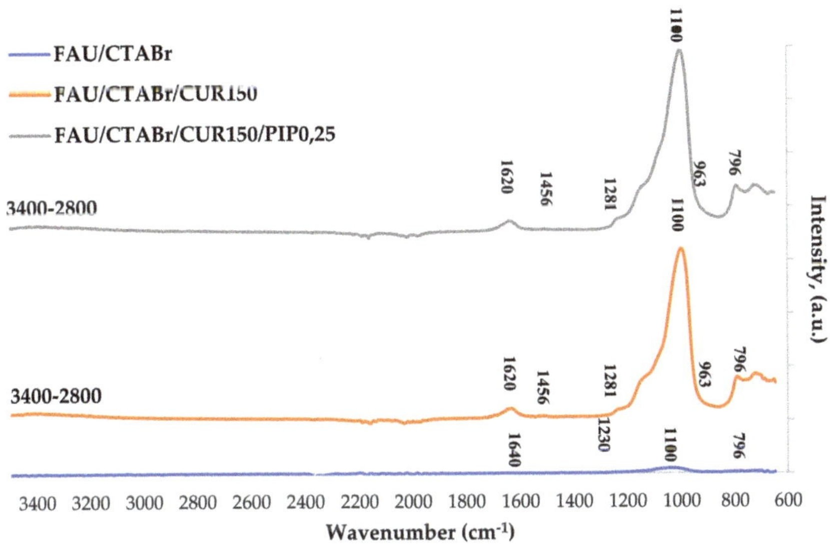

Figure 5. FT−IR spectra of hierarchical materials obtained from commercial FAU-type zeolite modified with curcumin.

Several most characteristic bands have been assigned to the spectrum of pure curcumin, including bands at 2917, 2849, and 1427 cm^{-1}, which can be attributed to HC stretching vibrations and deformation vibrations of methyl groups. The bands at 1627 and 1502 cm^{-1} correspond to C=C, C=O, and C-C stretching vibrations of the benzene ring, while olefinic C-H bending vibrations appear at 1438 cm^{-1} and aromatic C-O stretching vibrations at 1274 cm^{-1}. The bands at 1025 and 963 cm^{-1} can be assigned to C-O stretching and out-of-plane bending of aromatic C-H bonds, respectively [67].

Evidence for an interaction between curcumin molecules and the FAU zeolite backbone was also confirmed using Fourier transform infrared (FT-IR) spectroscopy. For the FAU/CUR150 and FAU/CUR150/PIP0.25 complexes, characteristic bands are found at approximately 1456 cm^{-1} (C=C) and 1620 cm^{-1} (C=N). For curcumin-loaded zeolite, the FT-IR spectrum shows bands in the 3000−2800 cm^{-1} region with two specific bands (2917 and 2849 cm^{-1}) that occur in the spectrum of pure curcumin but not in the spectrum of FAU zeolite. Therefore, these bands can be attributed to the presence of curcumin in the hierarchical zeolite. The characteristic bands of curcumin are also shifted from 1506 to 1512 cm^{-1}, from 1438 to 1442 cm^{-1}, and from 1274 to 1281 cm^{-1} due to the formed hydrogen bonding between the carrier's hydroxyl groups and the functional groups of curcumin such as phenolic hydroxyls, enolic C-OH and methoxyl groups [68].

2.4. Elemental Analysis

Elemental analysis was accomplished to obtain the exact composition of the zeolite samples studied. The results show the estimated carbon and hydrogen content (wt.%) in the tested materials (Table 2). The elemental analysis confirmed the effective modification of both the commercial FAU type material and the hierarchical zeolites with curcumin.

Elemental analysis of pure curcumin shows the presence of 30.26 wt.% C and 4.61 wt.% H, which can be attributed to the structure of the active substance (curcumin contains up to 21 carbon atoms and 20 hydrogen atoms). According to data from the literature, the theoretical values of carbon and hydrogen in curcumin are similar to those obtained experimentally [69]. While carbon % in the commercial zeolite FAU and hierarchical zeolite is very small (below 1%); this percentage is much higher after introducing curcumin into both commercial FAU-type zeolite and hierarchical zeolite, which confirms a successful incorporation of curcumin in the pores and on the surface of the materials studied.

Table 2. Elemental analysis of curcumin, commercial FAU zeolite, and hierarchical materials before and after curcumin incorporation.

Sample Name	%C	%H
Commercial zeolite FAU	0.74	0.35
Curcumin	30.26	4.61
FAU/CTABr	0.02	2.28
FAU/CUR150	5.69	2.64
FAU/CTABr/CUR150	17.30	3.24
FAU/CTABr/CUR150/PIP0.25	19.23	2.52

2.5. Thermogravimetric Studies

Figure 6 shows the TG (panel A in Figure 6) and DTG (panel B in Figure 6) profiles of commercial zeolite, hierarchical zeolites, active substances, and zeolites with incorporated curcumin and piperine.

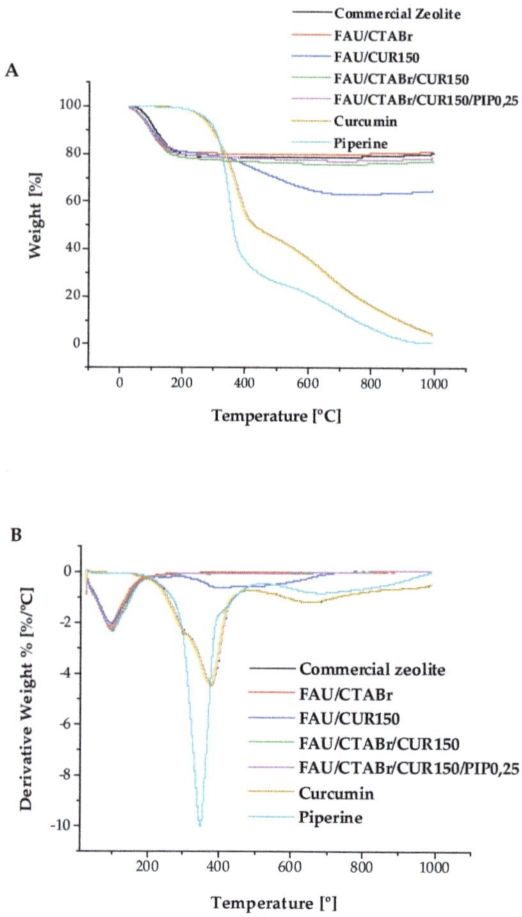

Figure 6. Thermograms of (**A**) TG and (**B**) DTG of the samples studied.

The TG profiles (panel A in Figure 6) for both commercial zeolite FAU and hierarchical zeolite with and without active substances (FAU/CTABr, FAU/CTABr/CUR150, FAU/CTABr/CUR150/PIP0,25) show that the mass loss starts at temperatures as low as ~50 °C, which is directly related to the dehydration of the samples [70]. The thermograms of the materials containing curcumin (FAU/CTABr/CUR150, FAU/CTABr/CUR150/PIP0,25) mainly show the mass loss associated with water loss, which may indicate the incorporation of curcumin into the pores of the zeolite. Two significant mass losses can be observed for the commercial FAU zeolite containing curcumin (FAU/CUR150). The first one that occurs at around 50 °C (a loss of ~5.00%) is directly related to the loss of water from the zeolite material. On the other hand, the second one (a mass loss of 25.00%) is observed at around 340 °C and is related to the mass loss of curcumin on the surface of the commercial zeolite [71]. The TG curve of pure curcumin shows the mass loss starting at about 200 °C. The observed mass loss is 100% at 1000 °C, which means that curcumin has been completely decomposed. Thermal decomposition of curcumin occurs in two stages. TG analysis suggests that stage (I) is due to the decomposition of functional groups in curcumin, and stage (II) is due to decomposition of two benzene rings in curcumin. Analyzing the TG profile of piperine, a mass loss is observed at approximately 210 °C. The TG curve of piperine shows the mass loss of about 38.00% at 344 °C and almost 65.00% at 500 °C [72,73].

The DTG profiles (panel B in Figure 6) indicate the presence of exothermic effects typical for the FAU zeolite and the materials synthesized on its basis, which do not cause structural changes. The curves obtained are the result of thermal dehydration, during which a given material releases physically and chemically bound water. The DTG curves for all synthesized materials show an intense peak at 100 °C, resulting from the onset of thermal dehydration. At about 300 °C, the stabilization of the system occurs, hence the lack of effect on the DTG curve [70]. A commercial zeolite containing curcumin is an exception, since its mass loss occurs in two steps as explained above, and stabilization of the system occurs only at about 800 °C. In the DTG thermogram, thermal decomposition of pure curcumin occurs in two stages. There is an inflection point (broad downward peak) at about 300 °C on the DTG curve, which can be considered as the end point of stage I. Thus, stage I, in which the mass loss is 50%, starts at about 200 °C and ends at about 450 °C. Stage II, whose mass loss exceeds 50%, begins at about 425 °C and gradually continues to 1000 °C, which is associated with a DTG peak at about 650 °C. The DTG curves for piperine show an intense peak at 350 °C, resulting from the onset of thermal dehydration [72,73]. Similar results were presented elsewhere [70–73].

2.6. Scanning Electron Microscopy (SEM)

Scanning electron microscopy (SEM) was used to determine the morphology of the zeolite materials before and after incorporation of curcumin. SEM images of the materials studied at 20 μm magnifications are shown in Figure 7.

Figure 7 shows SEM images of curcumin, piperine, pure hierarchical zeolite, and hierarchical zeolite with curcumin. The image of pure curcumin (panel A in Figure 7) shows irregularities in shape and average particle size. In the case of piperine (panel B in Figure 7), more regular and smoother structures are visible, having somewhat smaller particles than those of curcumin. The SEM microscopic image of the pure hierarchical zeolite (panel C in Figure 7) shows agglomerates with irregular shape and visible roughness. However, the SEM image of hierarchical zeolite with curcumin (panel D in Figure 7) shows no significant changes in the size and morphology of the curcumine after being introduced into zeolite. It can be concluded that curcumin can be adsorbed on the surface of the zeolite or in pores, and this process has no significant effect on the size and morphology of hierarchical zeolite [67,68,74–76]. It was previously stated that on the hydroxylated surfaces, such as zeolitic ones, curcumine is well dispersed [77].

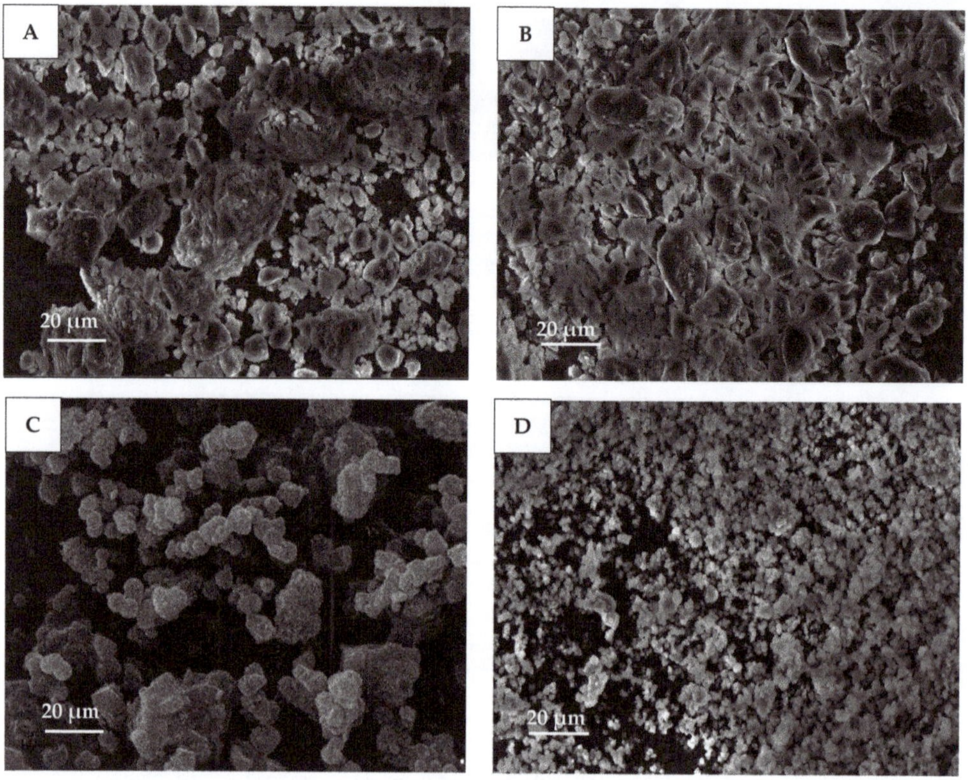

Figure 7. SEM mages of (**A**) curcumin, (**B**) piperine, (**C**) FAU/CTABr, (**D**) FAU/CTABr/CUR150.

2.7. Confocal Microscopy

Through confocal microscopy, the autofluorescence of pure curcumin was observed. By monitoring the sample, the incorporation of the active ingredient into the structure and onto the surfaces of the FAU-type hierarchical zeolite was confirmed (Figure 8).

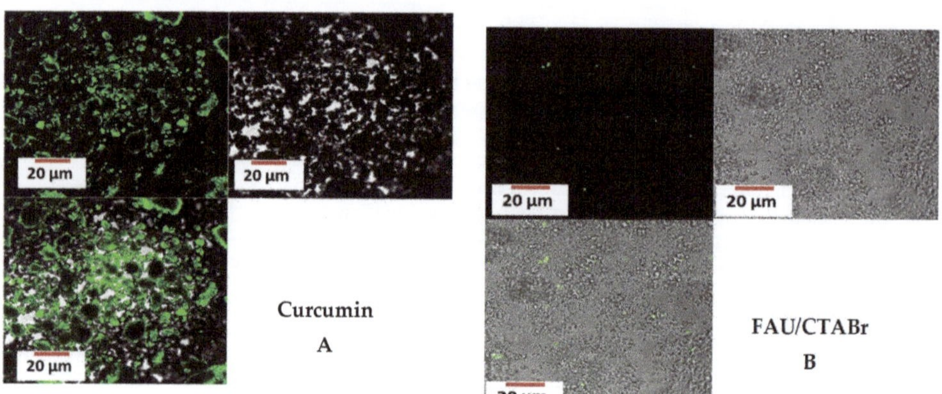

Figure 8. *Cont.*

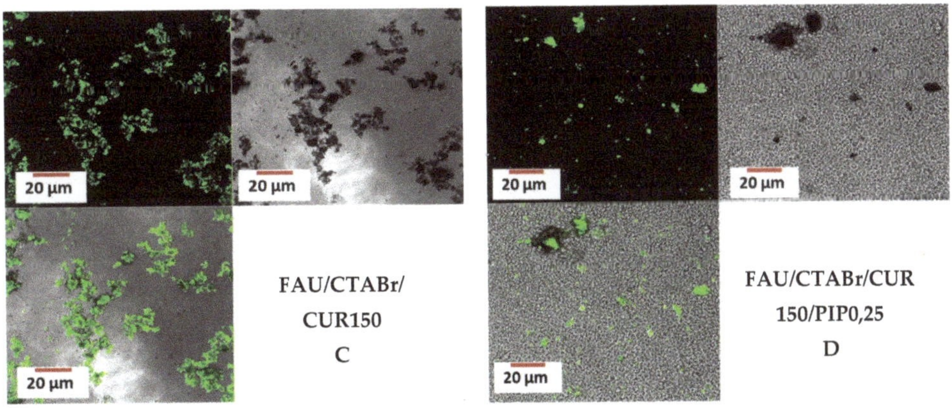

Figure 8. Confocal microscope images showing: (**A**) curcumin, (**B**) pure hierarchical zeolite, (**C**) hierarchical zeolite with incorporated curcumin, and (**D**) hierarchical zeolite with incorporated curcumin and piperine.

Curcumin has fluorescent properties; thus, the autofluorescence changes observed on the images help us locate curcumin in hierarchical zeolites. The highest autofluorescence intensity is visible for pure curcumin, and with the changes in curcumin autofluorescence shown above, we can assess the distribution of curcumin (localization) in the hierarchical zeolite. For hierarchical zeolite samples with incorporated curcumine, the active ingredient was found to be concentrated in the form of small clusters [78]. In addition, confocal microscopy studies show negligible autofluorescence properties of pure hierarchical zeolites. Observing confocal microscope images of the material with applied curcumin and piperine, one can see that the autofluorescence of the active ingredient is much lower (piperine was applied in smaller amounts, and thus, the distribution of this ingredient is much more homogeneous). It is presumed that this is due to the overshadowing of curcumin molecules by piperine molecules (piperine has autofluorescence properties as well).

2.8. Encapsulation Efficiency and Loading Capacity of Curcumin

Encapsulation efficiency (*EE*) and loading capacity (*LC*) are important factors to consider when evaluating the success of a DDS. The release of the active substance from hierarchical zeolites and the target effectiveness of the carriers containing curcumin depend on the *EE* and *LC* values [79]. The results of the encapsulation efficiency and loading capacity of curcumin into hierarchical zeolites are presented in Table 3.

Table 3. Encapsulation efficiency (*EE*) and loading capacity (*LC*) of curcumin into hierarchical zeolites.

Sample	Active Substance	*EE* (% ± SD)	*LC* (% ± SD)
FAU/CUR150	Curcumin	55.8 ± 0.62	9.7 ± 0.52
FAU/CTABr/CUR150	Curcumin	69.6 ± 0.12	14.7 ± 0.22
FAU/CTABr/CUR150/PIP0,25	Curcumin and Piperine	78.9 ± 0.45	20.4 ± 0.16

The encapsulation efficiency of curcumin is the highest in the case of FAU/CTABr/CUR150/PIP0,25 (78.9%) and loading capacity (20.4%). The lowest efficiency of curcumin encapsulation is for commercial FAU zeolite (FAU/CUR150; 55.8%), and the loading capacity is 9.7%. As expected, the best parameters are for hierarchical zeolite.

2.9. UV–Visible Spectroscopy

UV–vis spectroscopy makes it possible to determine the absorption properties of the tested compound. Each substance has a characteristic UV–vis spectrum and wavelength. According to literature data, pure curcumin has a wavelength of 415–430 nm [80]. Figure 9 shows the UV–vis spectra of pure curcumin in acetone and curcumin loaded on zeolite in acetone.

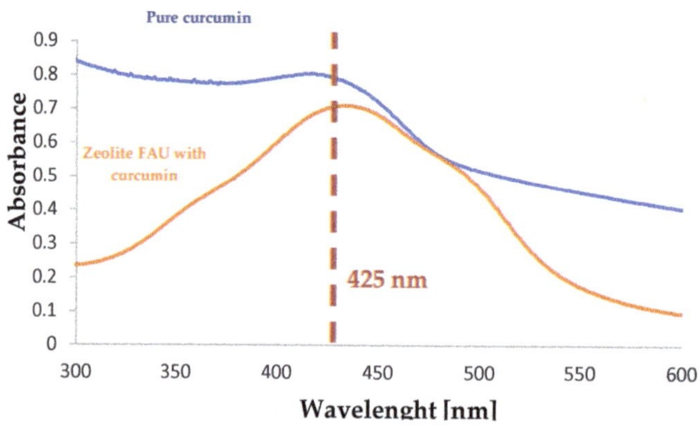

Figure 9. UV–vis spectra of pure curcumin solution and curcumin after adsorption on FAU zeolite.

After the loading process, solutions of pure curcumin and zeolite with loaded curcumin were separated by sedimentation and analyzed by UV–vis spectroscopy. In the UV–vis spectra (Figure 9), the presence of curcumin and curcumin residues when combined with zeolite FAU was confirmed by the appearance of an intense band at λ_{max} = 425 nm.

2.10. Open System Release Study of Curcumin

The controlled release of active substances from pharmaceutical compounds is important for evaluating their efficacy. The subject of this study was the release of curcumin from zeolite materials into the acceptor fluid, which was phosphate buffer (pH = 5.8) with glycerine. The use of UV–vis spectroscopy in release studies allows us to determine the release kinetics of the bioactive substance from the carrier to the acceptor medium. As a result of the release of curcumin from the hierarchical zeolite, there is a change in its concentration in the acceptor solution. Thus, the amount of substance released is calculated based on the observed change in absorbance over time according to the Lambert–Beer law. This measurement was carried out at the analytical wavelength of the test compound (425 nm). The results of the in vitro release kinetics of the active substance are presented in the form of a release profile showing the dependence of the % release of the active substance on time. The availability of the active substance is measured under laboratory conditions and determines the amount of substance that is released from the carrier and passes into the acceptor fluid. Active substance release studies were conducted for the FAU/CTABr/CUR150 system. Each measurement was repeated three times, and the mean value and standard deviation were calculated. Figure 10 shows the curcumin release profiles from hierarchical zeolites.

As shown in Figure 10, the curcumin release from hierarchical zeolites in simulated gastrointestinal media shows an initial burst-type release in the first 30 min. This release may be related to the surface-bound curcumin molecules on the surface of the zeolite materials [73]. Next, a continuous release is visible for 720 min (12 h). The curcumin release process from hierarchical zeolites under artificial and natural light (panel A in Figure 10) is rapid in 30 min, after which the process slows down, and the substance is gradually released. The carrier containing curcumin, in this case, shows a higher % release (about 25%) compared to hierarchical zeolite containing curcumin and piperine (about 11%). This fact may be related to the blocking of curcumin by larger piperine molecules. When the release of curcumin from the zeolite material is carried out in the absence of light, the release is much slower, and the active ingredient is released at a much lower rate. Analysis of data displayed in panel B of Figure 10 shows that % release of curcumin from the same material in the absence of light is twice as small. This is analogous to the carrier that contains curcumin and piperine. A completely different situation is encountered in the case of curcumin release with access to three types of light (panel C in Figure 10). The active ingredient in both cases (with and without piperine) is gradually released, and the final release rates are approximately 25% and 15%, respectively. The release of curcumin from hierarchical zeolite is limited due to its low solubility and variable or low light intensities. The whole process was carried out under daylight, room light, and fume hood [81].

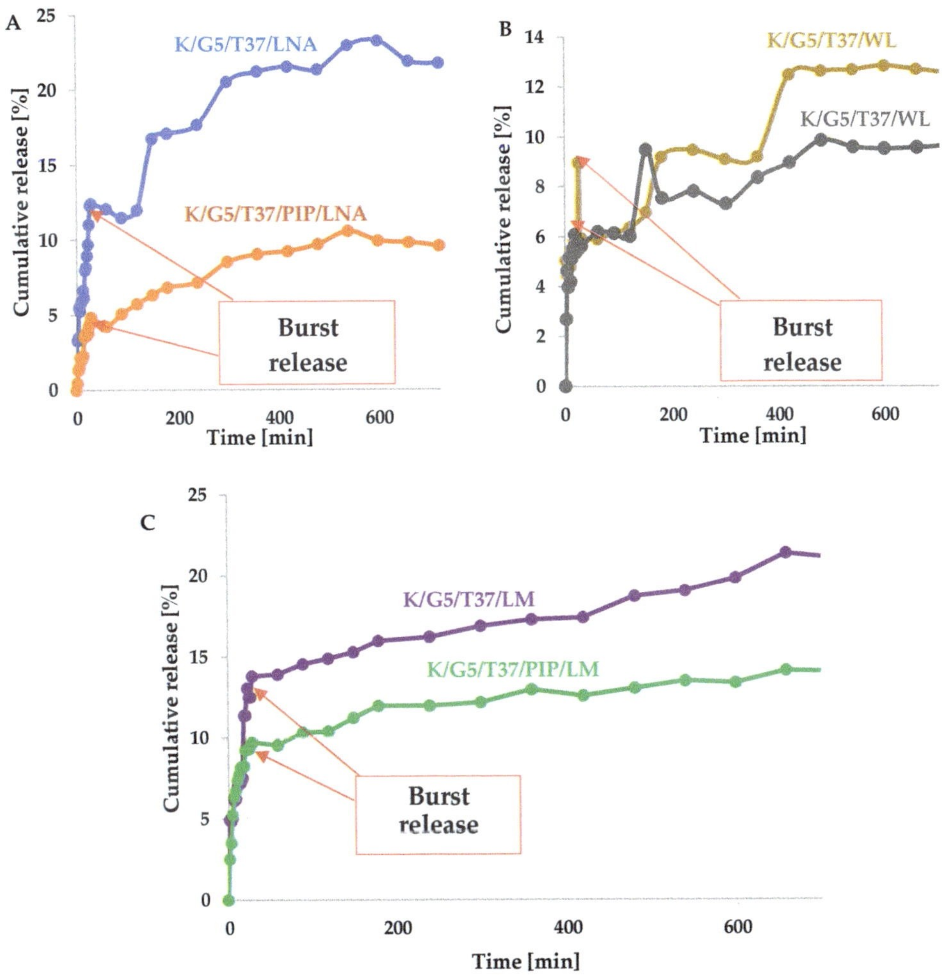

Figure 10. The light-induced release profile of curcumin from hierarchical zeolites; (**A**): the curcumin release process from hierarchical zeolites under artificial and natural light; (**B**): the curcumin release process from hierarchical zeolites in the absence of light; (**C**): the curcumin release process from hierarchical zeolites under mixed light (natural light, artificial light, and an additional light source of 36 W and luminous flux of 2900 lm).

2.11. *Study of Curcumin Release in a Closed System*

The photodynamic release of curcumin in a special reactor equipped with LEDs was carried out several times. The process of photodynamic release of active substance was carried out for FAU/CTABr/CUR150. The study started by checking the effect of the wavelength emitted by LEDs on curcumin release. Preliminary conclusions were obtained after testing all available types of LEDs (Table 4). Each of the experiments was repeated three times.

Table 4. Selecting the appropriate light length for the photodynamic release of curcumin from hierarchical zeolites.

Wavelength	Color	% Release	±SD
365–370	UV	12.64%	±0.70
450	Blue	18.28%	±0.27
500	Cyan	47.99%	±0.18
525	Green	36.98%	±0.43
595	Yellow	18.38%	±0.32
620–630	Red	25.11%	±0.26

The most optimal wavelength is 500 nm (CYAN). However, to further improve the release of curcumin, the testing was conducted by combining several wavelengths emitted by LEDs and by checking their effect on the release of the active substance (Table 5). Each of the experiments was repeated three times.

Table 5. Summary of combined wavelengths relevant to the release of curcumin from the zeolite carrier.

Color	Wavelength	% Release	±SD
Blue–Green	450 and 525 nm	45.89%	±0.21
Cyan–Green	500–525 nm	51.20%	±0.33
Blue–Cyan–Green	430–550	53.24%	±0.54
Yellow–Red	450–525 nm	31.07%	±0.12
Multi-colored	365–630 nm	23.48%	±0.61

Each test was conducted three times to verify the repeatability of the method. After a series of tests, the most favorable wavelength emitted by the LEDs was selected. The focus was on blue, cyan, and green, as well as their combinations. All results are summarized in graphs (Figure 11).

As can be seen from the graph, the most favorable wavelength for the photodynamic release of curcumin from the zeolite carrier is 430–550 nm (Blue–Cyan–Green). At this wavelength range, the active substance is released up to 53.24% ± 0.54 in about 3 h, indicating a good efficiency of its photodynamic release.

2.12. *Potential Mechanism of Photodynamic Light-Triggered Release of Curcumin from Hierarchical Zeolites*

The exact mechanism of photodynamic release of curcumin from the zeolite carrier is shown in Figure 12.

Curcumin loaded inside and on the surface of the FAU-type hierarchical zeolite was released with the help of light of the appropriate wavelength. During the laboratory tests, a series of tests were carried out using seven different types of light, and a combination of several of them was also checked. The conducted research gave a lot of information, due to which it was possible to choose a specific wavelength that will enable the best efficiency of the process. Due to the tests carried out, it is assumed that the curcumin anchored on the surface of the zeolite is detached first followed by the one that was built into the pores of the carrier. Curcumin is linked to hierarchical zeolite by hydrogen bonds. The use of light of the appropriate wavelength breaks the hydrogen bonds, which releases the curcumin into the phosphate buffer at pH = 5.8.

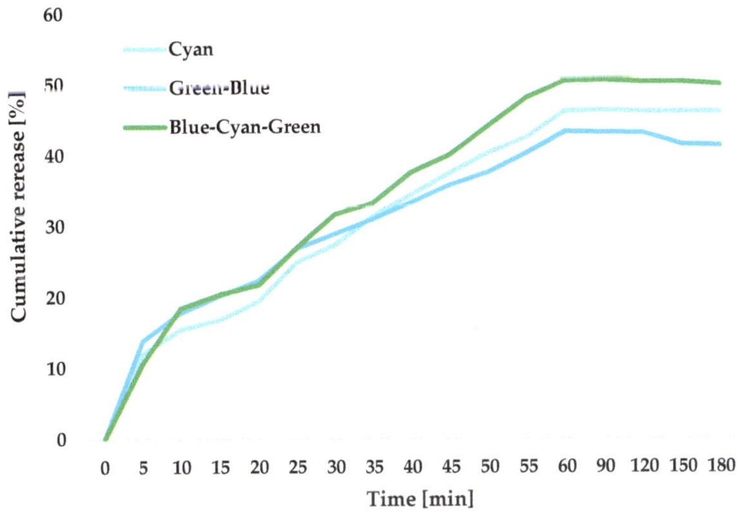

Figure 11. The release profile of curcumin from hierarchical zeolites using the photodynamic release method.

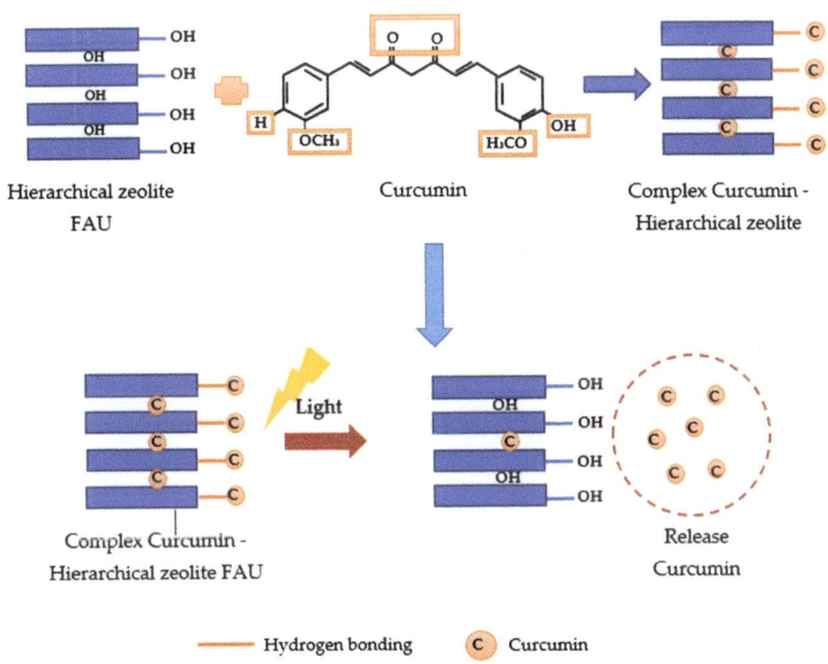

Figure 12. Scheme of photorelease of curcumin from the surface and interior of hierarchical FAU-type zeolite.

2.13. Study of the Kinetics of Curcumin Release from Hierarchical Zeolite

Another aspect of this study was evaluation of the selected kinetic models (zero-order, first-order, and Higuchi) by using experimental data for curcumin release from hierarchical zeolites. By analyzing the kinetics of the release of the active substance from the respective carriers, it is possible to assess the rate of release process. To evaluate the release kinetics of curcumin, the corresponding correlation plots presented in Table 5 were prepared, and then, the correlation coefficient R^2 was determined for the obtained release profiles. Table 5 summarizes the results for the open system release.

Based on the analysis of the values of the R^2 parameter calculated for the release profiles of curcumin from hierarchical zeolites, the release of the active substance from all the analyzed carriers follows the first-order model due to the highest R^2 correlation coefficient compared to the other kinetic models. The good fit of the empirical data to the kinetic model provides some information about the mechanism of diffusion of curcumin molecules from the zeolite carrier.

3. Materials and Methods

3.1. Materials

Commercial FAU zeolite was purchased from Alfa Aesar by Thermo Fisher Scientific (Kandel, Germany). Curcumin (purity > 90%) was purchased from ROTH, Poland. The piperine was purchased from TriMen Chemicals SA (Lodz, Poland). Substances needed for the preparation of the phosphate buffer solution (pH = 5.8) were purchased: sodium hydroxide czda-basic 98.8% purchased from POCH Basic (Gliwice, Poland), potassium diphosphate purchased from Sigma-Aldrich Sp. z.o.o (Poznań, Poland). Vegetable glycerin 99.5% was purchased from P. P. H. "STANLAB" Sp. J. (Lublin, Poland). Ethyl alcohol was bought from P. P. H. "STANLAB" Sp. J. (Lublin, Poland). Tween 80 was purchased from Sigma-Aldrich Sp. z.o.o (Poznań, Poland). All other chemicals and solvents were analytical grade and used without further purification.

3.2. Incorporation of Curcumin to Hierarchical Zeolite-Type FAU

To incorporate curcumin (CUR) to hierarchical zeolite type FAU, 250.00 mg of zeolite carrier was added into 10.00 mL of acetone. Next, 200.00 mg of curcumin was added to the resulting mixture, and the whole mixture was placed on a magnetic stirrer, where the active substance was applied at room temperature for 20 h. Next, the mixture was filtered, and the resulting product was dried in air at room temperature for 16 h to eliminate residual solvent. For comparison, the application process of curcumin along with piperine was carried out. The same amounts of carrier and active ingredient (CUR) were used, and 0.5 mg of piperine was added, and the whole process was identical as described above for incorporation of pure curcumin. After the process was completed, the percentage of curcumin to the selected carrier (%*LOAD*) was calculated. The efficiency of the application of curcumin to the hierarchical zeolite type FAU was determined using Equation (1).

$$\%\, load = \frac{final\ mass\ of\ curcumin\ fixed\ in\ the\ carrier\ [g]}{mass\ of\ the\ complex\ (curcumin + carrier)[g]} \times 100\% \qquad (1)$$

Equation (1). Expression to determine the curcumin % in zeolite carrier [31].

3.3. Optimization of the Curcumin Release from Hierarchical Zeolites

Both the optimization process and all experiments were conducted using hierarchical zeolite synthesized with the structure-directing agent CTABr (hexadecyltrimethylammonium bromide) with curcumin. A sample containing piperine was also tested. During optimization of the curcumin release (CUR) from a given carrier, the following factors were considered: the type of permeation promoter (glycerol, ethanol, Tween 80), the amount of permeation promoter (1.00 and 5.00 mL), the temperature of the curcumin release process (37 and 43 °C), the effect of light used (no light, artificial light, natural light, and simultaneous natural and artificial light) and whether the addition of piperine affects the release of the active substance (Table 6).

Table 6. Rate constants and half-life of curcumin release.

Material	Kinetic Model			Total Amount Released [%]
	0 Order	1st Order	Higuhi	
	R^2	R^2	R^2	
Open system				
K/G5/T37/LNA	0.7745	0.7953	0.7745	23.28
K/G5/T37/PIP/LNA	0.8103	0.8195	0.8103	10.57
K/G5/T37/WL	0.8151	0.8262	0.8144	12.84
K/G5/T37/PIP/WL	0.6675	0.6795	0.6675	9.65
K/G5/T37/LM	0.6693	0.6961	0.6693	21.35
K/G5/T37/PIP/LM	0.5836	0.6044	0.5836	14.13
Fotodynamic release				
Cyan	0.6063	0.6486	0.6063	46.26
Green-Blue	0.5360	0.5747	0.5360	41.29
Blue-Cyan-Green	0.5939	0.6398	0.5939	50.46

The materials were designated by their respective acronyms: C/X/TY/LY, where C—hierarchical zeolite/curcumin (CUR) complex (2.00 mg), X—type of permeation promoter (G: glycerol, E: ethyl alcohol, TW: Tween 80), TY—temperature used during the release (T37—release temperature of 37 °C, and T43—release temperature changed from 37 °C to 43 °C during the process), LY—type of light used (LN—natural light, LA—artificial light, LNA—natural and artificial light, LM—mixed light (natural light, artificial light, and an additional light source of 36 W and luminous flux of 2900 lm), WL—no light). Additionally, the effect of piperine addition on the curcumin release process was verified and designated as PIP.

Characteristic wavelength related to the maximum absorbance of curcumin in each solvent was observed at 425 nm.

The optimal conditions for the release of curcumin (CUR) from the hierarchical zeolite type FAU are: glycerol as the permeation promoter in the amount of 5.00 mL, temperature of 37 °C, the most convenient type of light is mixed light (natural light, artificial light, and additional light source of 36 W and a luminous flux of 2900 lm). The same conditions were also used for the experiment without piperine.

3.4. Curcumin Release from Hierarchical Zeolites

3.4.1. Curcumin Release in an Open System

A temperature-controlled magnetic stirrer and a Varian Cary 50 Bio UV–vis spectrophotometer were used to evaluate the release kinetics of curcumin from the hierarchical zeolite. Then, 2.00 mg of the carrier containing curcumin (0.68 mg of pure curcumin) was weighed in a glass bottle (dry sample), and 20.00 mL of phosphate buffer, pH 5.8, was added. Furthermore, a permeation promoter was introduced, in our case glycerol at 5.00 mL, to accelerate curcumin release. The permeation promoter was used as a stabilizer because curcumin is unstable in phosphate buffer and only releases minimally. The process temperature was maintained at 37 °C, and the release process was carried out for 24 h, with measurements every 2 min for the first 30 min, then every 30 min for 2.5 h, after which measurements were carried out every 1 h for 9 h. The release rate of curcumin was determined by UV–vis in the range of 200 to 800 nm (Figure 13). Each experiment was repeated three times.

At a predetermined time, approximately 3.00 mL of the reaction mixture was taken and centrifuged for 1 min. The centrifuged mixture was transferred into a measuring cell, which was then placed in a UV–vis spectrophotometer.

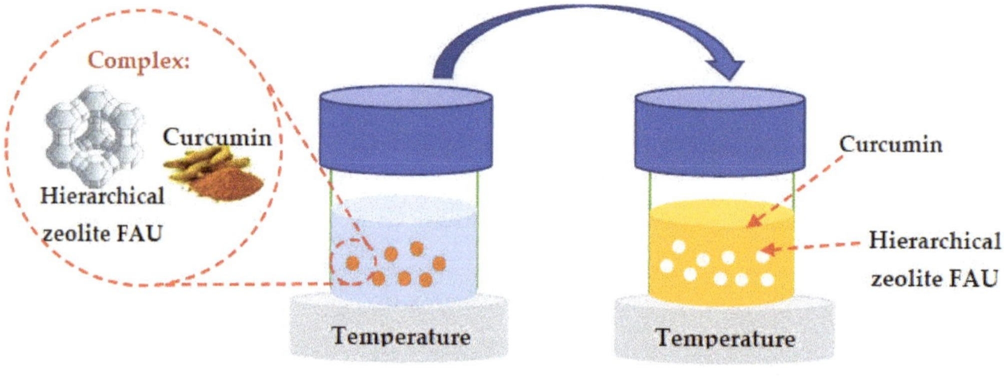

Figure 13. Schematic of the standard release of curcumin from the zeolite carrier.

Preparation Procedure for the Different Components of Curcumin Release

To release curcumin from the hierarchical zeolite-type FAU, the acceptor fluid and sample were prepared, respectively.

Preparation of the Acceptor Fluid

An acceptor solution that is a mixture of glycerol and phosphate buffer with a pH of 5.8 (1:4, v/v) was used to study the release kinetics of curcumin from hierarchical zeolites. To prepare the phosphate buffer, 17.60 g of potassium dihydrogen phosphate (V) (KH_2PO_4) was weighed in a glass beaker, and thę weight was dissolved in 1000.00 mL of demineralized water. In a separate vessel, 4.07 g of sodium hydroxide (NaOH) was dissolved in 1000.00 mL of demineralized water. The phosphate buffer was prepared by combining 732.00 mL of NaOH solution with the previously prepared KH_2PO_4 solution. After thorough mixing of the components of the buffer, its pH value was measured, which was 5.8.

Sample preparation

First, 2.00 mg of hierarchical zeolite containing curcumin and 5.00 mL of glycerin as the permeation promoter were weighed. The material was then placed in a glass bottle, and the material was poured with 20.00 mL of phosphate buffer. The entire bottle was secured with a screw cap and placed on a magnetic stirrer where the whole process was carried out. The process temperature was maintained at 37 °C. Curcumin release studies were carried out in phosphate buffer at pH 5.80. The process of evaluating curcumin release kinetics was carried out for 24 h. The release parameters were determined as shown in Tables 7 and 8.

The percentage of the release of active ingredients from hierarchical zeolites was calculated based on the following equation [81]:

$$\%release = \frac{A_p}{A_w}\left(\frac{m_w[\text{mg}] \times C_w}{V_w[\text{mL}]}\right)\left(\frac{1}{D_w}\right)\left(\frac{V_p[\text{mL}]}{m_p[\text{mg}]}\right) \times 100\% \quad (2)$$

where: A_p—absorbance of the sample (a.u.), A_w—absorbance of the standard (a.u.), m_w—mass of the standard (mg), m_p—mass of active substance contained in the sample (mg), C_w—purity of the standard (a.u.), D_w—dilution of the standard (a.u.), V_w—the volume of the standard solution (mL), V_p—the volume of phosphate buffer (mL).

Equation (2). The formula for calculating the percent release of curcumin from the carrier [82].

Table 7. Parameter optimization in the photodynamic release of curcumin.

Sample	Stabilizer	Stabilizer Quantity (mL)	Process Temperature (°C)	Type of Light
C/G1/T37/LNA	Glycerin	1	37	natural light artificial light
C/E1/T37/LNA	Ethyl alcohol	1	37	natural light artificial light
C/TW1/T37/LNA	Tween 80	1	37	natural light artificial light
C/G5/T37/LM	Glycerin	5	37	natural light artificial light additional light source
C/G5/T37/NL	Glycerin	5	37	no light
C/G5/T43/LNA	Glycerin	5	43	natural light artificial light
C/G5/T37/PIP/LM	Glycerin	5	37	natural light artificial light additional light source
C/G5/T37/PIP/NL	Glycerin	5	37	no light
C/G5/T37/PIP/LNA	Glycerin	5	43	natural light artificial light additional light source

Table 8. Parameters of the study of curcumin release from hierarchical zeolites.

Measurement Parameter	Parameter Value
Process temperature	37 °C
Agitator speed	200 RPM/min
Testing time	24 h
Media quantity	2.00 mg
Amount of promoter	5.00 mL
Buffer volume	20.00 mL

3.4.2. Curcumin Release in a Closed System

The photodynamic release of curcumin was carried out in a special reaction chamber equipped with a magnetic stirrer with heating function and LEDs, which were characterized by the corresponding emitted wavelength (Table 9).

Table 9. Spectral colors depend on the wavelength band.

Wavelength Range (nm)	Perceived Color
365–370	UV
395–405	UV
450	Blue
500	Cyan
525	Green
595	Yellow
620–630	Red

The reaction chamber was sealed to prevent access to any other light source. The test sample was irradiated with each type of light to see which has the greatest effect on the curcumin release process. For photodynamic release studies of curcumin, the same conditions were used for the standard release process of the active substance.

Figure 14 shows the excitation profiles of the light sources used during the study of curcumin release from hierarchical zeolites of the FAU type.

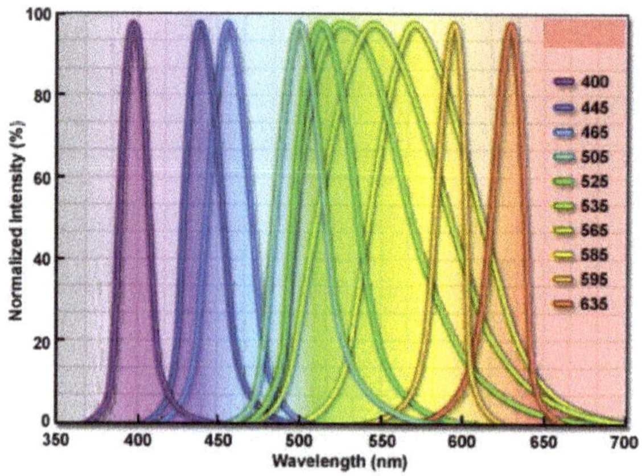

Figure 14. Excitation profiles of the light sources used (adapted with permission from ref. [83] 2023, Zeiss).

In addition, the effect of the overlap of several wavelengths emitted by the LEDs used was demonstrated:

- blue and green—450 and 525 nm;
- cyan and green—500–525 nm;
- blue and cyan—450–500 nm;
- blue, green and cyan—450–525 nm;
- yellow and red and yellow—595–630 nm.

3.5. Characterization of Materials

Curcumin-loaded zeolite carriers were characterized by using the following techniques:

- X-ray diffraction (XRD);
- Low-temperature N_2 adsorption/desorption;
- Fourier transform infrared spectroscopy (FT-IR);
- Elemental analysis;
- Thermogravimetric analysis (TG);
- Scanning electron microscopy (SEM);
- Confocal microscopy;
- Release of the active substance.

3.5.1. X-ray Diffraction

X-ray diffraction patterns for hierarchical zeolites with curcumin were measured on a Bruker D8 ADVANCE powder diffractometer with a Johansson monochromator (Bruker; X Cu-Kα = 0.15406 nm). These patterns were obtained in the range of small angles (SAXD: 2 = 0.6–0.8°) with a scan rate of 0.02°/3 s and the range of wide angles (WAXD: 2 = 6.0–60.0°) with a scan rate of 0.05°/1 s.

3.5.2. Low-Temperature Nitrogen Adsorption/Desorption Isotherms

Low-temperature nitrogen adsorption/desorption isotherms were performed using an ASAP 2420 MICROMERITICS instrument. The test samples were degassed under vacuum at 120 °C for 24 h before measurement. The isotherms were obtained at ca. −196 °C, in the relative pressure range p/p0 from 0.02 to 1.00.

3.5.3. Fourier Transform IR Spectroscopy

Characterization of the individual functional groups was performed by using Fourier transform IR spectroscopy. For this purpose, a Cary 630 FTIR spectrometer (Agilent Technologies) with an attenuated total reflection (ATR) element (ZnSe crystal) was employed. All spectra were recorded at room temperature in the range of 4000–650 cm^{-1}, with a resolution of 16 cm^{-1}. For each sample, 72 scans were recorded. In addition, 8 background scans were performed.

3.5.4. Elemental Analysis

Elemental composition analysis of hierarchical zeolites with incorporated curcumin was performed on a FLASH 2000, THERMO SCIENTIFIC apparatus. The apparatus for organic elemental analysis uses the method of dynamic complete combustion of samples (2–200 mg) in an electronically temperature-controlled reduction–oxidation furnace.

3.5.5. Thermogravimetric Analysis TG

Thermogravimetric analysis was performed using a Netzsch TG 209 Libra apparatus. The temperature range of the test was from 30 to 995 °C, with a standard rate of temperature change of 10 °C/min. The test consisted of heating a small amount of the sample placed in a measuring crucible made of aluminum oxide. During the analysis, the change (loss) of mass and temperature of the sample during a specified time in a nitrogen atmosphere was continuously recorded.

3.5.6. Scanning Electron Microscopy SEM

The microstructure analysis and morphology of the hierarchical zeolites loaded with curcumin were determined by scanning electron microscopy (SEM, Quanta FEG 250 (FEI)) under a low-pressure vacuum (70 Pa) and 10 kV beam acceleration voltage. Samples of zeolite materials were placed on a 400 mesh carbon-coated copper grid. Under the influence of the electron beam, the sample emitted various signals that were picked up by the detector and converted into an image of the sample.

3.5.7. Confocal Microscopy

Confocal microscopy was used to obtain high-quality images and reconstruct images in three dimensions. Before taking images using a confocal microscope (Zeiss Axiovert 200 M), the obtained zeolite materials were applied to a coverslip. Curcumin autofluorescence was excited using a 488 nm laser.

3.5.8. Encapsulation Efficiency and Loading Capacity of Curcumin

The encapsulation efficiency (*EE*) and loading capacity (*LC*) of the curcumin were evaluated based on the concentration of unencapsulated curcumin in the supernatant. The unencapsulated curcumin in the supernatant was previously dissolved in ethanol prior to the centrifugation process and then analyzed using a high-performance liquid chromatograph (Varian 920-LC, Agilent Technologies, Santa Clara, CA, USA) coupled with a UV–vis detector. The chromatographic determination of curcumin in the supernatant involved the quantification of the unincorporated active ingredient present in the fluid. The *EE* of curcumin was calculated by Equation (3) while the *LC* was calculated by Equation (4) [84].

$$EE\% = \frac{\text{Total amount of drug added} - \text{unentrapment drug}}{\text{Total amount of drug added}} \times 100\% \tag{3}$$

$$LC\% = \frac{\text{Total amount of drug added} - \text{unentrapment drug}}{\text{Total amount of drug added}} \times 100\% \tag{4}$$

3.5.9. UV–Visible Spectroscopy

UV–vis measurements were made with the Varian Cary 50 Bio UV–Vis spectrophotometer in the range of 300–600 nm (curcumin absorption peak λmax = 425 nm). Curcumin solutions were diluted with pure acetone and measured in quartz cells with an optical path length of 1 cm.

3.5.10. Curcumin Release from Hierarchical Zeolites

Hierarchical zeolite curcumin release tests were performed according to two techniques described in Sections 3.3 and 3.4.

4. Conclusions

Curcumin is a compound that has numerous advantages; however, due to its low water solubility and high photosensitivity, its clinical applications are limited. Therefore, it was important to select a suitable carrier, which turned out to be a hierarchical zeolite obtained by proper treatment of a commercial FAU-type zeolite. Studies on curcumin release from hierarchical zeolites were carried out under simulated gastrointestinal conditions using two methods. The first is the classical method of releasing the active substance in the open circulation. Using this method, curcumin is released much more slowly and to a much smaller extent. This is probably because curcumin is very sensitive to light. Therefore, the photodynamic release of curcumin from zeolite carriers proved to be a more advantageous and efficient method. Curcumin was released much faster, and the percentage of release was twice as high as that achieved by the classical release method. The results of our study show that the percentage of curcumin released from hierarchical zeolites is higher for the photodynamic release process. This observation was confirmed by XRD analysis, showing the presence of curcumin in the pores of the hierarchical zeolites, and confirming strong interactions of the zeolite with the active substance. SEM analysis showed a stable structure of the hierarchical zeolite, the integrity of which was preserved after curcumin loading. Using confocal microscopy, it was possible to locate the active substance in the carrier. These results confirmed the presence of curcumin in the FAU zeolite carrier and support the potential use of this porous material as a drug carrier.

Author Contributions: E.M.: methodology, visualization, data curation, writing—original draft; A.F.-G.: conceptualization, methodology, data curation, writing—editing; M.J.: data curation; writing—editing; I.N.: conceptualization, methodology; data curation; funding acquisition, project administration, writing—editing. All authors have read and agreed to the published version of the manuscript.

Funding: The work was carried out within the project "Advanced biocomposites for tomorrow's economy BIOG-NET", funded by the Foundation for Polish Science from the European Regional Development Fund (POIR.04.04.00-00-1792/18-00).

Data Availability Statement: The data presented in this study are available upon request from the authors.

Conflicts of Interest: The authors declare no conflict of interest.

References

1. Allen, T.M.; Cullis, P.R. Drug delivery systems: Entering the mainstream. *Science* **2004**, *303*, 1818–1822. [CrossRef] [PubMed]
2. Regi, M.V.; Balas, F.; Arcos, D. Mesoporous Materials for Drug Delivery. *Angew. Chem. Int. Ed.* **2007**, *46*, 7548–7558. [CrossRef] [PubMed]
3. Regi, M.V. Ordered Mesoporous Materials in the Context of Drug Delivery Systems and Bone Tissue Engineering. *Chem. Eur. J.* **2006**, *12*, 5934–5943. [CrossRef]
4. Rathbone, M.J.; Hadgraft, J.; Roberts, M.S.; Lane, M.E. *Modified-Release Drug Delivery Technology*, 2nd ed.; CRC Press: Boca Raton, FL, USA, 2008; pp. 303–321.
5. Rámila, A.; Muñoz, B.; Pérez-Pariente, J.; Vallet-Regí, M. Mesoporous MCM-41 as drug host system. *J. Sol-Gel Sci. Technol.* **2003**, *26*, 1199–1202. [CrossRef]

6. Horcajada, P.; Rámila, A.; Pérez-Pariente, J.; Vallet-Regí, M. Influence of Pore Size of MCM-41 Matrices on Drug Delivery Rate. *Microporous Mesoporous Mater.* **2004**, *68*, 105–109. [CrossRef]
7. Rouquerol, J.; Avnir, D.; Fairbridge, C.W.; Everett, D.H.; Haynes, J.M.; Pernicone, N.; Ramsay, J.D.F.; Sing, K.S.W.; Unger, K.K. Recommendations for the characterization of porous solids (Technical Report). *IUPAC Pure Appl. Chem.* **1994**, *66*, 1739–1758. [CrossRef]
8. Andersson, J.; Rosenhoim, J.; Areva, S.; Linden, M. Influences of Material Characteristics on Ibuprofen Drug Loading and Release Profiles from Ordered Micro- and Mesoporous Silica Matrices. *Chem. Mater.* **2004**, *16*, 4160–4167. [CrossRef]
9. Feliczak-Guzik, A. Hierarchical zeolites: Synthesis and catalytic properties. *Microporous Mesoporous Mater.* **2018**, *259*, 33–45. [CrossRef]
10. Jia, X.; Khan, W.; Wu, Z.; Choi, J.; Yip, A.C.K. Modern synthesis strategies for hierarchical zeolites: Bottom-up versus top-down strategies. *Adv. Powder Technol.* **2019**, *30*, 467–484. [CrossRef]
11. Verboekend, D.; Milina, M.; Mitchell, S.; Ramírez, J.P. Hierarchical Zeolites by Desilication: Occurrence and Catalytic Impact of Recrystallization and Restructuring. *Cryst. Growth Des.* **2013**, *13*, 5025–5035. [CrossRef]
12. Ding, K.; Corma, A.; Maciá-Agulló, J.A.; Hu, J.G.; Krämer, S.; Stair, P.C.; Stucky, G.D. Constructing hierarchical porous zeolites via kinetic regulation. *J. Am. Chem. Soc.* **2015**, *137*, 11238–11241. [CrossRef] [PubMed]
13. Baerlocher, C.; McCusker, L.B.; Olson, D.H. *Atlas of Zeolite Framework Types*, 6th ed.; Elsevier: Amsterdam, The Netherlands, 2007; pp. 140–141.
14. Bacakova, L.; Vandrovcova, M.; Kopova, I.; Jirka, I. Applications of zeolites in biotechnology and medicine–a review. *Biomater. Sci.* **2018**, *6*, 974–989. [CrossRef] [PubMed]
15. Hartmann, M.; Machoke, A.G.; Schwieger, W. Catalytic test reactions for the evaluation of hierarchical zeolites. *Chem. Soc. Rev.* **2016**, *45*, 3313–3330. [CrossRef] [PubMed]
16. Elsatar, A.G.A.; Farag, M.M.; Youssef, H.F.; Salih, S.A.; Mounier, M.M.; El-Meliegy, E. Different zeolite systems for colon cancer therapy: Monitoring of ion release, cytotoxicity and drug release behavior. *Prog. Biomater.* **2019**, *8*, 101–113. [CrossRef] [PubMed]
17. Kralj, M.; Pavelic, K. Medicine on a small scale: How molecular medicine can benefit from self-assembled and nanostructured materials. *EMBO Rep.* **2003**, *4*, 1008–1012. [CrossRef] [PubMed]
18. Thom, D.C.; Davies, J.E.; Santerre, J.P.; Friedman, S. The hemolytic and cytotoxic properties of a zeolite-containing root filling material in vitro. *Oral Surg. Oral Med. Oral Pathol. Oral Radiol. Endodontol.* **2003**, *95*, 101–108. [CrossRef]
19. Piasek, A.; Bartoszek, A.; Namieśnik, J. Substancje pochodzenia roślinnego przeciwdziałające kardiotoksyczności towarzyszącej chemioterapii nowotworów. *Adv. Hyg. Exp. Med.* **2009**, *63*, 142–158.
20. Goldman, P. Herbal medicines today and the roots of modern pharmacology. *Ann. Intern. Med.* **2001**, *135*, 594–600. [CrossRef]
21. Robinson, M.M.; Zhang, X. The World Medicines Situation 2011. In *Traditional Medicines: Global Situation, Issues and Challenges*, 3rd ed.; World Health Organization: Geneva, Switzerland, 2011.
22. Anand, P.; Thomas, S.G.; Kunnumakkara, A.B.; Sundaram, C.; Harikumar, K.B.; Sung, B.; Tharakan, S.T.; Misra, K.; Priyadarsini, I.K.; Rajasekharan, K.N.; et al. Biological Activities of Curcumin and its Analogues (Congeners) Made by Man and Mother Nature. *Biochem. Pharmacol.* **2008**, *76*, 1590–1611. [CrossRef]
23. Lakshmi, S.; Padmaja, G.; Remani, P. Antitumour Effects of Isocurcumenol Isolated from Curcuma zedoaria Rhizomes on Human and Murine Cancer Cells. *Int. J. Med. Chem.* **2011**, *2011*, 253962. [CrossRef]
24. Itokawa, H.; Shi, Q.; Akiyama, T.; Morris-Natschke, S.L.; Lee, K.H. Recent advances in the investigation of curcuminoids. *Chin. Med.* **2008**, *3*, 11. [CrossRef] [PubMed]
25. Lobo, R.; Prabhu, K.S.; Shirwaikar, A.; Shirwaikar, A. *Curcuma zedoaria Rosc.* (white turmeric): A review of its chemical, pharmacological and ethnomedicinal properties. *J. Pharm. Pharmacol.* **2009**, *61*, 13–21. [CrossRef] [PubMed]
26. Goel, A.; Kunnumakkara, A.B.; Aggarwal, B.B. Curcumin as "Curecumin": From kitchen to clinic. *Biochem. Pharmacol.* **2008**, *75*, 787–809. [CrossRef] [PubMed]
27. Nagahama, K.; Utsumi, T.; Kumano, T.; Maekawa, S.; Oyama, N.; Kawakami, J. Discovery of a new function of curcumin which enhances its anticancer therapeutic potency. *Sci. Rep.* **2016**, *60*, 30962. [CrossRef]
28. Jun, X.L.; Xiang-Fe, H.; Hang, L.Z. Anti-Cancer Agents. *Med. Chem.* **2012**, *12*, 210–211.
29. Anand, P.; Kunnumakkara, A.B.; Newman, R.A.; Aggarwal, B.B. Bioavailability of Curcumin: Problems and Promises. *Mol. Pharm.* **2007**, *4*, 807–818. [CrossRef]
30. Toden, S.; Goel, A. The Holy Grail of Curcumin and its Efficacy in Various Diseases: Is Bioavailability Truly a Big Concern? *J. Restorat. Med.* **2017**, *6*, 27–36. [CrossRef]
31. Han, H.K. The effects of black pepper on the intestinal absorption and hepatic metabolism of drugs. *Expert Opin. Drug Metab. Toxicol.* **2011**, *7*, 721–729. [CrossRef]
32. Shoba, G.; Joy, D.; Joseph, T.; Majeed, M.; Rajendran, R.; Srinivas, P.S. Influence of piperine on the pharmacokinetics of curcumin in animals and human volunteers. *Planta Med.* **1998**, *64*, 353–356. [CrossRef]
33. Photodynamic Therapy to Treat Cancer. Available online: http://www.cancer.gov/cancertopics/factsheet/Therapy/photodynamic (accessed on 10 January 2022).
34. Silva, L.P.; Magalhães, C.M.; Montenegro, A.N.; Ferreira, P.J.; Duarte, D.; Rodríguez-Borges, J.E.; Vale, N.; Esteves da Silva, J.C.G. Study of the combination of self-activating photodynamic therapy and chemotherapy for cancer treatment. *Biomolecules* **2019**, *9*, 384. [CrossRef]

35. Moret, F.; Reddi, E. Strategies for optimizing the delivery to tumors of macrocyclic photosensitizers used in photodynamic therapy (pdt). *J. Porphyr. Phthalocyanines* **2017**, *21*, 239–256. [CrossRef]
36. Cheng, Y.; Cheng, H.; Jiang, C.; Qiu, X.; Wang, K.; Huan, W.; Yuan, A.; Wu, J.; Hu, Y. Perfluorocarbon nanoparticles enhance reactive oxygen levels and tumour growth inhibition in photodynamic therapy. *Nat. Commun.* **2016**, *6*, 8785–8793. [CrossRef] [PubMed]
37. Agostinis, P.; Berg, K.; Cengel, K.A.; Foster, T.H.; Girotti, A.W.; Gollnick, S.O.; Hahn, S.M.; Hamblin, M.R.; Juzeniene, A.; Kessel, D. Photodynamic therapy of cancer: An update. *CA Cancer J. Clin.* **2011**, *61*, 250–281. [CrossRef] [PubMed]
38. Wu, W.; Shao, X.; Zhao, J.; Wu, M. Controllable photodynamic therapy implemented by regulating singlet oxygen efficiency. *Adv. Sci.* **2017**, *4*, 1700113. [CrossRef] [PubMed]
39. Yang, Q.Q.; Farha, A.K.; Kim, G.; Gul, K.; Gan, R.Y.; Corke, H. Antimicrobial and anticancer applications and related mechanisms of curcumin-mediated photodynamic treatments. *Trends Food Sci. Technol.* **2020**, *97*, 341–354. [CrossRef]
40. Tuanwei, L.; Lifeng, Y. Functional Polymer Nanocarries for Photodynamic Therapy. *Pharmaceuticals* **2018**, *11*, 2–4.
41. Lee, S.C.; Seong, Y.S.; Kim, S.S.; Koh, H.J.; Kwon, O.W. Photodynamic therapy with verteporfin for polypoidal choroidal vasculopathy of the macula. *Ophthalmologica* **2004**, *218*, 193–201. [CrossRef]
42. Choudhary, S.; Nouri, K.; Elsaie, M.L. Photodynamic therapy in dermatology: A review. *Lasers Med. Sci.* **2009**, *24*, 971–980. [CrossRef]
43. Chrepa, V.; Kotsakis, G.A.; Pagonis, T.C.; Hargreaves, K.M. The effect of photodynamic therapy in root canal disinfection: A systematic review. *J. Endod.* **2014**, *40*, 891–898. [CrossRef]
44. Cieplik, F.; Buchalla, W.; Hellwig, E.; Al-Ahmad, A.; Hiller, K.A.; Maisch, T.; Karygianni, L. Antimicrobial photodynamic therapy as an adjunct for treatment of deep carious lesions-a systematic review. *Photodiagn. Photodyn. Ther.* **2017**, *18*, 54–62. [CrossRef]
45. Marchal, S.; Dolivet, G.; Lassalle, H.P.; Guillemin, F.; Bezdetnaya, L. Targeted photodynamic therapy in head and neck squamous cell carcinoma: Heading into the future. *Lasers Med. Sci.* **2015**, *30*, 2381–2387. [CrossRef] [PubMed]
46. Vohra, F.; Al-Kheraif, A.A.; Qadri, T.; Hassan, M.I.; Ahmed, A.; Warnakulasuriya, S.; Javed, F. Efficacy of photodynamic therapy in the management of oral premalignant lesions. A systematic review. *Photodiagn. Photodyn. Ther.* **2015**, *12*, 150–159. [CrossRef] [PubMed]
47. Goldstein, M.; Heilweil, G.; Barak, A.; Loewenstein, A. Retinal pigment epithelial tear following photodynamic therapy for choroidal neovascularization secondary to AMD. *Eye* **2005**, *19*, 1315–1324. [CrossRef] [PubMed]
48. Silva, R.M.; Figueira, J.; Cachulo, M.L.; Duarte, L.; Faria de Abreu, J.R.; Cunha-Vaz, J.G. Polypoidal choroidal vasculopathy and photodynamic therapy with verteporfin. *Graefes Arch. Clin. Exp. Ophthalmol.* **2005**, *243*, 973–979. [CrossRef] [PubMed]
49. Akaza, E.; Yuzawa, M.; Mori, R. Three-year follow-up results of photodynamic therapy for polypoidal choroidal vasculopathy. *Jpn. J. Ophthalmol.* **2011**, *55*, 39–44. [CrossRef] [PubMed]
50. Muslubas, I.S.; Hocaoglu, M.; Arf, O.; Ozdemir, H.; Karacorlu, M. Treatment outcomes in patients with polypoidal choroidal vasculopathy. *Turk. J. Ophthalmol.* **2016**, *46*, 16–20. [CrossRef]
51. Kachynski, A.V.; Pliss, A.; Kuzmin, A.N.; Ohulchanskyy, T.Y.; Baev, A.; Qu, J.; Prasad, P.N. Photodynamic therapy by in situ nonlinear photon conversion. *Nat. Photon.* **2014**, *8*, 455–461. [CrossRef]
52. Wang, S.; Zhang, L.; Dong, C.; Su, L.; Wang, H.; Chang, J. Smart ph-responsive upconversion nanoparticles for enhanced tumor cellular internalization and near-infrared light-triggered photodynamic therapy. *Chem. Commun.* **2015**, *51*, 406–408. [CrossRef]
53. Wang, W.; Lin, L.; Ma, X.; Wang, B.; Liu, S.; Yan, X.; Li, S.; Tian, H.; Yu, X. Light-induced hypoxia-triggered living nanocarriers for synergistic cancer therapy. *ACS Appl. Mater. Interfaces* **2018**, *10*, 19398–19407. [CrossRef]
54. Wang, P.; Li, X.; Yao, C.; Wang, W.; Zhao, M.; El-Toni, A.M.; Zhang, F. Orthogonal near-infrared upconversion co-regulated site-specific o2 delivery and photodynamic therapy for hypoxia tumor by using red blood cell microcarriers. *Biomaterials* **2017**, *125*, 90–100. [CrossRef]
55. Li, J.; Meng, X.; Deng, J.; Lu, D.; Zhang, X.; Chen, Y.; Zhu, J.; Fan, A.; Ding, D.; Kong, D. Multifunctional micelles dually responsive to hypoxia and singlet oxygen: Enhanced photodynamic therapy via interactively triggered photosensitizer delivery. *ACS Appl. Mater. Interface* **2018**, *10*, 17117–17128. [CrossRef] [PubMed]
56. Lin, S.; Yang, L.; Shi, H.; Du, W.; Qi, Y.; Qiu, C.; Liang, X.; Shi, W.; Liu, J. Endoplasmic reticulum-targeting photosensitizer hypericin confers chemo-sensitization towards oxaliplatin through inducing pro-death autophagy. *Int. J. Biochem. Cell Biol.* **2017**, *87*, 54–68. [CrossRef] [PubMed]
57. Tian, X.; Zhu, Y.; Zhang, M.; Luo, L.; Wu, J.; Zhou, H.; Guan, L.; Battaglia, G.; Tian, Y. Localization matters: A nuclear targeting two-photon absorption iridium complex in photodynamic therapy. *Chem. Commun.* **2017**, *53*, 3303–3306. [CrossRef]
58. Shibu, E.S.; Hamada, M.; Murase, N.; Biju, V. Nanomaterials formulations for photothermal and photodynamic therapy of cancer. *J. Photochem. Photobiol. C Photochem. Rev.* **2013**, *15*, 53–72. [CrossRef]
59. Voon, S.H.; Kiew, L.V.; Lee, H.B.; Lim, S.H.; Noordin, M.I.; Kamkaew, A.; Burgess, K.; Chung, L.Y. In vivo studies of nanostructure-based photosensitizers for photodynamic cancer therapy. *Small* **2014**, *10*, 4993–5013. [CrossRef] [PubMed]
60. Musielak, E.; Feliczak-Guzik, A.; Jaroniec, M.; Nowak, I. Modification and Functionalization of Zeolites for Curcumin Uptake. *Materials* **2022**, *15*, 6316. [CrossRef] [PubMed]
61. Jyoti, K.; Baunthiyal, M.; Singh, A. Characterization of silver nanoparticles synthesized using *Urtica dioica Linn.* leaves and their synergistic effects with antibiotics. *J. Radiat. Res. Appl. Sci.* **2016**, *9*, 217–227. [CrossRef]

62. Maulvi, F.A.; Lakdawala, D.H.; Shaikh, A.A.; Desai, A.R.; Choksi, H.H.; Vaidya, R.J.; Ranch, K.M.; Koli, A.R.; Vyas, B.A.; Shah, D.O. In vitro and in vivo evaluation of novel implantation technology in hydrogel contact lenses for controlled drug delivery. *J. Control. Release* **2016**, *226*, 47–56. [CrossRef]
63. Thommes, M.; Kaneko, K.; Neimark, A.V.; Olivier, J.P.; Reinoso, F.R.; Rouquerol, J.; Sing, K.S.W. Physisorption of Gases, with Special Reference to the Evalua-tion of Surface Area and Pore Size Distribu-tion (IUPAC Technical Report). *Pure Appl. Chem.* **2015**, *87*, 1051–1069. [CrossRef]
64. Kim, S.; Philippot, S.; Fontanay, S.; Duval, R.E.; Lamouroux, E.; Canilho, N.; Pasc, A. pH-and glutathione-responsive release of curcumin from mesoporous silica nanoparticles coated using tannic acid–Fe(III) complex. *RSC Adv.* **2015**, *5*, 90550–90558. [CrossRef]
65. Moradi, P.; Hajjami, M.; Tahmasbi, B. Fabricated copper catalyst on biochar nanoparticles for the synthesis of tetrazoles as antimicrobial agents. *Polyhedron* **2020**, *175*, 114169–114180. [CrossRef]
66. Chaudhuri, H.; Dash, S.; Ghorai, S.; Pal, S.; Sarkar, A. SBA-16: Application for the removal of neutral, cationic, and anionic dyes from aqueous medium. *J. Environ. Chem. Eng.* **2016**, *4*, 157–166. [CrossRef]
67. Ameri, A.; Taghizadeh, T.; Kiakalaieh, A.T.; Forootanfar, H.; Mojtabavi, S.; Jahandar, H.; Tarighi, S.; Faramarzi, M.A. Bio-removal of phenol by the immobilized laccase on the fabricated parent and hierarchical NaY and ZSM-5 zeolites. *J. Taiwan Inst. Chem. Eng.* **2021**, *120*, 300–312. [CrossRef]
68. Preisig, D.; Haid, D.; Varum, F.J.O.; Bravo, R.; Alles, R.; Huwyler, J.; Puchkov, M. Drug loading into porous calcium carbonate microparticles by solvent evaporation. *Eur. J. Pharm. Biopharm.* **2014**, *87*, 548–558. [CrossRef]
69. Jayandran, M.; Haneefa, M.M.; Balasubramanian, V. Characterization and comparative studies of turmeric oleoresin derived from selected turmeric plants. *Asian J. Pharm. Sci. Technol.* **2015**, *5*, 18–21.
70. Ribeiro, S.P.S.; Martins, R.C.; Barbosa, G.M.; Rocha, M.; Landesmann, A.; Nascimento, M. Influence of the zeolite acidity on its synergistic action with a flame-retarding polymeric intumescent formulation. *J. Mat. Sci.* **2020**, *54*, 619–630. [CrossRef]
71. Karimi, M.; Habibizadeh, M.; Rostamizadeh, K.; Khatamian, M.; Divband, B. Preparation and characterization of nanocomposites based on different zeolite frameworks as carriers for anticancer drug: Zeolite Y versus ZSM-5. *Pol. Bull.* **2019**, *76*, 2233–2252. [CrossRef]
72. Sun, X.Z.; Williams, G.R.; Hou, X.X.; Zhu, L.M. Electrospun curcumin-loaded fibers with potential biomedical applications. *Carbohydr. Polym.* **2013**, *94*, 147–153. [CrossRef]
73. Chen, Z.; Xia, Y.; Liao, S.; Huang, Y.; Li, Y.; He, Y.; Tong, Z.; Li, B. Thermal degradation kinetics study of curcumin with nonlinear methods. *Food Chem.* **2014**, *155*, 81–86. [CrossRef]
74. Amosa, M.; AlKhatib, M.; Jami, M.; Jimat, D.; Uthman, O.; Muyibi, S. Morphological synthesis and environmental application of ZSM-5 zeolite crystals from combined low-water and fluoride syntheses routes. *Adv. Environ. Biol.* **2014**, *8*, 613–625.
75. Kolev, T.M.; Velcheva, E.A.; Stamboliyska, B.A.; Spiteller, M. DFT and experimental studies of the structure and vibrational spectra of Curcumin. *Int. J. Quantum Chem.* **2005**, *102*, 1069–1079. [CrossRef]
76. Li, X.; Yang, Q.; Ouyang, J.; Yang, H.; Chang, S. Chitosan modified halloysite nanotubes as emerging porous microspheres for drug carrier. *Appl. Clay Sci.* **2016**, *126*, 306–312. [CrossRef]
77. Ishtiaq, M.; Asghar, S.; Khan, I.U.; Iqbal, M.S.; Khalid, S.H. Development of the Amorphous Solid Dispersion of Curcumin: A Rational Selection of Polymers for Enhanced Solubility and Dissolution. *Crystals* **2022**, *12*, 1606. [CrossRef]
78. Tiwari, A.; Singh, A.; Garg, N.; Randhawa, J.K. Curcumin encapsulated zeolitic imidazolate frameworks as stimuli responsive drug delivery system and their interaction with biomimetic environment. *Sci. Rep.* **2017**, *7*, 12598–12610. [CrossRef]
79. Hudiyanti, D.; Al Khafiz, M.F.; Anam, K.; Siahaan, P.; Christa, S.M. In Vitro Evaluation of Curcumin Encapsulation in Gum Arabic Dispersions under Different Environments. *Molecules* **2022**, *27*, 3855. [CrossRef]
80. Kuźmińska, J.; Sobczak, A.; Wierzchowski, M.; Gośliński, T.; Jelińska, A. Historia i chemia kurkuminy. *Farmacja Współczesna* **2021**, *14*, 140–145.
81. Rezaei, A.; Fathi, M.; Jafari, S.M. Nanoencapsulation of hydrophobic and low-soluble food bioactive compounds within different nanocarriers. *Food Hydrocoll.* **2019**, *88*, 146–162. [CrossRef]
82. Prokopowicz, M.; Różycki, K.M. Innovation in cosmetics. *World Sci. News* **2017**, *72*, 448–456.
83. Education in microscopy and digital imaging. Available online: https://zeiss-campus.magnet.fsu.edu/articles/lightsources/leds.html (accessed on 29 January 2023).
84. Al Khafiz, M.F.; Hikmahwati, Y.; Anam, K.; Hudiyanti, D. Key conditions of alpha-tocopherol encapsulation in gum Arabic dispersions. *ScopeIndex* **2019**, *10*, 2622–2627.

catalysts

MDPI

Article

Photocatalytic Decomposition of Azo Dyes and Phenols Using Polymer Composites Containing Nanostructured Poly(Titanium Oxide) Doped with Gold or Silver Nanoparticles under Light Irradiation in a Wide Wavelength Range

Evgeniia Salomatina [1,*], Pavel Shelud'ko [1], Vsevolod Kuz'michev [2] and Larisa Smirnova [1]

[1] Department of Polymers and Colloidal Chemistry, Chemistry Faculty, N.I. Lobachevsky State University of Nizhny Novgorod, Gagarin Avenue 23, bld. 5, 603022 Nizhny Novgorod, Russia
[2] Collective Use Center "New Materials and Resource-Saving Technologies", N.I. Lobachevsky Research Institute of Chemistry, Gagarin Avenue 23, bld. 5, 603022 Nizhny Novgorod, Russia
* Correspondence: salomatina_ev@mail.ru; Tel.: +7-(831)4623235

Abstract: The photocatalytic properties of poly(titanium oxide) (PTO) dispersed in optically transparent polymeric matrices of different natures under the action of both UV and visible light on aqueous solutions of azo dyes and phenols were investigated. PTO in materials forms clusters of mixed polymorphic modification—anatase and rutile—with an average size ~12 nm. With a one-electron transition $Ti^{4+} + e^- \rightarrow Ti^{3+}$ accompanied by a reversible break of the Ti-O bond, the formation of electron-hole pairs and, consequently, active oxygen species occurs in PTO under UV irradiation. The PTO band gap in nanocomposites is 3.11–3.35 eV. Its doping with gold and silver nanoparticles with sizes from ~10 to ~30 nm reduces the PTO band gap by up to 2.11 eV, which leads to the operating wavelength range extension of the materials. It provides the enhancement of nanocomposites' photocatalytic properties under UV irradiation and is the reason for their high activity under visible light action. It was found that azo dyes decompose by ~90% in this case. A phenol and para-nitrophenol conversion of 80–90% was proven at ~60 min upon their aqueous solutions' visible-light irradiation at the nanocomposite concentration in a solution of 0.5 g/L.

Keywords: poly(titanium oxide); doping; gold and silver nanoparticles; organic polymeric matrix; nanocomposite; photocatalytic activity; UV- and visible-light irradiation; degradation; azo dyes; phenols

1. Introduction

Nowadays, titanium dioxide attracts special attention from researchers working on the creation of new photocatalytic materials. The interest in titanium dioxide first appeared in the 1970s, when a group of Japanese scientists [1] discovered the phenomenon of water splitting on an anode made of TiO_2 under UV irradiation that provoked an avalanche of research into its catalytic properties. It is thought [2–5] that the photocatalytic properties of TiO_2 is a result of the processing of the UV-induced one-electron transition of $Ti^{4+} + e^- \rightleftarrows Ti^{3+}$, accompanied by the reversible bond break of Ti-O and the formation of charge carriers—the holes and electrons on its surface [1–7]. The photogenerated holes from the valence band TiO_2 diffuse to its surface and react with adsorbed water molecules, forming hydroxyl radicals [6,7]. Electrons from the conduction band generally react with molecular oxygen in the air, forming superoxide anion radicals. These radicals provide for the strong oxidative capacity of titanium dioxide in light irradiation. The possible mechanism of the photocatalytic action of TiO_2 and its application fields are reflected in reviews [8–10]. It was found that titanium dioxide causes the decomposition of organic pollutants in water and air, including bacteria, and differs in terms of superhydrophilic properties, chemical stability, long service life, non-toxicity and low cost [11,12]. TiO_2 can be applied to various surfaces, such as glass, stainless steel fiber, inorganic materials, sand and activated carbon, allowing for its continuous and repeated use [1–3,8–10].

The photocatalytic activity of metal oxides is determined by a combination of factors, including particle size and crystal structure, photocatalyst surface area, its acidity, porosity, concentration of surface hydroxyl groups, the number and nature of surface charge traps, etc. [11–13]. The increase in the catalytic activity of materials containing TiO_2 can be achieved by increasing the dispersity of the oxide particles. It is known [14] that the TiO_2 particles of anatase structures with a size less than 50 nm have the highest photocatalytic activity due to their specifically developed surface area. At the same time, there are some studies showing that materials containing not only anatase TiO_2 but also its mix with rutile in proportion of 80–85% and 20–15%, respectively, show high photocatalytic activity [11–14]. Unfortunately, modern methods for the synthesis of TiO_2 particles do not allow for the full control of their size.

It should be noted that even increasing the dispersity of TiO_2 does not remove it from drawbacks such as a wide band gap of 3.2 eV [15,16], which limits the wavelength range to UV light only, causing charge separation and the generation of active oxygen forms and a high recombination rate of electrons and "holes", which in general reduces its efficiency as a photocatalyst. A consequence of these disadvantages is the relatively low quantum yield of the reaction $Ti^{4+} + e^- \rightleftarrows Ti^{3+}$, which is ~17–20% [17]. Systems with higher charge separation efficiencies of ~50% are poly(titanium oxide) (PTO) gels in alcoholic media [18,19]. It is proven that in alcoholic media, PTO forms cluster nanostructures [20,21]; however, gels are unstable in the air. With this connection, when creating photocatalytic materials with high activity based on titanium dioxide, it is necessary to eliminate the problems of the latter's narrow absorption spectrum and enhance the efficiency of the charge carrier formation and separation.

To a certain extent, the way to solve this problem was found in the 1990s, when the catalytic properties of gold and silver nanoparticles (NPs) were discovered. It was found that doping TiO_2 with gold and silver NPs leads to an increase in the flux of electrons and holes generated on its surface when exposed to visible light [22–24]. As a result, such materials enable a more efficient use of the full spectrum of solar radiation and significantly reduce the energy consumption of the photocatalytic process. The reason that the photocatalytic properties are strengthened is closely connected to the reduction of the band gap width of titanium dioxide. The authors of papers [25–29] believe that, firstly, there is an excitation of the collective vibration of electrons on Au or Ag NPs' surface, possessing plasmon resonance in the visible spectral region. This is because the Fermi levels of titanium dioxide and Au or Ag particles are close, and these electrons are transferred to the conduction band of TiO_2 that then leads to the generation of reactive oxygen species on its surface. These discoveries have stimulated a new wave of research into the photocatalytic properties of TiO_2 doped with NP metals and their oxides [30,31]. It has been shown [32–36] that the high photocatalytic activity of Au/TiO_2 powders, films or coatings already appears in the action of visible light when purifying polluted water (for example, catalytic oxidation of phenols, antibiotic residues, oil refining products, herbicides, insecticides, inhibition of bacterial growth, etc.).

Despite the success achieved, the task of developing new accessible and controllable methods of producing nanocomposite materials containing highly dispersed particles of TiO_2 with anatase (anatase and rutile) polymorphic modification doped with noble metal NPs remains relevant.

The aim of this work was to study the photocatalytic properties of composites containing nanostructured PTO doped with gold or silver NPs in optically transparent polymer matrices in the decomposition reactions of azo dyes and phenols in aqueous media under the action of both UV and visible light.

2. Results and Discussions

Polymeric nanocomposites are solid and optically transparent materials containing PTO chemically bonded to the organic polymer matrix. From the point of view of the practical application of materials, the structure of the PTO in them and its size is crucially important because the anatase form of TiO_2 exhibits the greatest charge separation ability in the reaction $Ti^{4+} + e^- \rightarrow Ti^{3+}$, and hence the photocatalytic activity. In this regard, in order to clearly understand the PTO organization in nanocomposites, they were exposed to thermolysis to exclude the effect of the polymer matrix on the inorganic component. XRD analysis showed that the ash left after the degradation of the organic part of the composites at 350 °C was a crystalline TiO_2 with an anatase polymorphic modification (space group I41/amd, code 94566), regardless of the nature of the polymer matrix (Figure 1, curves 4–6). The diffractograms of the powders left after the degradation of the composites containing Au or Ag NPs show maximums at angles characteristic of two allotropic modifications of TiO_2—anatase (75–95%) and rutile (5–25%) (Figure 1, curves 7–12, Table 1).

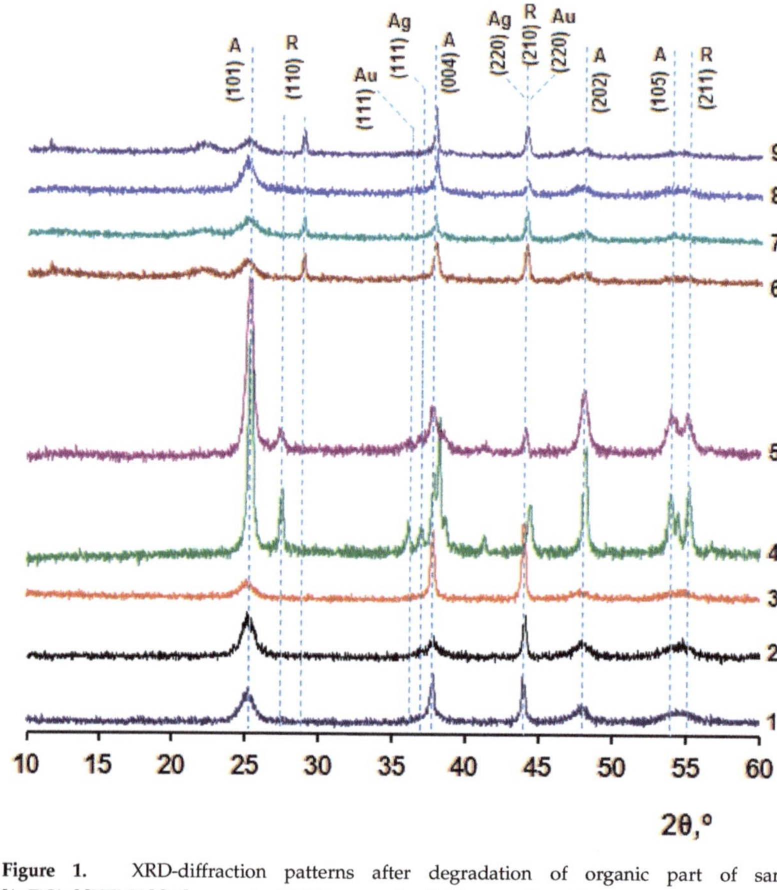

Figure 1. XRD-diffraction patterns after degradation of organic part of samples $[(\equiv TiO)_n]$:[HEMA]:[M]: curve 1—HEMA, curve 2—BMA, curve 3—AN, curve 4—HEMA + Ag NPs, curve 5—HEMA + НЧ Au,curve 6—BMA + Ag NPs; curve 7—BMA + Au NPs; curve 8—AN + Ag NPs; curve 9—AN + Au NPs. A—anatase, R—rutile, Au—gold, Ag—silver.

Table 1. Particle sizes in powders after degradation of the organic part of nanocomposites of composition [$(\equiv TiO)_n$]:[HEMA]:[M], containing Au or Ag NPs.

[M]	Type of NPs	Particle Diameter in the Powder after Degradation of the Organic Part of the Composites, nm			Anatase Content, %	E_g
		TiO_2	Au	Ag		
HEMA	without NPs	7	-	-	100.0	3.35
	Au	6	-	9	95.0	2.42
	Ag	6	12	-	81.2	2.75
BMA	without NPs	6	-	-	100.0	3.11
	Au	9	-	20	87.1	2.11
	Ag	12	27	-	75.1	2.49
AN	without NPs	8	-	-	100.0	3.20
	Au	11	-	23	93.8	2.48
	Ag	10	31	-	85.3	2.65

The character of the peaks in the diffractograms (their width from 2° to 5°) indicates the small size of TiO_2 particles in powders. The application of the Scherrer equation allowed for the calculating of the approximate diameter of the TiO_2 particles in the ash, which did not exceed 12 nm, regardless of the nanocomposite composition in all cases (Table 1). There can be noted only a slight increase in particle diameter during the transition from materials of binary composition to materials containing, in addition to PTO and HEMA, AN and BMA units. It is also seen that in ternary composites containing AN or BMA units, larger NPs form in comparison with the materials with binary composition—from ~20 to ~30 nm versus 10–12 nm (Table 1). Apparently, this effect is related to the formation of closer-packing macromolecules in the nanocomposites of ternary compositions, which facilitates the assembly processes of gold or silver atoms in NPs.

Thus, on the basis of obtaining results, it can be assumed that during the synthesis of amorphous polymer composites, PTO chains self-organize into nanocluster structures similar to photocatalytically active anatase with particle size ~10–12 nm. We have studied the structure of nanocomposites in detail in [37,38]. It should be noted that PTO clusters are cross-linking centers of organic (co)polymer macromolecules, so the composites are not soluble in organic solvents (dimethylformamide, tetrahydrofuran, ethyl alcohol, acetone and benzene) and water—they only swell in them.

Scanning electron microscopy with X-ray fluorescence analysis helped to answer the question of whether the metal particles are located in the volume of polymer matrix or near the PTO (Figure 2).

It was found that the surface of the composites, regardless of polymer matrix nature, consists of many small objects of sizes ~5 µm, with a uniform distribution of Ti, O and C atoms on it (Figure 2a). The titanium content on the surface was ~15–17%, which corresponds to the calculated value based on the composition of the charging mixture. A different picture can be observed in the study of nanocomposites with Au and Ag nanoparticles. In this case, finer inhomogeneities with sizes ranging from 0.5 to 1 micron were observed on the surface. It was found that the atoms of all the elements in the nanocomposites were also uniformly distributed on their surface. Moreover, the share of titanium on it increased to 23–27%. The content of gold and silver on the nanocomposite's surface was ~5 and 14%, respectively. The elemental analysis of the nanocomposite's surface confirmed the arrangement of metal particles on the surface of PTO. It, apparently, is caused by the assembling of metal atoms during NP synthesis near PTO clusters with high specific surface area. By computer simulation using DFT and TD-DFT (B3LYP) in previous works, there has been shown the coordination of gold nanoparticles (Au_8) on oxygen atoms in the PTO cluster and carbonyl groups of oxoethyl methacryate fragments chemically bound to PTO [39].

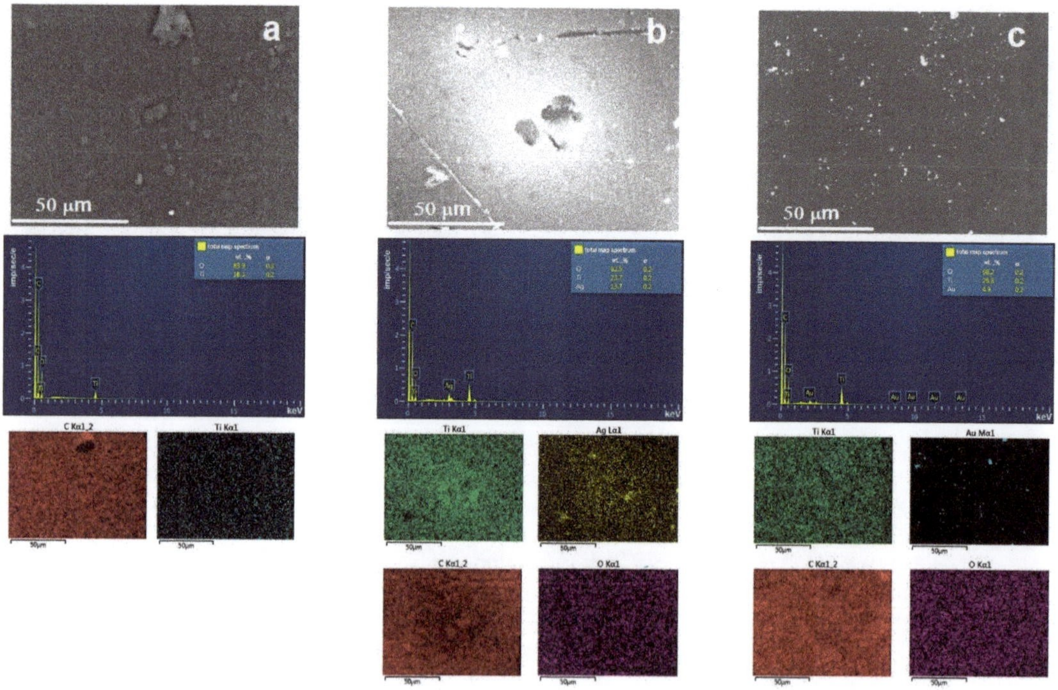

Figure 2. Elemental surface composition of nanocomposites $[(\equiv TiO)_n]$:[HEMA]:[AN]: (**a**) without nanoparticles; (**b**) with Ag NPs; (**c**) with Au NPs.

During UV- irradiation of composites, in their spectra in the region of 600–700 nm, an absorption band appears and grows, which is accompanied by a reversible darkening of the samples (Figure 3a). This phenomenon provides evidence for the occurrence of the one-electron transition $Ti^{4+} + e^- \rightarrow Ti^{3+}$ in composites, wherein in order to happen, as in TiO_2, there must occur the generation of reactive oxygen species on the material surfaces. The speed and depth of the process depend on the nature of the organic polymer matrix (Figure 3b). Binary materials are characterized by the greatest depth of obscuration to 70% in 180 min.

The question of how the $Ti^{4+} + e^- \rightarrow Ti^{3+}$ transition occurs in the obtained copolymers under UV irradiation is the subject of a separate study and is not considered in this work. Previously, using EPR spectroscopy, the authors [40–43] showed the formation of paramagnetic ions Ti^{3+} both on the surface and inside the TiO_2 particles in its colloidal solutions under UV irradiation. These ions are responsible for the dark blue coloration of the samples. The appearance of Ti^{3+} ions in photo-irradiated samples of PTO gels in alcoholic media has also been confirmed by EPR experiments [19]. In the gel samples, the strong resonance due to Ti^{3+} leads to the following values of the g-factor: $g_\perp = 1.924$ and $g_{II} = 1.878$. All this allowed the authors [18] to write the following mechanism of formation of Ti^{3+} centers in PTO gels in alcoholic media:

$$gel + hv \overset{\sigma_1}{\rightarrow} e^-_{CB} + h^+_{VB} \tag{1}$$

$$\ldots\text{-}O^{2-}-\ldots + h^+_{VB} \rightarrow \ldots\text{-}O^{2-}-\ldots + radicals \tag{2}$$

$$\ldots Ti^{4+}-\ldots + e^-_{CB} \rightarrow \ldots Ti^{3+}-\ldots \tag{3}$$

$$\ldots Ti^{3+}-\ldots + hv \overset{\sigma_2}{\rightarrow} \ldots Ti^{4+}-\ldots + e^-_{CB} \tag{4}$$

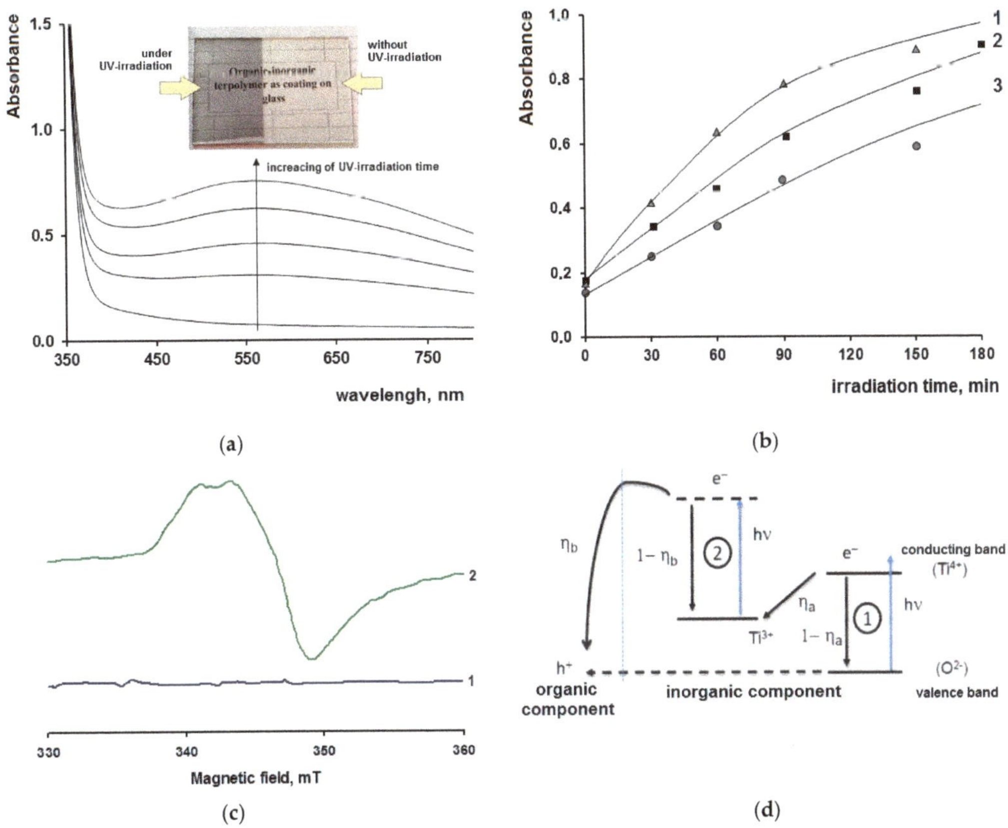

Figure 3. Optical properties of materials of different compositions: (**a**) typical change in the spectra of copolymers upon UV irradiation; (**b**) dependence of the optical density of composites $[(\equiv TiO)_n]$:[HEMA]:[M] at 600 nm on the time of UV irradiation: curve 1—M = HEMA, curve 2—M = BMA, curve 3—M = AN; (**c**) EPR spectra of material with composition $[(\equiv TiO)_n]$:[HEMA]:[M] = 1:5:1: curve 1—without UV irradiation, curve 2—UV irradiated sample; (**d**) scheme of possible processes occurring in PTO gels in alcoholic media or in organic–inorganic copolymers under UV irradiation.

The EPR spectra of the organic–inorganic copolymer based on titanium isopropoxide and HEMA was considered in [44,45]. It demonstrates the characteristic magnetic parameters of the paramagnetic Ti^{3+} centers in the distorted rhombic field of the oxygen ligand: $g_x = 1.973$, $g_y = 1.919$ and $g_z = 1.855$. The EPR spectra of organic–inorganic copolymers of ternary composition that contain AN or BMA units, in addition to PTO and HEMA, have a similar appearance (Figure 3c). It can be assumed that the formation of Ti^{3+} centers ("electrons") and "holes" in the hybrid organic–inorganic copolymers under UV irradiation occurs along a path similar to that of PTO gels in alcoholic media (Figure 3d) [46] and has a multistep character. The authors suggest that in the first step, Ti^{3+} centers ("electrons") are formed during photon absorption as a result of the electron transfer from the valence band with *2p*-orbitals of the oxygen atom O^{2-} ("hole") to the conduction band to the *3d*-orbital of Ti^{4+}. The photogenerated "holes" then rapidly transfer to the organic polymer matrix (poly(HEMA) or poly(HEMA-co-AN), poly(HEMA-co-BMA)), and the electrons remain "trapped" in the conduction band in the PTO (Figure 3d, quantum efficiency of the process is η_a). The maximum concentration of Ti^{3+} centers is limited by the internal photoeffect, which is associated with the excitation of electrons "trapped" in the conduction band in the PTO particle (Figure 3d). Secondarily excited electrons can leave the inorganic component and recombine with "holes" localized in the organic component of the copolymer (Figure 3d, this process is characterized by quantum efficiency η_b).

Thus, the combination of obtained results on the structure of polymer nanocomposites and the PTO in them, its doping with metal nanoparticles, and the optical properties of materials are prerequisites for nanocomposites that exhibit high photocatalytic activity under the action of both UV and visible radiation.

The photocatalytic properties of nanocomposites were studied in the decomposition reactions of azo dyes—methylene orange (MO) and congo red (CR)—and phenols in aqueous solutions. The choice of objects was due to the fact that they are widespread pollutants in wastewater in respective industries and are non-biodegradable substances. Their disposal is very important from the point of view of environmental protection, and the photocatalytic decomposition method in aqueous solutions seems to be the most cost effective on an industrial scale. MO and CR are widely used in the textile industry [47]. In addition, azo dyes are popular model substances for assessing the photocatalytic activity of solution discoloration materials recommended by ISO 10678:2010 [48]. According to the World Environment Organization, phenol and para-nitrophenol (p-NP) are two of 114 dangerous biologically and chemically persistent pollutants in wastewater, forming during the production of dyes, pharmaceuticals, fungicides, insecticides and herbicides [49]. The maximum permissible concentration of p-NP in water is 20 µg/L. Studies on animals have shown that p-NP is capable of producing various toxic effects on the body, such as irritation and inflammation of the eyes, skin and respiratory tract, which may lead to cyanosis, confusion and loss of consciousness; stomach pain and vomiting; blood diseases; and destruction of the endocrine system [50]. Thus, the obtaining of new materials to develop an effective technology for purifying water from these pollutants attracts considerable attention from the scientific community.

Let us consider the photocatalytic activity of polymer composites in powder form in the reaction of MO decomposition in aqueous solution under the action of UV light. There have been proposed several schemes of MO decomposition on TiO_2 under UV irradiation, one of which suggests a deep decomposition of MO to CO_2 and water [51–53]; at the same time, the spectrum of dye solutions shows a change of the absorption band intensity at 470 nm, which is typical for basic forms of MO (relating to the presence of -N=N- bonds in MO) [54]. Also, during the first few stages, they can decompose without discoloration by the N-demethylation mechanism [55,56], according to the scheme below (Scheme 1).

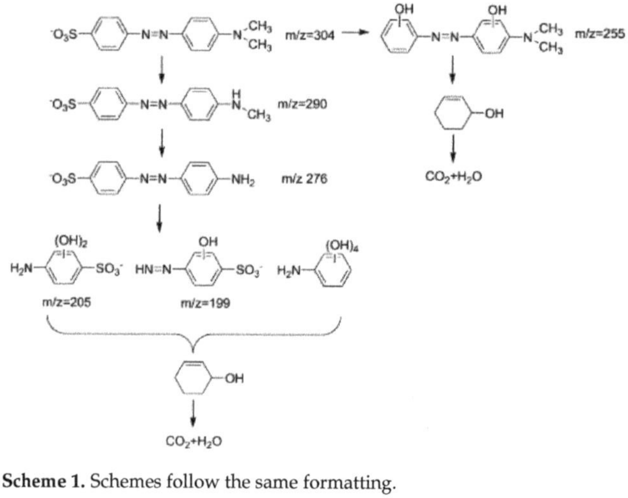

Scheme 1. Schemes follow the same formatting.

It was the wavelength at 470 nm that was chosen to control the depth of MO decomposition in the solution. Figure 4a shows the change in the spectra of MO solutions containing nanocomposites as heterogeneous photocatalysts during the UV irradiation of the purified system. It is seen that when the MO solution is UV-irradiated in the absence of a catalyst, the dye concentration remains practically unchanged for 10 h («empty» experiment). With the use of composites as photocatalysts, discoloration of the MO solution occurs from the first minutes of light irradiation. A comparison of the conversion depth of dye for 400 min of UV irradiation showed that the use of a binary composite reduces the MO concentration in the solution by ~60% (Figure 4b, curve 2), the material with BMA units by ~75% (Figure 4c, curve 2), and AN by ~80% (Figure 4d, curve 2). The doping of the PTO with Au or Ag nanoparticles provides a deeper MO decomposition—up to 90–95%. The best results are demonstrated by the material of composition [$(\equiv TiO)_n$]:[HEMA]:[AN] = 1:5:1, doped with silver NPs (Figure 4d, curve 4).

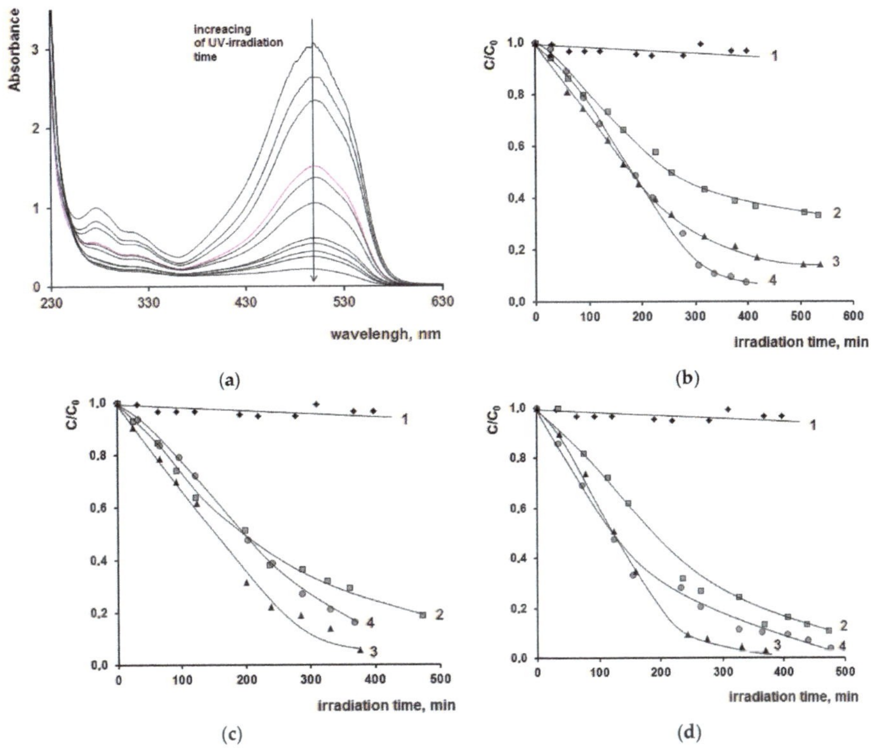

Figure 4. Photocatalytic properties of composites: (**a**) change in the absorption spectrum of MO aqueous solution with increasing time of its UV irradiation with the usage of copolymer of original composition [$(\equiv TiO)_n$]:[HEMA]:[M] + NPs of Ag as a catalyst; (**b–d**) a change in the MO concentration in the solution in time during its UV irradiation in the presence of composition materials with HEMA (**b**), BMA (**c**); and AN (**d**) units. Curve 1—«empty» experiment; curve 2—material with PTO without NPs; curve 3—PTO in the material is doped with gold NPs; and curve 4—PTO in the material is doped with silver NPs.

In contrast to MO, solutions of CR at UV irradiation start to decolorize already in the absence of catalysts, as evidenced by the reduced intensity of its characteristic absorption band at 500 nm (also related to the chromophoric azo group). However, the CR concentration decreases only by 25% during 3 h of UV irradiation in the absence of nanocomposite; the nanocomposites used lead to CR decomposition by 75% during the same period of time (Figure 5). The nanocomposite containing AN units and PTO doped with silver NPs (Figure 5b, curve 9) demonstrates the highest results. It is reported in [57,58] that the degradation of CR can be related both to the destruction of the chromophore azo group and to the detachment of the auxochrome NH_2 group. In addition, the nitro and hydroxyl groups of biphenyl and naphthalene intermediates formed during the destruction of the azo group can also be oxidized by reactive oxygen species generated on the composite surface with the formation of aromatic products.

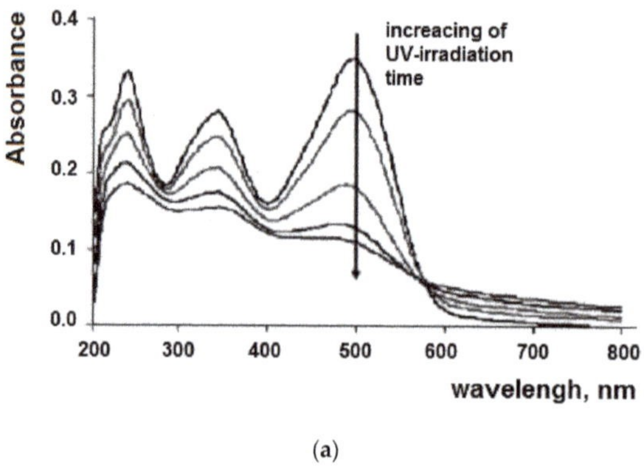

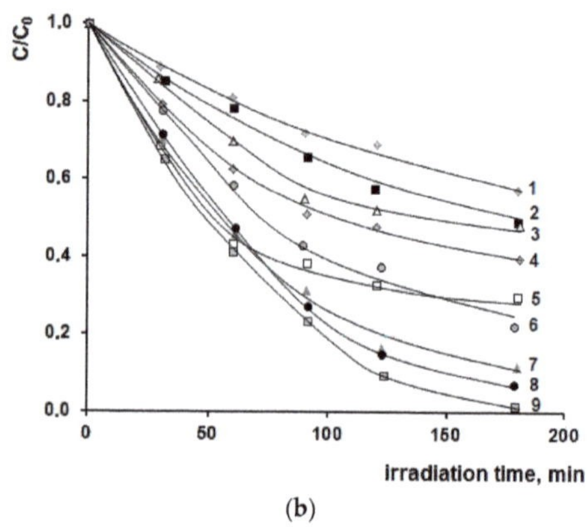

Figure 5. (**a**) Typical change in the spectrum of aqueous CR solution under UV irradiation in the presence of nanocomposites; (**b**) dependence of the relative concentration of aqueous CR solution under light irradiation in the presence of nanocomposites of different composition [$(\equiv TiO)_n$]:[HEMA]:[M]: curve 1—«empty» experiment, curve 2—HEMA, curve 3—BMA, curve 4—AN, curve 5—HEMA + NPs Au, curve 6—BMA + NPs Au, curve 7—HEMA + NPs Ag, curve 8—AN + NPs Au, curve 9—AN + NPs Ag.

It can be assumed that the CR degradation in aqueous solution in the nanocomposites' presence occurs by a similar mechanism, because the absorption bands at 1250–1020 cm^{-1} related to the C-N bond vibrations, and at 1520–1500 cm^{-1}, related to the nitrogen group [59,60], gradually disappear in the IR spectra of the product's CR decomposition (Figure 6).

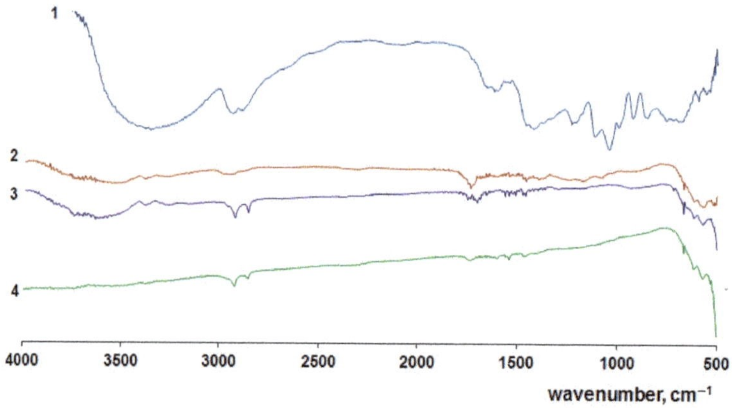

Figure 6. IR spectra of CR (curve 1) and its decomposition products under the UV irradiation of its solutions in the presence of nanocomposites [$(\equiv TiO)_n$]:[HEMA] + Ag NPs (curve 2), [$(\equiv TiO)_n$]:[HEMA]:[BMA] + Ag NPs (curve 3), [$(\equiv TiO)_n$]:[HEMA]:[AN] + Ag NPs (curve 4).

Nitro and hydroxyl groups of biphenyl and naphthalene intermediates formed during the destruction of the azo group can also be oxidized by reactive oxygen species generated on the composites' surface to form aromatic products. The presence of aromatic compounds in the products of the CR decomposition in solution when using nanocomposites under UV irradiation is reflected in the presence in their IR spectra of absorption bands at 2930–2850 cm^{-1}, relating to the vibrations of the benzene ring.

The combination of results obtained on the application of PTO-containing nanocomposites for the decomposition of azo dyes indicates their high photocatalytic activity. This is most clearly manifested in the materials where the PTO is doped with silver NPs. In this regard, the photocatalytic properties of such nanocomposites were investigated in the decomposition reactions of phenol and para-nitrophenol (p-NP) under the action of both UV and visible light. Figure 7a shows the change in the spectra of p-NP solutions under UV irradiation in the presence of photocatalysts. It can be seen that there occurs a decrease in the optical density value at wavelength λ = 318 nm, which is typical of the NO_2 group of p-NP [61,62]. On the basis of the spectral data using the calibration relation, the relative change of concentration of p-NP in solution under UV irradiation was calculated (Figure 7b).

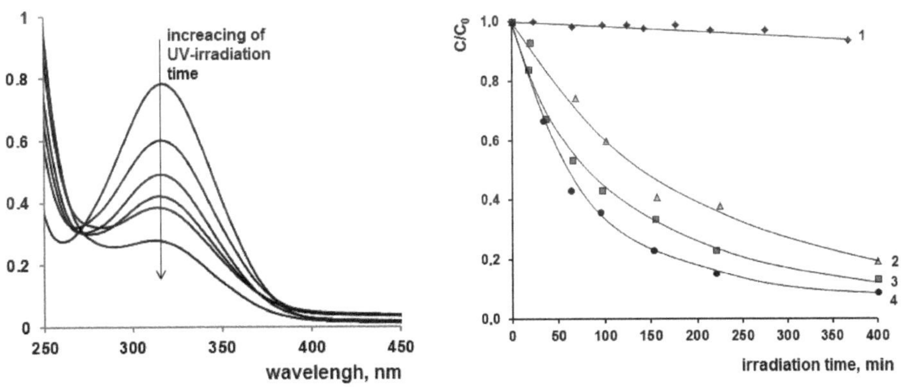

(a) (b)

Figure 7. Photocatalytic activity of copolymers under UV irradiation: (**a**) the change of absorption spectrum of p-NP solution in the presence of photocatalyst [(≡TiO)$_n$]:[HEMA]:[M]; (**b**) the change of p-NP concentration in solution: curve 1—«empty» experiment, curve 2—BMA + NPs Ag, curve 3—HEMA + NPs Ag; curve 4—AN + NPs Ag.

The maximum possible conversion of p-NP for 400 min of UV irradiation of its solutions—by ~85%—is observed when using a composite with AN and PTO units doped with Ag NPs. In the absence of a catalyst, the change of p-NP concentration in solution does not exceed 5%.

It turns out that during the studying of the photocatalytic activity of nanocomposites under the action of visible light in the p-NP degradation reaction, their efficiency appeared to be higher (Figure 8). In the absence of nanocomposites, the p-NP concentration in the solution did not change, while when nanocomposites were used, an acceleration of the reaction and an order-of-magnitude shorter p-NP conversion time compared to UV irradiation was observed.

UV irradiation in the presence of nanocomposites with AN units and Ag NPs caused the conversion of p-NP in the solution by a maximum of 85% for 400 min; then, in the case of visible light, the depth of p-NP decomposition had already been 94% for 12 min. Calculations show that the photocatalytic decomposition of p-NP at the early stages can be described by the first order equation. The rate constants of p-NP decomposition upon the irradiation of its solutions with visible light in the presence of the system $[(\equiv TiO)_n]$:[HEMA]:[AN] + Ag NPs and $[(\equiv TiO)_n]$:[HEMA] + Ag NPs are $k = (20.51 \pm 0.16) \cdot 10^{-2}\ c^{-1}$ and $k = (7.65 \pm 0.32) \cdot 10^{-2}\ c^{-1}$, respectively.

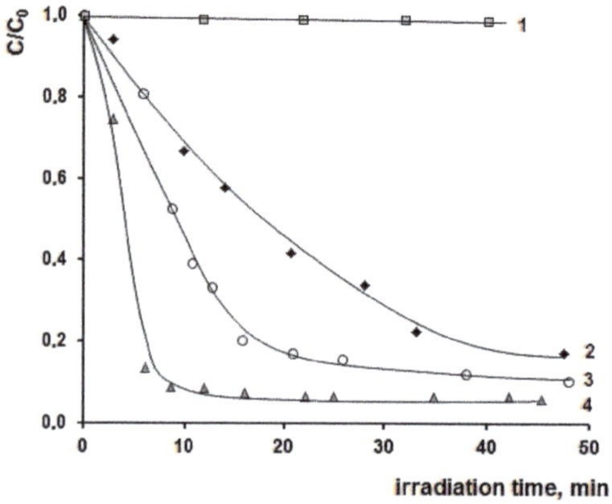

Figure 8. Changes in concentration of p-NP in solutions during their irradiation with light > 410 nm (light filter «ZhS-11») in the presence of nanocomposites $[(\equiv TiO)_n]$:[HEMA] [M]: curve 1—«empty» experiment; curve 2—BMA + Ag NPs, curve 3—HEMA + Ag NPs; curve 4—AN + Ag NPs.

According to [63,64], during the UV irradiation of p-NP in the presence of titanium dioxide, there occurs the detachment of the—NO_2 group, which is replaced by the OH group by electrophilic substitution. The hydroxylation products hydroquinone and 1,2,4-trihydroxybenzene are formed as a result. The compound 4-nitrocatechol [65] is also found as an oxidation intermediate of p-NP. Further oxidation of the intermediates leads to their complete mineralization into p-NP molecules to carbon dioxide and water [66].

IR analysis of the p-NP conversion products in our case showed that in p-NP spectra under UV irradiation and in the presence of nanocomposite, the bands typical for NO_2-groups (1500–1560 cm^{-1}) disappeared and the bands corresponding to the stretch vibration of carbonyl (-C=O) group appeared (1720–1790 cm^{-1}) (Figure 9). It can be inferred that when using nanocomposites as intermediate reaction products, a mixture of hydroquinone, benzoquinone, and N-oxime are formed, which corresponds to the data of [67]. In the spectra of the reaction products, when the system was irradiated with visible light, only bands related to small inorganic molecules had already been detected after 40 min of irradiation, which suggests the complete oxidation of p-NP to CO_2, water and nitrile, and nitrate ions occur, as described in papers [7,68].

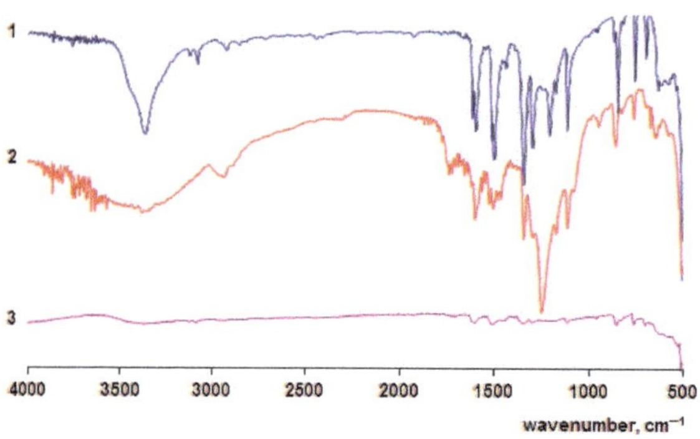

Figure 9. IR spectra of p-NP original solution (curve 1) and intermediate products of its decomposition under UV irradiation for 200 min (curve 2) and at visible light irradiation for 40 min (curve 3) in the presence of nanocomposite $[(\equiv TiO)_n]$:[HEMA]:[AN] + Ag NPs.

The investigation results of the photocatalytic activity of nanocomposites containing nanostructured PTO doped with Ag NPs under the action of both UV and visible light on phenol solutions are shown in Figures 10 and 11. The conversion of phenol in aqueous solution (C = 19 mmol/L) was monitored with the spectrophotometric method of changing the optical density of its solutions at wavelengths of 270 nm and 400 nm. According to previous research [69,70], these bands reflect the presence of OH groups in phenol.

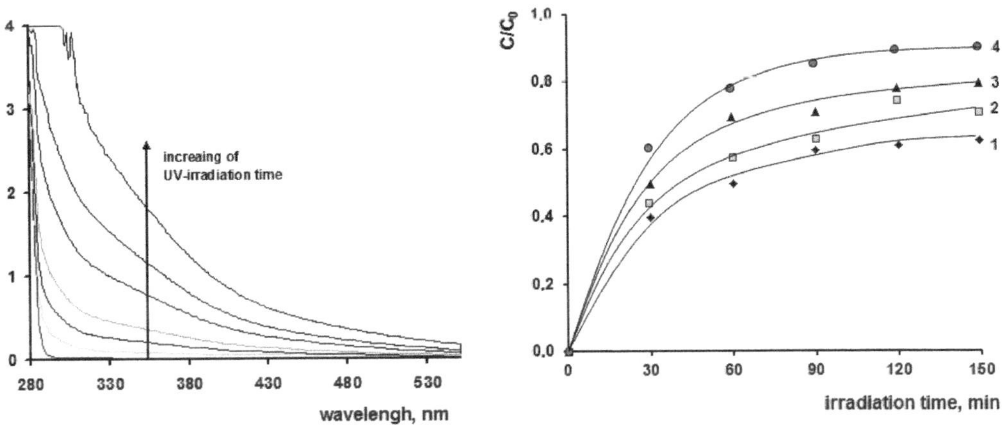

(a) (b)

Figure 10. (**a**) The change of spectrum character of phenol aqueous solution under UV irradiation in the presence of nanocomposites (C = 19 mmol/L); (**b**) relative optical density dependence of phenol solution at 400 nm under UV irradiation in the presence of nanocomposites: curve 1—«empty» experiment, curve 2—BMA + Ag NPs, curve 3—HEMA + Ag NPs, curve 4—AN + Ag NPs.

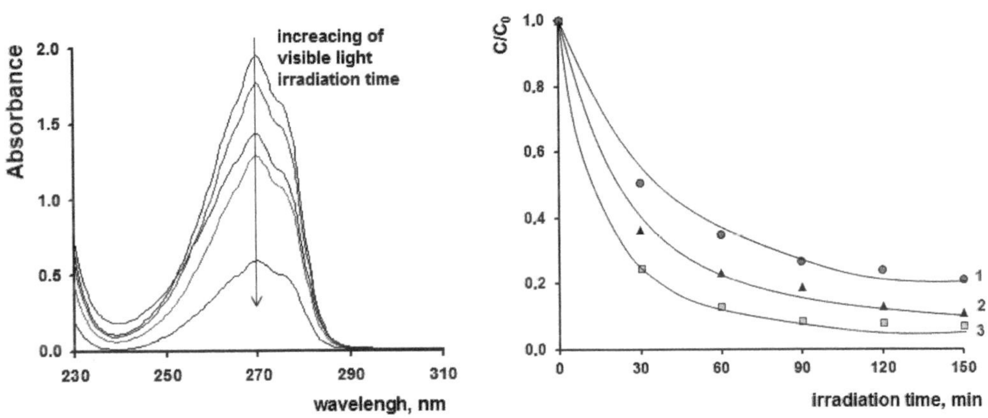

(a) (b)

Figure 11. (**a**) The change in the nature of the aqueous solution spectrum of phenol at its irradiation with visible light in the presence of nanocomposites; (**b**) the dependence of the relative concentration of phenol in the solution under irradiation of visible light in the presence of nanocomposites of different composition: curve 1—BMA + Ag NPs, curve 2—HEMA + NPs Ag, curve 3—AN + Ag NPs.

The UV irradiation of phenol solutions leads to an increase in the absorption band of its solutions at a wavelength of 400 nm, even in the absence of nanocomposites in the system (Figure 10a). Thus, the oxidation of phenol by ~40% for 2.5 h can occur without the use of nanocomposites.

However, the application of nanocomposites [(≡TiO)$_n$]:[HEMA]:[BMA], [(≡TiO)$_n$]:[HEMA] and [(≡TiO)$_n$]:[HEMA]:[BMA] modified with silver NPs under UV irradiation leads to a much deeper transformation of phenol by 65, 78 and 89% during the same time, respectively.

Under the action of visible light on phenol solutions in the presence of nanocomposites, a change in the optical density at λ = 270 nm was observed in the spectrum of the solution (Figure 11a). In this case exactly, this band was chosen as the indicator that would be responsible for phenol concentration in the solution. The dependence of the relative concentration of phenol in solution on time were constructed from spectral data (Figure 11b).

It is seen from the figure that for 1 h of irradiation in the presence of nanocomposite with AN and PTO doped with Ag NPs, the conversion of ~90% of phenol occurs.

A different picture of changes in the spectra of the phenol solutions during its photocatalytic transformation under UV and visible light seems to be connected with the formation of different intermediate products. Studies from the previous years have been carried out using chromatography-mass spectrometry. For this purpose, at 260 °C, corresponding to the phenol melting point, the reaction products that are obtained after 1 h of light irradiation to the phenol solutions were investigated. The chromatograms are shown in Figure 12.

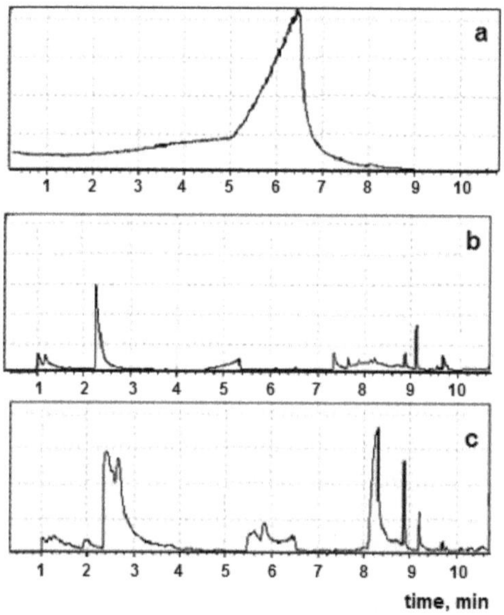

Figure 12. Chromatograms of original phenol (**a**), intermediate products of phenol decomposition at UV-irradiation (**b**) or visible light irradiation (**c**) to solutions in presence of composite [(≡TiO)$_n$]:[HEMA]:[AN] + Ag NPs.

The chromatogram of the initial phenol shows only one peak at 6 min, which disappears in the chromatograms of its products. According to the integrated library of the chromatography-mass spectrometer, the intermediate products of the phenol transformation are:

- In the case of UV light—CO_2, 2-butanone, 2-propionic acid;
- In the case of visible light—CO_2, acetone, 2-propanol, butanediol—1,4, 2,4-dimethylpentanone-3, butyric acid butyl ester, 2-methyl-2-(2-hydroxyethoxy)acetic acid ethyl ester.

Not even trace amounts of phenol are observed in the IR spectra of the products of the complete transformation of phenol, which indicates its complete mineralization (Figure 13).

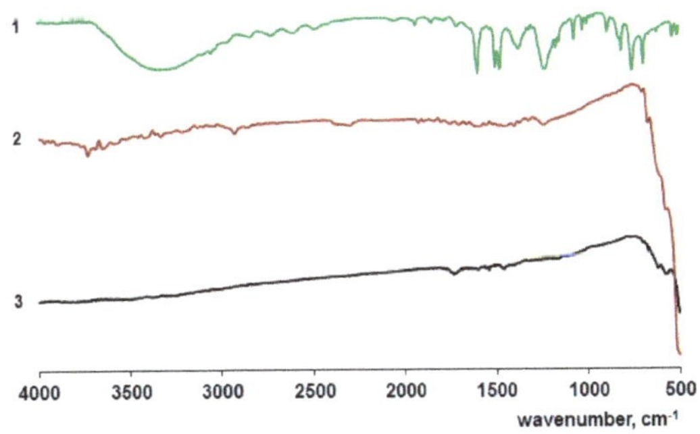

Figure 13. IR spectra of initial phenol (curve 1), its decomposition products in the presence of composite [(≡TiO)$_n$]:[HEMA]:[AN] + Ag NPs under UV-irradiation (curve 2), under visible light irradiation (curve 3). Time of light irradiation is 400 min.

The mechanism of phenol phototransformation under the action of UV light in the presence of TiO_2 has been reviewed in detail by authors [71,72]. According to spectral data, during the UV irradiation of phenol solutions, apparently, their oxidation occurs due to the replacement of hydrogen atoms in the benzene ring by OH groups, due to the generation of large numbers of OH radicals in solutions in the presence of photocatalysts [73,74]. It can be assumed that the complete mineralization of phenol occurs through the formation of several reaction intermediates, such as p-benzoquinone, o-benzoquinone and hydroquinone, and/or maleic and other carboxylic acid compounds [75].

We can suggest that phenol transformation begins with the eliminating of the hydroxyl group from the benzene ring when phenol solutions are irradiated with visible light, as described in [76], the consequence of which is the decrease in the intensity of its solution band at 270 nm.

Another piece of evidence for the photocatalytic decomposition of pollutants in water is the results of the analysis of the total carbon content in the solutions. Table 1 shows the values of total, organic and inorganic carbon in the water samples.

The data in Table 2 show that the application of the developed nanocomposites as photocatalysts under the action of both UV and visible light reduces the concentration of pollutants in water.

Table 2. Total organic carbon content in water before and after the photocatalytic decomposition of its organic pollutants when using the polymeric nanocomposite [(≡TiO)$_n$]:[HEMA]:[AN] = 1:5:1 + Ag NPs.

ExperimentConditions	TOC *, mg/L	NPOC **, mg/L	TC ***, mg/L	IC ****, mg/L
MO + UV light, without nanocomposite	80.5 ± 1.9	78.5 ± 1.9	91 ± 2.2	10.5 ± 0.2
MO + UV light,with nanocomposite	14.5 ± 1.5	14.8 ± 1.5	14.7 ± 1.5	0.12 ± 0.1
Phenol + UV light, without nanocomposite	95.5 ± 2.3	99.5± 2.4	101.0 ± 2.4	5.5 ± 0.1
Phenol + UV light, with nanocomposite	7.5 ± 0.7	7.4 ± 0.7	7.9 ± 0.7	0.4 ± 0.1
Phenol + visible light, with nanocomposite	9.9 ± 1.0	10.0 ± 1.0	10.9 ± 1.0	1.0 ± 0.2

* Total organic carbon method, TOC = | TC − IC |. ** Organic carbon method, NPOC. *** Total carbon method, TC. **** Inorganic carbon method, IC.

The enhancement of the photocatalytic activity of the nanocomposites where the PTO is doped with gold or silver NPs under light irradiation in a wide wavelength range seems to be related to the change in the band gap of the PTO (Table 1). The results of diffuse reflection and calculations using the Kubelka–Munk absorption function (Figure 14) showed that the introduction of a third monomer, BMA or AN, into the materials had almost no effect on the value of the band gap of PTO, which remained at 3.2–3.3 eV. PTO doping with gold and silver NPs reduced this parameter to 2.49 eV and 2.11 eV, respectively (Table 1). The metal NPs presence's effect on the PTO band gap can be attributed to:

- Charge transfer transitions between the d-electrons of the metal and the conduction band or valence band of the PTO;
- Formation of impurity levels in band gap of the PTO; if these energy levels are located close to the edges of band gap, then they can overlap it, thus its width decreases [77];
- The appearance of allowed energy states in the band gap of TiO_2, as a consequence of the presence of segregated clusters of M_xO_y on its surface [78], which can transfer excited state energy from NPs into the conduction band of the PTO according to the scheme proposed earlier for powdered titanium dioxide [79]. At the same time, the recombination rate of the electron-hole pair will decrease, and its lifetime will increase.

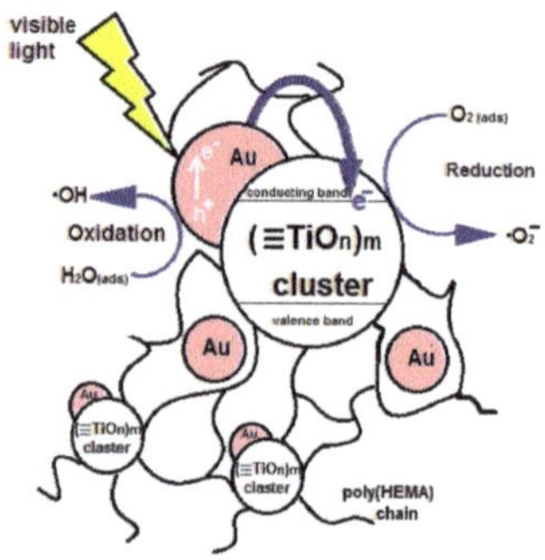

Figure 14. Schematic diagram of possible processes occurring in organic–inorganic copolymers containing nanostructured PTO modified with Au NPs under the action of visible light.

We used literature data to describe the mechanism of photocatalytic processes occurring on the surface of the polymer nanocomposites containing PTO. We believe that Au (Ag) NPs transfer the absorbed excited state energy to the conduction band of the PTO NPs, as it was previously proposed for TiO_2-anatase [80] (Figure 14).

It can be assumed that the reason for the increased photocatalytic activity of the copolymers doped with Au (Ag) NPs is related to the difference in the Fermi levels of PTO particles and Au (Ag) NPs [67]. The Fermi level of the Au (Ag) NPs is higher than that of TiO_2 and, consequently, that of PTO clusters [25]. The Au and Ag particles in the copolymer structure are located in close proximity to the PTO clusters [39]. In this regard, surface plasmons of metal NPs, collectively oscillating under the action of visible radiation, immediately migrate to the conduction band of PTO. As a result, the Ti–O bond is excited, accompanied by its rupture, and the one-electron transition $Ti^{4+} + e \rightarrow Ti^{3+}$ takes place. This leads to the formation of O_2^- ions, superoxide and hydroxyl radicals, which are involved in the oxidation of organic substances in water.

3. Materials and Methods

3.1. Synthesis of Nanocomposites

Polymeric nanocomposites containing nanostructured poly(titanium oxide) were synthesized from titanium isopropoxide $Ti(OPr^i)_4$ (ACROS Organics, Geel, Belgium, base substance content 98%, without additional purification), hydroxyethyl methacrylate (HEMA) (Sigma Aldrich, Darmstadt, Germany, base substance content 99.9%, without further purification) as main components, and acrylonitrile (AN) or butyl methacrylate (BMA) as third monomers, purified by vacuum distillation according to conventional procedures. The synthesis of the samples involved the polycondensation of titanium alkoxide in organic monomer medium and the subsequent radical polymerization of the system at 70 °C, and is described in detail in our previous papers [37,38]. Molar ratio of the components in all cases was $[Ti(OPr^i)_4]$:[HEMA]:[M] = 1:5:1 (where M is HEMA, AN or BMA).

Into the reaction system precursors, $AgNO_3$ (JSC «Uralskiy zavod chimicheskikh reagents», Moscow, Russia) and $HAuCl_4$ (gold content not less than 48%, JSC "Aurat", Moscow, Russia) were additionally introduced at the stage when the components mixed in the concentration with 1 wt. % by $Ti(OPr^i)_4$ mass for obtaining materials with PTO doped with Au or Ag NPs. In all cases, a single batching up of all reagents into reaction mixture was carried out.

The result of synthesis is solid polymer composites with a molar ratio of organic and inorganic component units $[(\equiv TiO_n)_m]$:[HEMA]:[M] = 1:5:1 (where M is HEMA, AN or BMA, $(\equiv TiO_n)_m$ is poly(titanium oxide)) as thin films and bulk samples. Gold or silver nanoparticles were formed by UV irradiation of solid samples after their synthesis. The UV radiation source was a mercury lamp DRT-230 (power 1600 mW/m^2).

3.2. Investigation of Nanocomposites' Optical Properties

The formation of Ti^{3+} centers in the materials during UV irradiation conditioned by the one-electron transition $Ti^{4+} + e^- \rightarrow Ti^{3+}$ with the breakage of the Ti–O bond was controlled by the spectrophotometric method of increasing the absorption band in the region of 600–700 nm. The gold and silver NPs' formation in the materials was monitored by the appearance and growth of plasmon resonance bands in the regions of 390–420 nm (Ag NPs) and 500–600 nm (Au NPs). Spectrophotometer SHIMADZU—UV-1650 PC (Tokyo, Japan) has been used in previously mentioned research.

The band gap of the PTO in polymeric composites was determined on the basis of their diffuse reflectance results. Diffuse reflection spectra of the samples were obtained using an optical system that included a halogen lamp (2–4.5 eV), a monochromator, a monochromatic beam profile formation system, an integral sphere and a photomultiplier tube. The samples were placed in the integrated sphere perpendicular to the emitting light. The detector was placed at a 90° angle to the line between the entrance hole of the sphere and the sample. The band gap width was determined using the Kubelka–Munk function [81]:

$$F(R_\infty) = \frac{(1 - R_\infty)^2}{2R_\infty} = \frac{k}{s} \tag{5}$$

where $R_\infty = \frac{I_{sample}}{I_{standard}}$—relative diffuse reflectance; k—absorption coefficient; and s—scattering coefficient.

The Kubelka–Munk function is proportional to the absorption coefficient α:

$$(F(R_\infty))^2 = C'(hv - E_g) \tag{6}$$

where v is the frequency, h is Planck's constant, and C' is the coefficient of proportionality.

The linear part of the dependence $F(R) = f(hv)$ was extrapolated up to the intersection with the abscissa axis, which allowed for the determining of the value of the band gap energy E_g of PTO in the material. As an example, Figure 15 shows the Kubelka–Munk function graphs of some samples and marks the linear sections by which the band gap energy of PTO in polymer nanocomposites was determined.

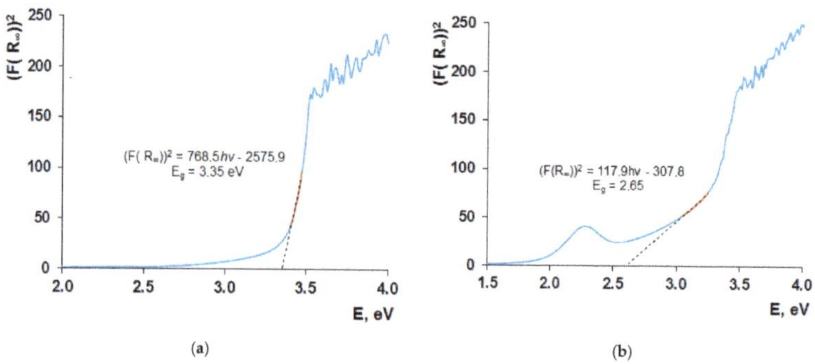

Figure 15. Kubelka-Munk function of samples with compositions [(≡TiO)$_n$]:[HEMA] = 1:6 (**a**) and [(≡TiO)$_n$]:[HEMA]:[AN] = 1:5:1 + Ag NPs (**b**).

3.3. Study of the Structure of Materials Based on Organic-Inorganic Copolymers

The structure of synthesized materials was studied by X-ray diffraction analysis (XRD). X-ray diffractograms were recorded on an XRD-6000 diffractometer by Shimadzu (Tokyo, Japan, CuK$_\alpha$-radiation, reflection imaging geometry) with scanning step 0.02 in the range of 2θ 10–60°. The accuracy of diffraction angles (2θ) was ±0.02°. The intensity of diffraction peaks was estimated by their height (100-ball scale), and the inorganic phase in the polymer was identified with the «Find It» database.

The study was conducted for powders remaining after the degradation of the composites' organic part. For this purpose, samples were heat-treated in a muffle furnace at a temperature of 300 to 400 °C when the temperature of the organic polymer matrix destruction was reached but the polymorphic transition of anatase into rutile was excluded. Crystallite size—*D* (in angstroms)—in powders was determined by the Scherrer formula given in Equation (3):

$$D = \frac{K\lambda}{\left(\frac{a}{2}\cos\theta\right)} \quad (7)$$

where *K* is Scherrer coefficient equal to 0.94; λ—wavelength equal to 0.15418 nm for Cu-Kα; *a*—half-width; and θ—diffraction angle of the transducer.

The morphology and elemental analysis of the surface of polymer composite films were characterized using a JEOL JSM-IT300LV scanning electron microscope (JEOL Inc., Peabody, MA, USA) equipped with energy and wave dispersion elemental analyzers.

3.4. Investigation of Photocatalytic Activity

Photocatalytic properties of obtained nanocomposites were studied in aqueous solutions in the reactions of the azo dyes' (methylene orange (MO), congo red (CR)) decomposition under UV irradiation and phenol and para-nitrophenol (p-NP) under the action of both UV and visible light. Concentrations of the substances in solutions are given in Table 3.

Table 3. Component ratios in photocatalytic cells depending on the type of water pollutant.

Type of Water Pollutant	Characteristic Wavelengths in the Absorption Spectrum, nm	Concentration of Pollutant, mmol/L	Content of Nanocomposite in Cleaning Solution, g/L
MO	470	3.06	
CR	500	14.00	
p-NP	318 and 400	0.07	0.5 *
phenol	350 and 270	19.00	

* 0.05 g/L PTO content in the solution, based on the composition of the nanocomposite.

For the experiment, there were selected aliquots of these solutions in which powdered polymer composites were placed as photocatalysts. The ratios of the components in the photocatalytic cells are shown in Table 2. The thickness of the absorbing layer in each cell was 2 cm.

The obtained suspensions were incubated to achieve adsorption–desorption equilibrium in the heterogeneous system. The photocatalytic cells were covered with quartz glass when experiments were conducted with UV light, while when using visible light, they were covered with a light filter «ZhS-11» which emits light with a wavelength greater than 410 nm, cutting off the UV part of the spectrum. A DRT-230 lamp was used as a source of UV light (distance from the lamp—17 cm, power—1600 mW/cm^2, the most intense emission bands—313 nm, 330 nm, 366 nm), and lamp KGM-24-150 (illumination in the plane of the cell with the solution was 24 kLk when irradiated with full lamp light) was used as visible light source.

Spectra of azo dyes and phenols solutions were recorded for every 30 min of irradiation in order to control their decomposition by changes in intensity of characteristic absorption bands indicated in Table 1. Absorption spectra were recorded using a SHIMADZU—UV-1650 PC spectrophotometer. At the end of the process, the dependences of changes in the concentration of azo dyes and phenol solutions on the time of their irradiation were plotted: C/C_0 = f(t) (C—current concentration, mmol/L, C_0—concentration of original solution of the substance, mmol/L). For this purpose, calibration dependences of the optical density of the characteristic absorption band maximum on the water pollutant concentration were preliminarily plotted.

Products of photocatalytic transformation of substances in solutions were studied by infrared spectroscopy on Infralum FT 801 (Simex, Novosibirsk, Russia) and pyrolysis-gas chromatography mass spectroscopy using GCMS-QP2010 Ultra "SHIMADZU" (Tokio, Japan).

3.5. Total Organic Carbon Analysis

Determination of total organic carbon was carried out according to the certified method M-02-2405-13 of LLC "ANALIT PRODUCTS".

Measurement of carbon in the samples was carried out in parallel by two methods—differential TOC, as the difference of total carbon and inorganic dissolved carbon dioxide, and measurement of NPOC organic carbon not removed by purging. The analysis settings for both methods were set to 2 M hydrochloric acid (+0.5%). The NPOC flow and purge times were set to 100 mL/min and 60 s, respectively.

4. Conclusions

The combination of the results obtained allows us to conclude that composites containing nanostructured PTO in optically transparent polymeric matrices exhibit high photocatalytic properties under UV irradiation during the decomposition of organic pollutants in water at a PTO concentration in the volume of the purified system of 0.05 g/L. The PTO doping with gold or silver NPs increases the activity of the materials by 1.5 times under the action of UV light, and it leads to the manifestation of their high efficiency under visible light action on the solutions of persistent organic pollutants. It was found that phenol and para-nitrophenol in this case turn into harmless substances for the surrounding environment.

The advantage of the nanocomposites developed is also the simplicity of their synthesis. It involves a single batching up of all components, including precursors of the gold and silver NPs $HAuCl_4$ and $AgNO_3$, and a simultaneous hydrolytic polycondensation and radical polymerization of the system. At the end of the synthesis, the optically transparent nanocomposites containing PTO in the form of clusters of predominantly anatase polymorphic modification of an average size of ~12 nm and corresponding precursors were obtained. Subsequent UV irradiation of the composites for 2 h provided the PTO doping of gold or silver nanoparticles with sizes ranging from 12 to 31 nm and 9 to 23 nm, respectively.

Author Contributions: Conceptualization, E.S. and L.S.; methodology, E.S. and P.S.; investigation, E.S., P.S. and V.K.; data curation, L.S.; writing—E.S. and L.S.; writing—review and editing, E.S. and V.K.; visualization, E.S.; supervision, L.S.; project administration, E.S.; funding acquisition, E.S. All authors have read and agreed to the published version of the manuscript.

Funding: This research was funded by a grant from the President of Russian Federation, grant number MK-2195.2021.1.3.

Data Availability Statement: Data is contained within the article.

Acknowledgments: This work was carried out using the equipment of the Collective Use Center "New Materials and Resource-saving Technologies" (N.I. Lobachevsky State University of Nizhny Novgorod). The authors express their deep gratitude to Evgeny Bulanov, associate professor in the Chemistry Faculty of N.I. Lobachevsky State University of Nizhny Novgorod, for the X-ray diffraction analysis, and to Aleksander Piskunov, Vice Director of the Institute of Organometallic Chemistry of the Russian Academy of Sciences (Nizhny Novgorod), for EPR measurements.

Conflicts of Interest: The authors declare no conflict of interest.

References

1. Fujishima, A.; Honda, K. Electrochemical Photolysis of Water at a Semiconductor Electrode. *Nature* **1972**, *238*, 37–38. [CrossRef] [PubMed]
2. Fujishima, A.; Zhang, X.; Tryk, D. Heterogeneous photocatalysis: From water photolysis to applications in environmental cleanups. *Int. J. Hydrog. Ener.* **2007**, *14*, 2664–2672. [CrossRef]
3. Zhu, S.; Wang, D. Photocatalysis: Basic principles, diverse forms of implementations, and emerging scientific opportunities. *Adv. Energy Mater.* **2017**, *7*, 1700841. [CrossRef]
4. Marcì, G.; Palmisano, L. *Heterogeneous Photocatalysis. Relationships with Heterogeneous Catalysis and Perspectives*, 1st ed.; Elsevier: Amsterdam, The Netherlands, 2019; pp. 1–24.
5. Tsang, C.H.A.; Li, K.; Zeng, Y.; Zhao, W.; Zhang, T.; Zhan, Y.; Xie, R.; Leung, D.Y.C.; Huang, H. Titanium oxide-based photocatalytic materials development and their role in air pollutants degradation: Overview and forecast. *Environ. Int.* **2019**, *125*, 200–228. [CrossRef]
6. Fujishima, A.; Rao, T.N.; Tryk, D.A. Titanium dioxide photocatalysis. *J. Photochem. Photobiol. C Photochem. Rev.* **2000**, *1*, 1–21. [CrossRef]
7. Schneider, J.; Matsuoka, M.; Takeuchi, M.; Zhang, J.; Horiuchi, Y.; Anpo, M.; Bahnemann, D.W. Understanding TiO_2 Photocatalysis: Mechanisms and Materials. *Chem. Rev.* **2014**, *114*, 9919–9986. [CrossRef]
8. Reddy, P.V.G.; Reddy, B.R.P.; Reddy, M.V.K.; Reddy, K.R.; Shetti, N.P.; Saleh, T.A.; Aminabhavi, T.M. A review on multicomponent reactions catalyzed by zero-dimensional/one-dimensional titanium dioxide (TiO_2) nanomaterials: Promising green methodologies in organic chemistry. *J. Environ. Manag.* **2021**, *279*, 111603. [CrossRef]
9. Chen, D.; Cheng, Y.; Zhou, N.; Chen, P.; Wang, Y.; Li, K.; Huo, S.; Cheng, P.; Peng, P.; Zhang, R.; et al. Photocatalytic degradation of organic pollutants using TiO_2-based photocatalysts: A review. *J. Clean. Prod.* **2020**, *268*, 121725. [CrossRef]
10. Dharma, H.N.C.; Jaafar, J.; Widiastuti, N.; Matsuyama, H.; Rajabsadeh, S.; Othman, M.H.D.; Rahman, M.A.; Jafri, N.N.M.; Suhaimin, N.S.; Nasir, A.M.; et al. A Review of Titanium Dioxide (TiO_2)-Based Photocatalyst for Oilfield-Produced Water Treatment. *Membranes* **2022**, *12*, 345. [CrossRef]
11. Arun, J.; Nachiappan, S.; Rangarajan, G.; Alagappan, R.P.; Gopinath, K.P.; Lichtfouse, E. Synthesis and application of titanium dioxide photocatalysis for energy, decontamination and viral disinfection: A review. *Environ. Chem. Lett.* **2023**, *21*, 362–399. [CrossRef]
12. Prakash, J.; Cho, J.; Mishra, Y.K. Photocatalytic TiO_2 nanomaterials as potential antimicrobial and antiviral agents: Scope against blocking the SARS-COV-2 spread. *Micro Nano Eng.* **2022**, *14*, 100100. [CrossRef]
13. Qi, Y.; Xiang, B.; Zhang, J. Effect of titanium dioxide (TiO_2) with different crystal forms and surface modifications on cooling property and surface wettability of cool roofing materials. *Sol. Energy Mater. Sol. Cells* **2017**, *172*, 34–43. [CrossRef]
14. Liu, H.Y.; Hsu, Y.L.; Su, H.Y.; Huang, R.C.; Hou, F.Y.; Tu, G.C.; Liu, W.H. A comparative study of amorphous, anatase, rutile, and mixed-phase TiO_2 films by mist chemical vapor deposition and ultraviolet photodetectors applications. *IEEE Sens. J.* **2018**, *18*, 4022–4029. [CrossRef]
15. Zhang, J.; Zhou, P.; Liu, J.; Yu, J. New understanding of the difference of photocatalytic activity among anatase, rutile, and brookite TiO_2. *Phys. Chem. Chem. Phys.* **2014**, *16*, 20382–20386. [CrossRef]
16. Nair, R.V.; Gummaluri, V.S.; Matham M., V.; Vijayan, C. A review on optical bandgap engineering in TiO_2 nanostructures via doping and intrinsic vacancy modulation towards visible light applications. *J. Phys. D App. Phys.* **2022**, *55*, 313003. [CrossRef]
17. Xiong, L.-B.; Li, J.-L.; Yang, B.; Yu, Y. Ti^{3+} in the surface of titanium dioxide: Generation, properties, and photocatalytic application. *J. Nanomat.* **2012**, *2012*, 831524. [CrossRef]
18. Kuznetsov, A.; Kameneva, O.; Alexandrov, A.; Bityurin, N.; Chhor, K.; Kanaev, A. Chemical Activity of Photoinduced Ti^{3+} Centers in Titanium Oxide Gels. *J. Phys. Chem. B* **2006**, *110*, 435–441. [CrossRef] [PubMed]
19. Kuznetsov, A.; Kameneva, O.; Rozes, L.; Sanchez, C.; Bityurin, N.; Kanaev, A. Extinction of photo-induced Ti^{3+} centres in titanium oxide gels and gel-based oxo-PHEMA hybrids. *Chem. Phys. Let.* **2006**, *429*, 523–527. [CrossRef]
20. Rozes, L.; Sanchez, C. Titanium oxo-clusters: Precursors for a Lego-like construction of nanostructured hybrid materials. *Chem. Soc. Rev.* **2011**, *40*, 1006–1030. [CrossRef] [PubMed]

21. Museur, L.; Gorbovyi, P.; Traore, M.; Kanaev, A.; Rozes, L.; Sanchez, C. Luminescence properties of pHEMA-TiO_2 gels based hybrids materials. *J. Lumin.* **2012**, *132*, 1192–1199. [CrossRef]
22. Li, R.; Zhou, Q. Impact of Titanium Dioxide (TiO_2) Modification on Its Application to Pollution Treatment—A Review. *Catalysts* **2020**, *10*, 804. [CrossRef]
23. Sanzone, G.; Zimbone, M.; Cacciato, G.; Ruffino, F.; Carles, R.; Privitera, V.; Grimaldi, M.G. Ag/TiO_2 nanocomposite for visible light-driven photocatalysis. *Superlattices Microstruct.* **2018**, *123*, 394–402. [CrossRef]
24. Sornalingam, K.; McDonagh, A.; Zhou, J.L.; Johira, M.A.H.; Ahmed, M.B. Photocatalysis of estrone in water and wastewater: Comparison between Au-TiO2 nanocomposite and TiO2, and degradation by-products. *Sci. Total Environ.* **2018**, *610–611*, 521–530. [CrossRef] [PubMed]
25. Sakthivel, S.; Shankar, M.V.; Palanichamy, M.; Arabindoo, B.; Bahnemann, D.W.; Murugesan, V. Enhancement of photocatalytic activity by metal deposition: Characterization and photonic efficiency of Pt, Au, and Pd deposited on TiO_2. *Water Res.* **2004**, *38*, 3001–3008. [CrossRef] [PubMed]
26. Ismael, M. A review and recent advances in solar-to-hydrogen energy conversion based on photocatalytic water splitting overdoped-TiO_2 nanoparticles. *Solar Energy* **2020**, *211*, 522–546. [CrossRef]
27. Tang, K.Y.; Chen, J.X.; Legaspi, E.D.R.; Owh, C.; Lin, M.; Tee, I.S.Y.; Kai, D.; Loh, X.J.; Li, Z.; Regulacio, M.D.; et al. Gold-decorated TiO_2 nanofibrous hybrid for improved solar-driven photocatalytic pollutant degradation. *Chemosphere* **2021**, *265*, 129114. [CrossRef]
28. Yu, Y.; Wen, W.; Qian, X.Y.; Liu, J.B.; Wu, J.M. UV and visible light photocatalytic activity of Au/TiO_2 nanoforests with Anatase/Rutile phase junctions and controlled Au locations. *Sci. Rep.* **2017**, *7*, 41253. [CrossRef]
29. Pandiyaraj, K.N.; Vasu, D.; Ghobeira, R.; Tabaei, P.S.E.; Geyter, N.D.; Morent, R.; Pichumani, M.; Padmanabhanan, P.V.A.; Deshmukh, R.R. Dye wastewater degradation by the synergetic effect of an atmospheric pressure plasma treatment and the photocatalytic activity of plasma-functionalized Cu-TiO_2 nanoparticles. *J. Hazard. Mat.* **2020**, *405*, 124264. [CrossRef]
30. Kozlov, D.A.; Lebedev, V.A.; Polyakov, A.Y.; Khazova, K.M.; Garshev, A.V. The microstructure effect on the Au/TiO_2 and Ag/TiO_2 nanocomposites photocatalytic activity. *Nanosyst. Phys. Chem. Math.* **2018**, *9*, 266–278. [CrossRef]
31. Gao, Y.; Zhang, W.; Liu, P. Enhanced Photocatalytic Efficiency of TiO_2 Membrane Decorated with Ag and Au Nanoparticles. *Appl. Sci.* **2018**, *8*, 945. [CrossRef]
32. Moslah, C.; Kandyla, M.; Mousdis, G.A.; Petropoulou, G.; Ksibi, M. Photocatalytic Properties of Titanium Dioxide Thin Films Doped with Noble Metals (Ag, Au, Pd, and Pt). *Phys. Status Solidi (A)* **2018**, *215*, 1800023. [CrossRef]
33. Malik, A.S.; Liu, T.; Rittiruam, M.; Saelee, T.; Da Silva, J.L.; Praserthdam, S.; Praserthdam, P. On a high photocatalytic activity of high-noble alloys Au–Ag/TiO_2 catalysts during oxygen evolution reaction of water oxidation. *Sci. Rep.* **2022**, *12*, 2604. [CrossRef] [PubMed]
34. Yoshiiri, K.; Wang, K.; Kowalska, E. TiO_2/Au/TiO_2 Plasmonic Photocatalysts: The Influence of Titania Matrix and Gold Properties. *Inventions* **2022**, *7*, 54. [CrossRef]
35. Yang, X.; Wang, Y.; Zhang, L.; Fu, H.; He, P.; Han, D.; Lawson, T.; An, X. The Use of Tunable Optical Absorption Plasmonic Au and Ag Decorated TiO_2 Structures as Efficient Visible Light Photocatalysts. *Catalysts* **2020**, *10*, 139. [CrossRef]
36. Varapragasam, S.J.P.; Mia, S.; Wieting, C.; Balasanthiran, C.; Hossan, M.Y.; Baride, A.; Rioux, R.M.; Hoefelmeyer, J.D. Ag–TiO_2 Hybrid Nanocrystal Photocatalyst: Hydrogen Evolution under UV Irradiation but Not under Visible-Light Irradiation. *ACS Appl. Energy Mater.* **2019**, *2*, 8274–8282. [CrossRef]
37. Salomatina, E.; Bityurin, N.; Gulenova, M.; Gracheva, T.; Drozdov, M.; Knyazev, A.; Kir'yanov, K.; Markin, A.; Smirnova, L. Synthesis, structure, and properties of organic-inorganic (co)polymers containing poly(titanium oxide). *J. Mater. Chem. C.* **2013**, *1*, 6375–6385. [CrossRef]
38. Salomatina, E.V.; Moskvichev, A.N.; Knyazev, A.V.; Smirnova, L.A. Effect of kinetic features in synthesis of hybrid copolymers based on $Ti(OPr^i)_4$ and hydroxyethyl methacrylate on their structure and properties. *Rus. J. App. Chem.* **2015**, *88*, 197–207. [CrossRef]
39. Loginova, A.S.; Ignatov, S.K.; Chukhmanov, E.P.; Salomatina, E.V.; Smirnova, L.A. Structure, spectra, and photoinduced electron-redistribution properties of TiO_2/organic copolymers with gold nanoparticles. A DFT study. *Comput. Theor. Chem.* **2017**, *1118*, 1–15. [CrossRef]
40. Howe, R.F.; Gratzel, M. EPR observation of trapped electrons in colloidal titanium dioxide. *J. Phys. Chem.* **1985**, *89*, 4495–4499. [CrossRef]
41. Gratzel, M.; Howe, R.F. Electron paramagnetic resonance studies of doped titanium dioxide colloids. *J. Phys. Chem.* **1990**, *94*, 2566–2572. [CrossRef]
42. Anpo, M.; Aikawa, N.; Kubokawa, Y. Photocatalytic hydrogenation of alkynes and alkenes with water over titanium dioxide. Platinum loading effect on the primary processes. *J. Phys. Chem.* **1984**, *88*, 3998–4000. [CrossRef]
43. Micic, O.I.; Zhang, Y.; Cromack, K.R.; Trifunac, A.D.; Thurnauer, M.C. Photoinduced hole transfer from titanium dioxide to methanol molecules in aqueous solution studied by electron paramagnetic resonance. *J. Phys. Chem.* **1993**, *97*, 13284–13288. [CrossRef]
44. Kuznetsov, A.I.; Kameneva, O.; Bityurin, N.; Rozes, L.; Sanchez, C.; Kanaev, A. Laser-induced photopatterning of organic-inorganic TiO_2-based hybrid materials with tunable interfacial electron transfer. *Phys. Chem. Chem. Phys.* **2009**, *11*, 1248–1257. [CrossRef] [PubMed]

45. Kameneva, O.; Kuznestov, A.I.; Smirnova, L.A.; Rozes, L.; Sanchez, C.; Aleksandrov, A.; Bityurin, N.; Chhor, K.; Kanaev, A. New photoactive hybrid organic-inorganic materials based on titanium-oxo-PHEMA nanocomposites exhibiting mixed valence properties. *J. Mater. Chem.* **2005**, *15*, 3380–3383. [CrossRef]
46. Uklein, A.; Gorbovyi, P.; Traore, M.; Museur, L.; Kanaev, A. Photo-induced refraction of nanoparticulate organic-inorganic TiO_2-pHEMA hybrids. *Opt. Mater. Express* **2013**, *3*, 533–545. [CrossRef]
47. Hamed, M.M.; Ahmed, I.M.; Metwally, S.S. Adsorptive removal of methylene blue as an organic pollutant by marble dust as eco-friendly sorbent. *J. Ind. Eng. Chem.* **2014**, *20*, 2370–2377. [CrossRef]
48. Mills, A.; Hill, C.; Robertson, P.K.J. Overview of the current {ISO} tests for photocatalytic materials. *J. Photochem. Photobiol. A Chem.* **2012**, *237*, 7–23. [CrossRef]
49. Krishnan, S.; Shriwastav, A. Application of TiO_2 nanoparticles sensitized with natural chlorophyll pigments as catalyst for visible light photocatalytic degradation of methylene blue. *J. Environ. Chem. Eng.* **2021**, *9*, 104699. [CrossRef]
50. Tichapondwa, S.M.; Newman, J.P.; Kubheka, O. Effect of TiO_2 phase on the photocatalytic degradation of methylene blue dye. *Phys. Chem. Earth Parts A/B/C* **2020**, *118–119*, 102900. [CrossRef]
51. Petrella, A.; Spasiano, D.; Cosma, P.; Rizzi, V.; Race, M.; Mascolo, M.C.; Ranieri, E. Methyl Orange Photo-Degradation by TiO_2 in a Pilot Unit under Different Chemical, Physical, and Hydraulic Conditions. *Processes* **2021**, *9*, 205. [CrossRef]
52. Zha, R.; Nadimicherla, R.; Guo, X. Ultraviolet photocatalytic degradation of methyl orange by nanostructured TiO_2/ZnO heterojunctions. *J. Mater. Chem. A* **2015**, *3*, 6565–6574. [CrossRef]
53. Regraguy, B.; Rahmani, M.; Mabrouki, J.; Drhimer, F.; Ellouzi, I.; Mahmou, C.; Dahchour, A.; El Mrabet, M.; El Hajjaji, S. Photocatalytic degradation of methyl orange in the presence of nanoparticles $NiSO_4/TiO_2$. *Nanotechnol. Environ. Eng.* **2022**, *7*, 157–171. [CrossRef]
54. He, Y.; Grieser, F.; Ashokkumar, M. The mechanism of sonophotocatalytic degradation of methyl orange and its products in aqueous solutions. *Ultrason. Sonochem.* **2011**, *18*, 974–980. [CrossRef] [PubMed]
55. Zhang, T.; Oyama, T.; Horikoshi, S.; Hidaka, H.; Zhao, J.; Serpone, N. Photocatalyzed N-demethylation and degradation of methylene blue in titania dispersions exposed to concentrated sunlight. *Solar Ener. Mater. Solar Cells* **2002**, *73*, 287–303. [CrossRef]
56. Dai, K.; Chen, H.; Peng, T.; Ke, D.; Yi, H. Photocatalytic degradation of methyl orange in aqueous suspension of mesoporous titania nanoparticles. *Chemosphere* **2007**, *69*, 1361–1367. [CrossRef]
57. Wang, S.; Luo, C.; Tan, F.; Cheng, X.; Ma, Q.; Wu, D.; Li, P.; Zhang, F.; Ma, J. Degradation of Congo red by UV photolysis of nitrate: Kinetics and degradation mechanism. *Sep. Purif. Technol.* **2021**, *262*, 118276. [CrossRef]
58. Ullah, I.; Haider, A.; Khalid, N.; Ali, S.; Ahmed, S.; Khan, Y.; Ahmed, N.; Zubair, M. Tuning the band gap of TiO_2 by tungsten doping for efficient UV and visible photodegradation of Congo red dye. *Spectrochim. Acta Part A Mol. Biomol. Spectrosc.* **2018**, *204*, 150–157. [CrossRef]
59. Qin, L.; Liang, F.; Li, Y.; Wu, J.; Guan, S.; Wu, M.; Xie, S.; Luo, M.; Ma, D. A 2D Porous Zinc-Organic Framework Platform for Loading of 5-Fluorouracil. *Inorganics* **2022**, *10*, 202. [CrossRef]
60. Qin, L.; Li, Y.; Liang, F.; Li, L.; Lan, Y.; Li, Z.; Lu, X.; Yang, M.; Ma, D. A microporous 2D cobalt-based MOF with pyridyl sites and open metal sites for selective adsorption of CO_2. *Microporous Mesoporous Mater.* **2022**, *341*, 112098. [CrossRef]
61. Yang, L.; Luo, S.; Li, Y.; Xiao, Y.; Kang, Q.; Cai, Q. High Efficient Photocatalytic Degradation of p-Nitrophenol on a Unique Cu_2O/TiO_2 p-n Heterojunction Network Catalyst. *Environ. Sci. Technol.* **2010**, *44*, 7641–7646. [CrossRef]
62. Yadav, V.; Verma, P.; Sharma, H.; Tripathy, S.; Saini, V.K. Photodegradation of 4-nitrophenol over B-doped TiO2 nanostructure: Effect of dopant concentration, kinetics, and mechanism. *Environ. Sci. Pollut. Res.* **2020**, *27*, 10966–10980. [CrossRef] [PubMed]
63. Moctezuma, E.; Leyva, E.; Aguilar, C.A.; Luna, R.A.; Montalvo, C. Photocatalytic degradation of paracetamol: Intermediates and total reaction mechanism. *J. Hazard. Mater.* **2012**, *243*, 130–138. [CrossRef] [PubMed]
64. Ahn, W.-Y.; Sheeley, S.A.; Rajh, T.; Cropek, D.M. Photocatalytic reduction of 4-nitrophenol with arginine-modified titanium dioxide nanoparticles. *Appl. Catal. B Environ.* **2007**, *74*, 103–110. [CrossRef]
65. Zhao, S.; Ma, H.; Wang, M.; Cao, C.; Yao, S. Study on the role of hydroperoxyl radical in degradation of p-nitrophenol attacked by hydroxyl radical using photolytical technique. *J. Photochem. Photobiol. A Chem.* **2013**, *259*, 17–24. [CrossRef]
66. Di Paola, A.; Augugliaro, V.; Palmisano, L.; Pantaleo, G.; Savinov, E. Heterogeneous photocatalytic degradation of nitrophenols. *J. Photochem. Photobiol. A Chem.* **2003**, *155*, 207–214. [CrossRef]
67. Kalvert, G.; Pitts, G. *Photochemistry*, 1st ed.; John Wiley and Sons, Inc.: New York, NY, USA, 1966; 899p.
68. Nakata, A.; Fujishima, A. TiO_2 photocatalysis: Design and applications. *J. Photochem. Photobiol C Photochem. Rev.* **2012**, *13*, 169–189. [CrossRef]
69. Belekbir, S.; El Azzouzi, M.; El Hamidi, A.; Rodríguez-Lorenzo, L.; Santaballa, J.A.; Canle, M. Improved Photocatalyzed Degradation of Phenol, as a Model Pollutant, over Metal-Impregnated Nanosized TiO_2. *Nanomaterials* **2020**, *10*, 996. [CrossRef]
70. Gadirova, E.M. Photochemical degradation of phenol in the presence of titanium dioxide nanoparticles. *Proceed. Univ. Appl. Chem. Biotechnol.* **2019**, *9*, 179–182.
71. Dang, T.T.T.; Le, S.T.T.; Channei, D.; Khanitchaidecha, W.; Nakaruk, A. Photodegradation mechanisms of phenol in the photocatalytic process. *Res. Chem. Intermed.* **2016**, *42*, 5961–5974. [CrossRef]
72. Grabowska, E.; Reszczyńska, J.; Zaleska, A. RETRACTED: Mechanism of phenol photodegradation in the presence of pure and modified-TiO_2: A review. *Water Res.* **2012**, *46*, 5453–5471. [CrossRef]

73. Dong, X.; Li, D.; Li, Y.; Sakiyama, H.; Muddassir, M.; Pan, Y.; Srivastava, D.; Kumar, A. A 3,8-connected Cd(II)-based metal–organic framework as an appropriate luminescent sensor for the antibiotic sulfasalazine. *CrystEngComm* **2022**, *24*, 7157–7165. [CrossRef]
74. Rao, C.; Zhou, L.; Pan, Y.; Lu, C.; Qin, X.; Sakiyama, H.; Muddassir, M.; Liu, J. The extra-large calixarene-based MOFs-derived hierarchical composites for photocatalysis of dye: Facile syntheses and contribution of carbon species. *J. Alloys Compd.* **2022**, *897*, 163178. [CrossRef]
75. Dobrosz-Gómez, I.; Gómez-García, M.Á.; Zamora, S.M.L.; GilPavas, E.; Bojarska, J.; Kozanecki, M.; Rynkowski, J.M. Transition metal loaded TiO_2 for phenol photo-degradation. *Comptes Rendus Chim.* **2015**, *18*, 1170–1182. [CrossRef]
76. Chowdhury, P.; Moreira, J.; Gomaa, H.; Ray, A.K. Visible-Solar-Light-Driven Photocatalytic Degradation of Phenol with Dye-Sensitized TiO_2: Parametric and Kinetic Study. *Ind. Eng. Chem. Res.* **2012**, *51*, 4523–4532. [CrossRef]
77. Rehman, S.; Ullah, R.; Butt, A.M.; Gohar, N.D. Strategies of making TiO_2 and ZnO visible light active. *J. Hazard. Mat.* **2009**, *170*, 560–569. [CrossRef]
78. Carp, O.; Huisman, C.L.; Reller, A. Photoinduced Reactivity of Titanium Dioxide. *Prog. Solid State Chem.* **2004**, *32*, 33–177. [CrossRef]
79. Saih, Y.; Segawa, K. Ultradeep hydrodesulfurization of dibenzothiophene (DBT) derivatives over Mo-sulfide catalysts supported on TiO_2-coated alumina composite (Review). *Catal. Surv. Asia* **2003**, *7*, 235–249. [CrossRef]
80. Suyoulema, L.J.; Sarina, W.W. A study of photodegradation of sulforhodamine B on Au–TiO_2/bentonite under UV and visible light irradiation. *Solid State Sci.* **2009**, *11*, 2037–2043.
81. Kortum, G.F.A. *Reflectance Spectroscopy: Principles, Methods, Applications*; Springer Science & Business Media: New York, NY, USA, 1969.

catalysts

MDPI

Article

Application of BiOX Photocatalyst to Activate Peroxydisulfate Ion-Investigation of a Combined Process for the Removal of Organic Pollutants from Water

Tünde Alapi [1,*], Bence Veres [1], Máté Náfrádi [1], Luca Farkas [1], Zsolt Pap [2] and Anett Covic [1]

1 Department of Inorganic and Analytical Chemistry, University of Szeged, Dóm Square 7, H-6720 Szeged, Hungary
2 Applied & Environmental Chemistry Department, University of Szeged, Rerrich Béla Square 1, H-6720 Szeged, Hungary
* Correspondence: alapi@chem.u-szeged.hu

Abstract: The persulfate-based advanced oxidation processes employing heterogeneous photocatalysts to generate sulfate radicals ($SO_4^{\bullet-}$) from peroxydisulfate ion (PDS, $S_2O_8^{2-}$) have been extensively investigated to remove organic pollutants. In this work, BiOX (X = Cl, Br, and I) photocatalysts were investigated to activate PDS and enhance the transformation rate of various organic substances under UV (398 nm) and Vis (400–700 nm) radiation. For BiOCl and BiOBr, in addition to excitability, the light-induced oxygen vacancies are decisive in the activity. Although without organic substances, the BiOI efficiency highly exceeds that of BiOBr and BiOCl for PDS activation (for BiOI, 15–20%, while for BiOBr and BiOCl, only 3–4% of the PDS transformed); each BiOX catalyst showed enhanced activity for 1,4-hydroquinone (HQ) transformation due to the semiquinone radical-initiated PDS activation. For sulfamethoxypyridazine (SMP), the transformation is driven by direct charge transfer, and the effect of PDS was less manifested. BiOI proved efficient for transforming various organic substances even under Vis radiation. The efficiency was enhanced by PDS addition (HQ is wholly transformed within 20 min, and SMP conversion increased from 40% to 90%) without damaging the catalyst; its activity did change over three consecutive cycles. Results related to the well-adsorbed trimethoprim (TRIM) and application of biologically treated domestic wastewater as a matrix highlighted the limiting factors of the method and visible light active photocatalyst, BiOI.

Keywords: benzoquinone; sulfonamides; sulfate radical ion; semiquinone; matrix effect; BiOX

1. Introduction

Nowadays, environmental pollution, including water pollution, has received particular attention, and treating water pollutants has become an urgent task. Some biologically active organic contaminants, such as endocrine-disrupting chemicals, pharmaceuticals, and personal care products, are non-biodegradable and cannot be removed entirely using conventional water treatment methods. Therefore, the effective elimination of these contaminants has become a leading research topic in environmental sciences. Moreover, the world has faced the pandemic caused by COVID-19, which resulted in the growing global use of antibiotics and their concentration in wastewater [1–3]. The concentration of antibiotics in influent wastewater is typically in the order of nanograms per liter and often reaches several milligrams per liter. During wastewater treatment, their concentration is usually reduced, although a significant part of them and some of their metabolites can often remain in the effluent and reach the surface waters. This further increases the probability of developing antibiotic-resistant bacterial strains, having unpredictable consequences [4].

Heterogeneous photocatalysis is one of the Advanced Oxidation processes (AOP), which is suitable for the economical removal of non-biodegradable water pollutants, especially by using photocatalysts that can be excited by visible light. In a photocatalyst excited, the electron (e_{cb}^-) in the conduction band and hole (h_{vb}^-) in the valence band, created by photoinitiated charge separation, must avoid recombination and reach the surface to react with the organic compound or generate reactive species (radicals and radical ions) to degrade it. The widely studied first-generation photocatalysts, such as TiO_2 and ZnO, can be excited in the UV range, and the transformation of the organic substances is generally initiated by hydroxyl radical ($^•OH$) based reactions [5]. In recent decades, the number of publications related to synthesizing and testing visible light-excitable photocatalysts with lower bandgap energy has increased significantly [6–8]. Another important goal is to achieve a longer lifetime of the photogenerated charges to enhance the oxidation and reduction processes realized through direct charge transfer on the surface and increase the efficiency of the transformation of organic and inorganic components in this way [5,9].

Bismuth-containing photocatalysts are intensively researched materials; most publications focus on bismuth oxohalides (BiOX, X = F, Cl, Br, I) due to their highly unique structure and beneficial optical and catalytic properties. Furthermore, their advantages are chemical stability, easy preparation, good adsorption properties, and superior photocatalytic activity [10–15]. BiOF and BiOCl are active in the UV range (band gap: ~3.6 eV for BiOF and ~3.2 eV for BiOCl), while BiOBr, and especially BiOI, are active in the visible range (band gap: ~2.6 eV for BiOBr and ~1.8–2.1 eV for BiOI) [14,16]. In addition to optical properties and widely different bandgap values, the adsorption capacity and surface properties of BiOX photocatalysts also differ. Photoinduced oxygen vacancies (OVs) are easily formed on their surface (especially for BiOBr and BiOCl), which can change their color, absorption, and adsorption properties. All of these can be beneficial in terms of photocatalytic activity [17,18]. The number and quality of reactive species formed on the surface of the excited catalyst are decisive for the efficiency and transformation of the organic target substances. The relative oxidation power of photogenerated h_{VB}^+ changes in the order of BiOCl > BiOBr > BiOI. The $^•OH$ forms with low efficiency in the case of BiOBr, while the $O_2^{•-}$ formation is characteristic for each BiOX [19]. Several attempts have been made to enhance the efficiency of BiOX photocatalysts, including the preparation of composite catalysts, such as BiOCl/BiOBr [20], BiOCl/BiOI, [21–23], and TiO_2/BiOX [24], with improved stability and enhanced visible-light activity.

The efficiency of heterogeneous photocatalysis can be enhanced via the addition of substances that are effective electron scavengers and the source of highly reactive radicals simultaneously, such as ozone [25] and H_2O_2 [26]. In recent years, persulfate (PS)-based AOPs have been considered a promising method to eliminate organic pollutants in water due to the favorable properties of sulfate radical ion ($SO_4^{•-}$), which is a strong nonselective oxidant with high redox potential (2.6–3.1 V), similar to the $^•OH$ (1.9–2.7 V) [27–32]. The source of $SO_4^{•-}$ can be peroxomonosulfate (HSO_5^-, PMS) or peroxydisulfate ion ($S_2O_8^{2-}$, PDS); their activation can occur via energy or electron transfer. In the case of the combination PS process with heterogeneous photocatalysis, the excited photocatalysts act as electron donors, and PS can capture photogenerated e_{cb}^- to generate $SO_4^{•-}$. In addition to $SO_4^{•-}$, other reactive species, such as singlet oxygen ($^1O_2$), also form and are responsible for the enhanced transformation and mineralization rate of target substances. Simultaneously, the photocatalytic process can be promoted due to the improved separation of charge carriers.

Recently intensive studies were conducted to investigate the combination of heterogeneous photocatalysis with the PS process [33–37], using PDS [38,39] and PMS [40–42] to enhance the efficiency. Some publications have reported the efficient application of BiOBr [43] or BiOI composite to activate PDS [44–46]. However, to our knowledge, the direct comparison of pure BiOI, BiOBr, and BiOCl for activation of PDS under UV and visible light has not yet been published. In this work, we compare the efficiency of the BiOX photocatalysts for the activation of PDS and the conversion of different organic compounds (1,4-benzoquinone (BQ), 1,4-hydroquinone (HQ), sulfamethoxypyridazine (SMP), and trimethoprim (TRIM)). The reusability of the BiOX catalyst showing the best activity and the investigation of its use in biologically treated municipal wastewater as a matrix was also studied.

2. Results and Discussion

2.1. Characterization of the BiOX Photocatalyst

The synthesis method optimized and published by Bárdos et al. [15] was used to prepare the BiOX photocatalysts, having spherical hierarchical microcrystals composed of individual nanoplates. The synthesized catalysts were characterized using powder X-ray diffractometry (XRD). The specific surface area (using measurements based on N_2 adsorption), the reflectance, and bandgap values (using DRS measurements) were determined. The results (Figure 1 and Table 1) provided data consistent with the data in the literature [13,47]. According to the bandgap value, BiOBr can be excited with the photons emitted by UV-LED, while BiOI can be excited by both LEDs. However, a significant part of the photons (λ > 512 nm) emitted by the Vis-LED is unsuitable for creating the photogenerated e_{cb}^{-}–h_{vb}^{+} pair. The energy of 398 nm photons is inadequate for the excitation of BiOCl, having a 3.26 eV bandgap value.

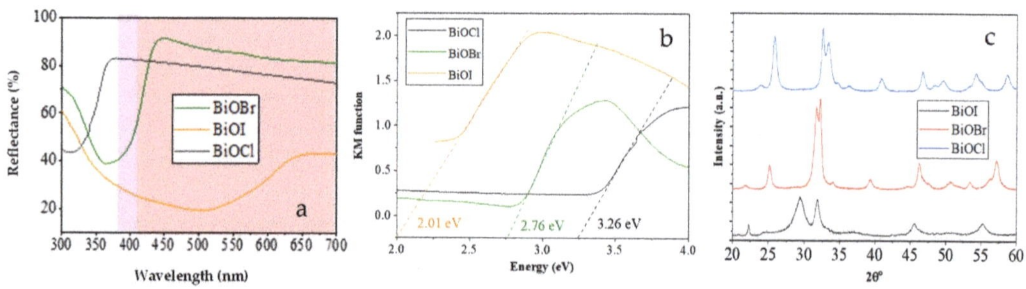

Figure 1. UV–Vis DRS spectra (**a**) (shadow indicates the emission range of LED (violet: UV-LED, red: Vis-LED)), Kubelka Munk Curves (**b**), and XRD patterns (**c**) of BiOCl, BiOBr, and BiOI.

Table 1. The specific surface areas and bandgap values of the photocatalysts.

	Bandgap	Surface Area	Primary Particle Size
BiOCl	3.26 eV (363 nm)	16 m^2 g^{-1}	31 nm
BiOBr	2.76 eV (435 nm)	18 m^2 g^{-1}	39 nm
BiOI	2.01 eV (512 nm)	45 m^2 g^{-1}	21 nm

2.2. Transformation of the Target Substances Using BiOX Photocatalysts

The efficiency of the photocatalytic elimination of organic substances depends on the quality and quantity of the reactive species formed on the surface of the excited photocatalyst and the reactivity of the target substances towards them. The formation of the non-selective $^{\bullet}OH$ is unlikely in the case of the irradiated BiOX [13,19,48]; the transformation of organic substances occurs mainly via reactions with other, more selective species, such as singlet oxygen ($^1O_2$) or $O_2^{\bullet-}/HO_2^{\bullet}$ [19]. Another option for the transformation of organic substances is the direct charge transfer [13,48]. In this work, four organic substances were chosen to determine and compare the activity of BiOX photocatalyst and investigate the relative contribution of the various processes to the transformation of organic substances. The target compounds were carefully chosen to characterize the activity of catalysts from different points of view.

2.2.1. Transformation of BQ

Fónagy et al. [49] demonstrated that the transformation of BQ in O_2-free suspension occurs via direct charge transfer; the reaction with e_{CB}^{-} results in the formation of HQ [50]:

$$BQ + 2\,e^{-} \rightarrow HQ\ k = 3.9 \times 10^{10}\ M^{-1}\ s^{-1} \quad (1)$$

consequently, the transformation rate of BQ and formation rate of HQ is primarily determined by the density of the photogenerated e_{cb}^- on the surface, which depends on the absorption property, excitability of the catalyst, and the recombination rate of photogenerated charges at a given photon flux and wavelength.

Figure 2 shows the BQ transformation in O_2-free suspensions of BiOX photocatalysts using UV LED (398 nm) and Vis LED (400–700 nm). BQ is also efficiently transformed by direct photolysis, even under Vis radiation. Although the transformation of BQ by both direct charge transfer and photolysis results in HQ, there is a characteristic difference between the two processes regarding HQ conversion. In the case of direct charge transfer, BQ completely transforms to HQ [49], while in the case of direct photolysis besides HQ, other aromatic and ring-opening products form [51].

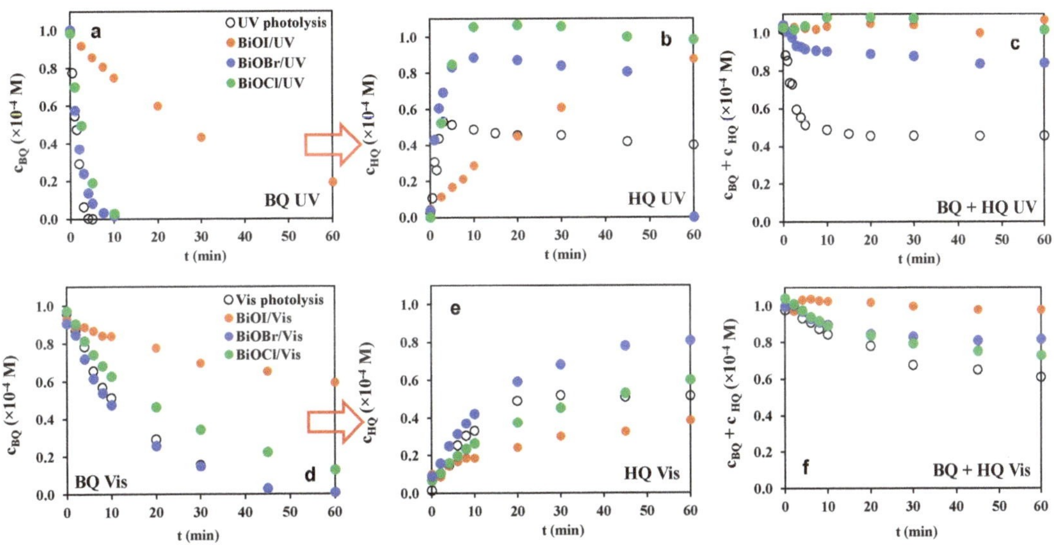

Figure 2. Transformation of BQ (**a**,**d**) and formation of HQ (**b**,**e**) and the sum of the BQ and HQ concentration (**c**,**f**) in O_2-free BiOX suspensions, under UV (**a**–**c**) and Vis (**d**–**f**) radiation (**a**,**d**).

Based on the bandgap energy (Table 1), BiOI and BiOBr can be excited with 398 nm photons, but not BiOCl, while only BiOI can be excited using a Vis LED. Nevertheless, in the case of UV LED, the transformation of BQ occurred at a rate similar to that of photolysis in the case of BiOCl and BiOBr, and transformation was observed even under visible light radiation. Both catalysts were blackened under irradiation; this change was reversible, and both photocatalysts changed their color back in the dark. A similar observation has been published and interpreted with photoinitiated oxygen/halogen vacancies, which may result in better absorption properties, band gap changes, and enhanced photocatalytic activity [18]. The oxygen vacancy-mediated mechanism of CO_2 photoreduction was suggested for BiOX [17], especially for BiOBr. In the case of BiOCl and BiOBr under UV light, the conversion of HQ from BQ was significantly higher (85 and 100%) than in the case of photolysis (less than 50%), which suggests that the heterogeneous photocatalytic transformation contributes to the BQ transformation (Figure 2). However, the direct photolysis of BQ must be addressed, especially for BiOCl and BiOBr. Within the emission range of the Vis LED, BiOCl, and BiOBr show high reflectance, and HQ conversion is closer to that measured during photolysis, indicating that it contributes significantly to the conversion.

The BiOI catalyst should be discussed separately. The BiOI can be excited by both LEDs and, opposite to the BiOCl and BiOI, efficiently absorb the emitted photons even in the emission range of Vis LED (Figure 1 and Table 1). The color change of BiOI was not observed under irradiation. The transformation of BQ in BiOI suspension was much slower than in solution via photolysis, and BQ completely transformed into HQ (Figure 2). Presumably, in the case of BiOI suspension, the photons are absorbed by the catalyst and the dominant way of BQ transformation is the direct charge transfer, beside that the role of photolysis is negligible.

2.2.2. Transformation of HQ

The HQ transformation was investigated in aerated suspensions (Figure 3). The concentration of the HQ does not change in irradiated and aerated solutions or O_2-free suspensions and is poorly (<2%) adsorbed on the surface of photocatalysts. The transformation requires the simultaneous presence of BiOX photocatalyst, irradiation, and dissolved O_2 [50] and occurs due to the reaction with $O_2^{\bullet-}$ [52] resulting in BQ. In this case, the HQ transformation rate directly refers to the $O_2^{\bullet-}$ formation rate [49].

$$O_2 + e^- \rightarrow O_2^{\bullet-} \qquad k = 2.3 \times 10^{10}\ M^{-1}\ s^{-1} \tag{2}$$

$$HQ + O_2^{\bullet-} \rightarrow HBQ^{\bullet} + HO_2^- \tag{3}$$

$$HBQ^{\bullet} \rightarrow BQ^{\bullet-} + H^+ \tag{4}$$

$$BQ^{\bullet-} + O_2 \rightarrow BQ + O_2^{\bullet-} \tag{5}$$

$$BQ + O_2^{\bullet-} \rightarrow BQ^{\bullet-} + O_2 \qquad k = 9.8 \times 10^{8}\ M^{-1}\ s^{-1} \tag{6}$$

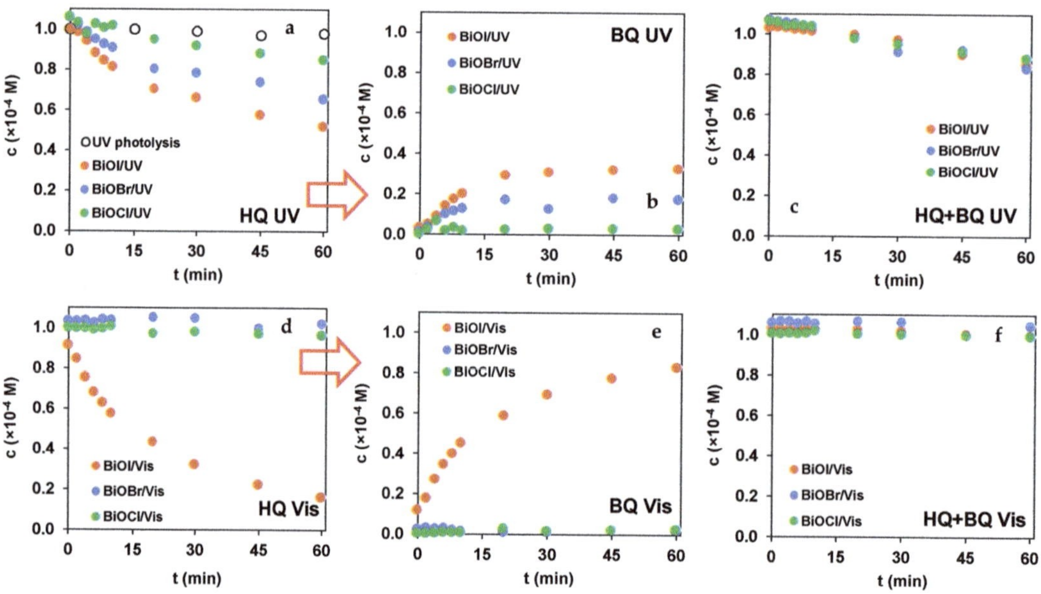

Figure 3. Transformation of HQ (**a**,**d**) and formation of BQ (**b**,**e**) and the sum of the HQ and BQ concentration (**c**,**f**) in aerated BiOX suspensions under UV (**a**–**c**) and Vis (**d**–**f**) radiation.

The formation rate of $O_2^{\bullet-}$ determined by Lv et al. [19] under simulated sunlight and changed in the order BiOCl > BiOBr ≅ BiOI, while the relative oxidation power of h_{vb}^+ was in the order of BiOCl > BiOBr > BiOI. However, in our case, the transformation rate of HQ changed in the order BiOI > BiOBr > BiOCl, reflecting the optical properties and excitability of the catalysts (Figures 1 and 3, Table 1). Under Vis radiation, the activity of BiOCl and BiOBr was negligible, while more than 80% of HQ was converted to BQ within 60 min in the BiOI suspension. The results are consistent with the bandgap values, based on which BiOCl and BiOBr cannot be excited by visible light, or only to a small extent, while BiOI can be well excited even by Vis light. The blackening of BiOCl and BiOBr was observed in suspensions containing O_2 and was less intense under Vis radiation than under UV. The sum of the HQ and BQ concentration decreased slowly in UV, while did not change in Vis irradiated suspensions (Figure 3). The slight decrease in the sum of HQ and BQ concentration suggests that further BQ transformation occurs, resulting in the ring-opening process. For BiOI, the application of Vis LED proved to be more efficient than UV LED at the same electric power (4.6 W), even though the UV photon flux is twice the photon flux in the visible range.

2.2.3. Transformation of SMP and TRIM

SMP and TRIM, as widely used, hardly biodegradable antibiotics, often detected in surface water, were used as environmentally relevant target substances. A significant difference between the substrates is that SMP is poorly adsorbed on BiOX photocatalysts, while the adsorption of TRIM is substantial, particularly in the case of BiOI.

The SMP transformation using BiOI photocatalysts was investigated by Náfrádi et al. [48], and the direct charge transfer was reported as the primary transformation way. The indirect photodegradation [53,54] and reaction with e_{cb}^- cannot be excluded partly because of the reactivity of pyridazine moiety towards e_{aq}^-. However, in the case of TiO_2 excited with 398 nm light, direct energy transfer [55] was observed, and a reaction with $^1O_2$ [56] was also assumed. Similar to HQ, direct photolysis of SMP is negligible in O_2-free irradiated suspension. The transformation surprisingly occurred at similar rates for all three catalysts using UV LED and was much faster than HQ transformation under the same experimental conditions (Figure 4). The observed activity cannot be interpreted solely based on excitability, especially in the case of BiOCl [57]. Most probably, the photoinitiated change of the catalysts properties, the formed oxygen vacancies, and the surficial Bi^{5+} and Bi^0 play a significant role in the efficiency and highly favorable to the redox capacity for the photocatalysts and transformation of SMP by direct charge transfer. The synergism between Bi^0 and Bi^{5+} was demonstrated and was proved to be the reason for the enhanced degradation of organic substances [52–55]. The decomposition was a hole-driven process; moreover, $O_2^{\bullet -}$ has an additive role. However, under visible light illumination, only the BiOI catalyst showed activity; the transformation efficiency was half that observed under UV radiation (Figure 4), similar to the BQ transformation also driven by direct charge transfer (Figure 2).

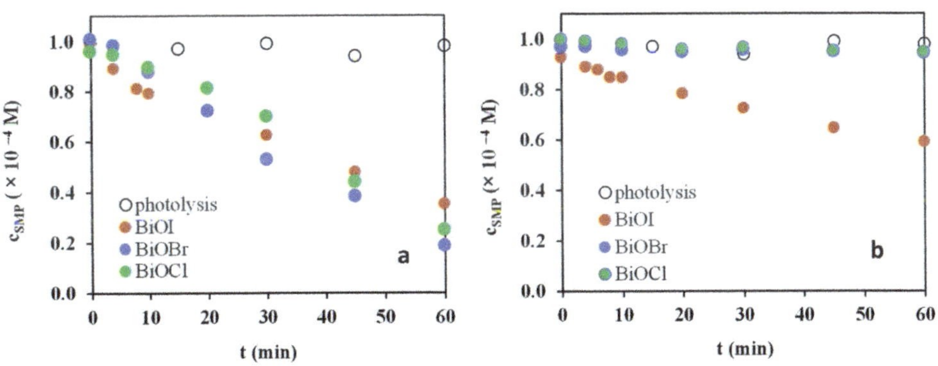

Figure 4. Transformation of SMP in aerated BiOX suspensions under UV (**a**) and Vis (**b**) irradiation.

Opposite the HQ and SMP, the relative adsorbed amount of TRIM was 39% for BiOI. Lower values, 6% for BiOBr, and less than 2% for BiOCl were determined. To distinguish the amount of adsorbed and transformed TRIM; NaF solution was added to the samples for TRIM desorption before analysis [48]. Opposite to the good adsorption properties, no more than 6% of TRIM transformed in BiOX-containing suspension, even in the case of

BiOI (Figure S1). The reason could be the low reactivity of TRIM towards less reactive ROS, such as $O_2^{\bullet-}$ and $^1O_2$. The transformation by charge transfer is most probably ruled out in this case. This result clearly shows the limitation of the applicability of BiOX photocatalysts because of the lack of the effective formation of highly oxidizing non-selective species, such as $^{\bullet}OH$.

2.3. *Effect of PDS*

The effects of PDS on the BiOX Structure and Properties, and the Transformation of PDS without Organic Substances

The effect of 1.0×10^{-3} M PDS on the photocatalyst stability was investigated under UV and Vis radiation. XRD and DRS results proved that there is no significant change in their structure, optical properties, and excitability of the photocatalysts (DRS spectra and band gap values have not changed) (Figure 5).

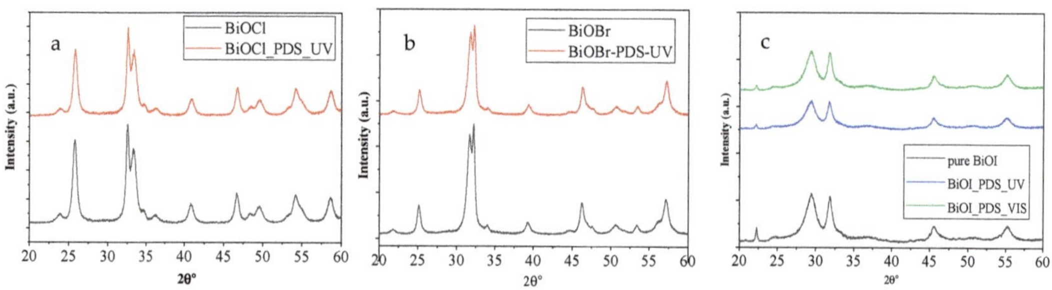

Figure 5. The Effect of PDS on the XRD diffractogram of BiOX ((**a**): BiOCl; (**b**): BiOBr; (**c**): BiOI) photocatalysts under UV and Vis radiation in aerated suspensions.

In an aerated BiOI suspension containing no organic matter, the transformation of 1.0×10^{-3} M PDS results in the formation of 3.0×10^{-4} M and 3.7×10^{-4} M SO_4^{2-} under UV and Vis radiation, respectively, i.e., 15–20% of PDS was transformed in 60 min. However, in the case of BiOBr and BiOCl, only 3–4% PDS was converted even under UV radiation (Figure 6). That means BiOI is much more efficient for PDS reduction than BiOBr or BiOCl.

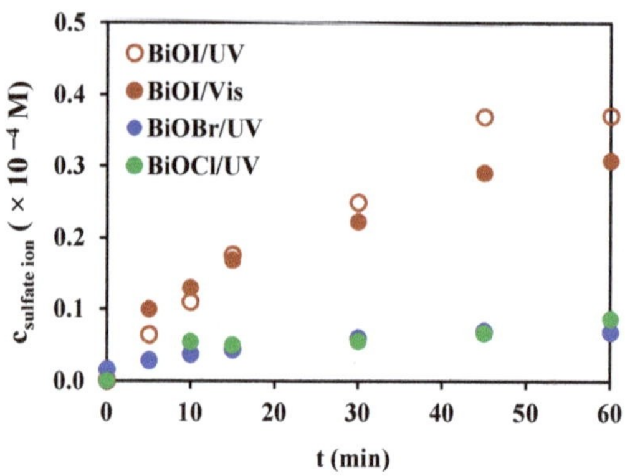

Figure 6. The concentration of sulfate ion results in PDS (1.0×10^{-3} M) transformation in UV and Vis irradiated, aerated BiOX suspensions without organic substances.

The ineffectiveness of BiOCl is not surprising given its excitation and absorption properties (Table 1); however, better efficiency of UV-irradiated BiOBr was expected, as publications reported for activation of PDS [43,58] or PMS [59], even under visible light radiation. In the case of PDS, the enhanced Bisphenol-A transformation was interpreted by oxygen vacancies mediated PDS activation process resulting in $^1O_2$ as the main reactive species under alkaline suspension of BiOBr [43]. A likely explanation of our results could be the competition between O_2 and PDS for e_{cb}^-. The e_{cb}^- capture by O_2 can be more favorable for BiOBr than BiOI due to the different conduction band potentials [19]; thus, O_2 can effectively compete with PDS for e_{cb}^-. Another important factor could be the surface charge, which is negative for all BiOX to varying degrees depending on the halogen atom [22,60] and probably less favorable for the surface reactions of negatively charged ions, such as PDS for BiOCl and BiOBr than BiOI. The different surface properties of BiOX were also well reflected in the amount of adsorbed SMP.

2.4. Effect of PDS on the Transformation of the Target Substances

Without irradiation, the organic substances do not transform; the direct oxidation with PDS is negligible in each case. The positive effect of PDS was pronounced in the case of HQ not only for BiOI but also for BiOCl and BiOBr photocatalysts (Figure 7); however, the PDS transformation was efficient without an organic substrate only for BiOI (Figure 6). In the case of BiOI, the enhanced transformation is most likely caused by the formed $SO_4^{\bullet-}$. However, this interpretation needs to be revised for BiOBr and BiOCl.

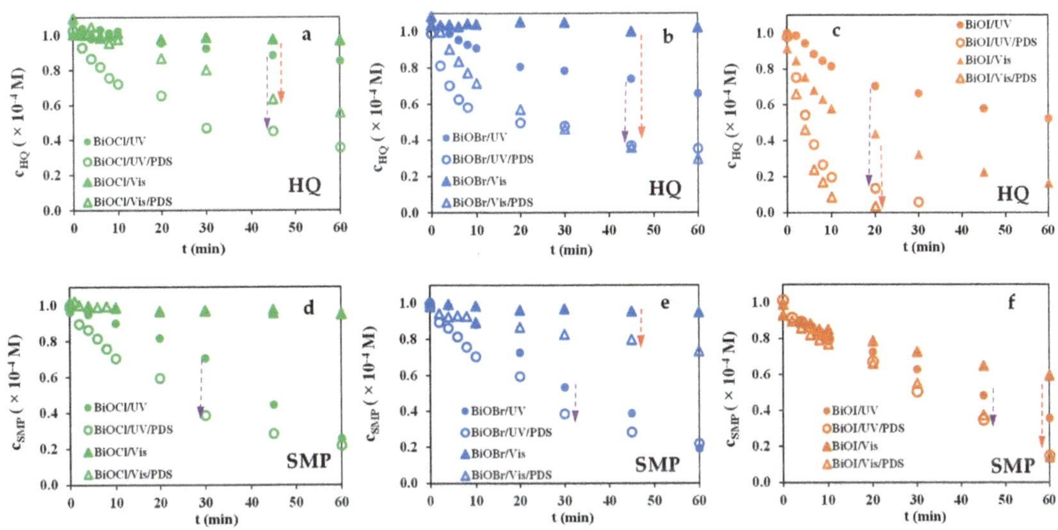

Figure 7. The effect of 1.0×10^{-3} M PDS on the transformation of HQ (**a**–**c**) and SMP (**d**–**f**) in UV and Vis irradiated, aerated BiOCl (**a**,**d**), BiOBr (**b**,**e**), and BiOI (**c**,**f**) suspensions. (The narrows show the effect of PDS (purple: UV; red: Vis).

The transformation of HQ results in BQ via semiquinone radical ($HBQ^\bullet$), which also can form via the comproportionation between BQ and HQ. The PDS can undergo reductive transformation into $SO_4^{\bullet-}$ by $HBQ^\bullet$ and enhance the transformation rate in this way [61–63]. However, the authors interpret the transformation of various organic substances in the presence of PDS or PMS differently. Fang et al. [62] reported that the $SO_4^{\bullet-}$ is responsible for the enhanced 2,4,4′-trichlorobiphenyl transformation in the case of $HBQ^\bullet$ activated PDS. Similarly, Zhou et al. [61] activated PMS by $HBQ^\bullet$ for the degradation of sulfamethoxazole. The $^\bullet OH$ or $SO_4^{\bullet-}$ were not detected, and $^1O_2$ was found to be responsible for sulfamethoxazole transformation. Similarly, Bu et al. [43] reported that the photon-initiated oxygen vacancies, Bi^{5+} and Bi^0, are responsible for the PDS transformation and cause enhanced activity of BiOBr in the presence of PDS. The Bisphenol-A transformation was justified by the formation of $^1O_2$, but the contribution of $SO_4^{\bullet-}$ was not confirmed.

Comparing the effect of PDS on HQ degradation for BiOX catalysts, BiOI is the most advantageous; the HQ is wholly transformed within 20 min. For BiOBr and BiOCl, 75% conversion was reached during 60 min even under UV radiation. Their efficiency was far exceeded by BiOI, even under Vis radiation. Most likely, in the case of BiOCl and BiOBr, mainly $HBQ^\bullet$ initiated the transformation of PDS is responsible for the enhanced efficiency. In addition, the photon-initiated oxygen vacancies, the surficial Bi^{5+} and Bi^0 may have a role. However, estimating the contribution of individual effects is beyond the scope of our work and requires several additional measurements.

The effect of PDS on SMP conversion is moderated compared to that observed for HQ (Figure 7), where the $HBQ^\bullet$-induced transformation of PDS also contributes. For BiOCl and BiOBr catalysts, the increased efficiency was observed only under UV radiation, confirming the crucial role of oxygen vacancies in activating PDS. The SMP transformation was enhanced at the beginning of the treatment; after that, the transformation slowed down. For BiOI, the positive effect of PDS was pronounced under UV and Vis radiation. After 60 min of Vis irradiation, the 40% conversion increased to more than 90% (Figure 7), while the SMP conversion increased by 10% under UV radiation. However, the conversion efficiency of SMP was still much lower than that of HQ when the $HBQ^\bullet$-initiated process also contributed to the enhanced efficiency.

The effect of PDS on TRIM transformation was investigated using BiOI under Vis radiation, and a slight increase in efficiency was observed; nearly 24% of TRIM was transformed (Figure S1). However, this measurement also confirms that the activity of BiOX catalysts is highly dependent on the compound to be converted, even in the presence of PDS. Moreover, the well-adsorbed TRIM can partially inhibit PDS activation on the surface.

Although $SO_4^{\bullet-}$ is effective for mineralization, significant TOC changes were not observed during the treatments. TOC did not decrease during the 60-min treatment without PDS, while a maximum decrease of 10% was measured in the presence of PDS for HQ and SMP. The pH of BiOX suspension was about 6.0–6.5, the addition of target substances did not significantly change that. During treatment, the pH decreased slightly (to 5.0–4.0 without PDS, and to 4.5–3.7 in the presence of PDS), likely because of the formed organic acids. Probably, pH change has no significant effect on PDS transformation and $SO_4^{\bullet-}$ initiated reactions [64]; however, it can affect the surface properties of the photocatalysts and the $HO_2/O_2^{\bullet-}$ ratio during treatment.

2.5. *Effect of Dissolved Oxygen*

PDS can affect the conversion rate of organic compounds in heterogeneous photocatalysis in many ways, such as forming new reactive species ($SO_4^{\bullet-}$, $^1O_2$) and enhancing the charge separation due to the inhibition of the recombination of photogenerated charges [34,36,65]. To investigate this phenomenon, measurements were performed in O_2-free suspensions (Figure 8).

While the conversion rates of BQ and SMP in O_2-free suspensions without PDS were negligible, the degradation proceeded efficiently in the presence of PDS (Figure 8). A significant difference is that in the case of HQ, the positive effect of dissolved O_2 is well manifested even in the presence of PDS. In contrast, for SMP, the transformation rate is not affected by O_2. In the case of HQ, O_2 plays an important role, as $O_2^{\bullet-}$ is highly efficient in its conversion, while for SMP, the direct charge transfer was assumed as the primary process. The PDS probably partially takes over the electron capture role of O_2 in the case of BiOI. Still, the positive effect of the formed $SO_4^{\bullet-}$ overcompensates this adverse effect, and both reactive species can contribute significantly to the HQ transformation in aerated suspension. In the case of SMP, the lack of an O_2 effect suggests that there is no contribution of $^1O_2$ or $O_2^{\bullet-}$ to the transformation in the presence of PDS and confirms the additive role of the highly reactive $SO_4^{\bullet-}$, the formation of which is not related to the dissolved O_2.

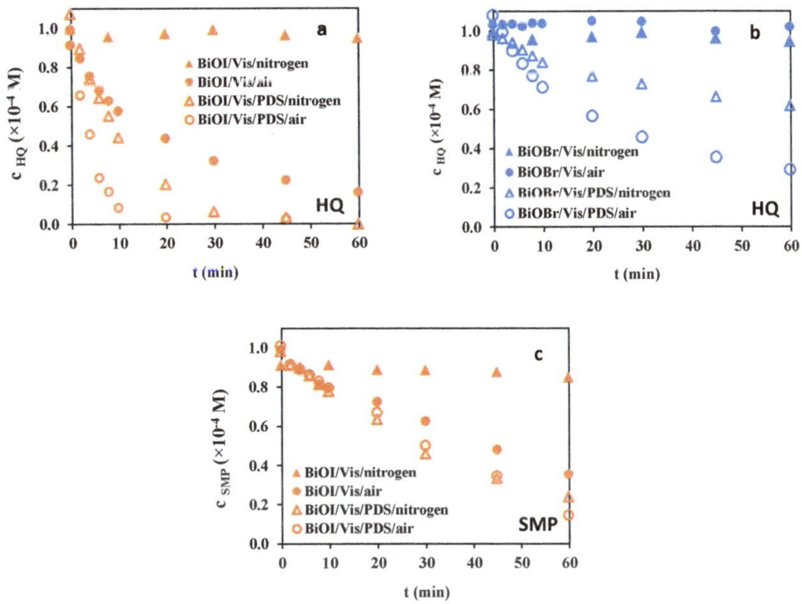

Figure 8. The Effect of dissolved O_2 on the transformation of HQ (**a**,**b**) and SMP (**c**).

2.6. Reusability of the BiOI Photocatalyst

The study of the stability and reusability of the catalyst has a crucial role in practical application. BiOI, the most efficient photocatalyst under Vis radiation, was chosen to investigate reusability. Another reason was that the efficient transformation of PDS without organic substances was proved only in the case of BiOI.

Degradation of $1.0 \times 10-4$ M HQ and $5.0 \times 10-5$ M SMP was monitored for three consecutive cycles (Figure 9), under Vis radiation, using $5.0 \times 10-4$ M PDS concentration. After the end of a cycle, the concentration was adjusted to $1.0 \times 10-4$ M for HQ and $5.0 \times$ 10-5 M for SMP by adding a small volume of concentrated solution. $5.0 \times 10-4$ M PDS was added to the suspension at the beginning of each run.

The transformation rate of both target substances did not change during the three consecutive cycles (Figure 9). Slightly decreased activity may be due to products' accumulation and competition with HQ and SMP for reactive species. There is no significant change in the XRD patterns and DRS spectra after the third cycle (Figures S2 and S3); thus, there is probably no change in the photocatalyst structure. Based on the ion chromatography measurements, the SO_4^{2-} concentration at the end of the cycle was 2.77×10^{-4} M, 5.23×10^{-4} M, and 7.80×10^{-4} M, so the extent of PDS decomposition was almost constant (1.4×10^{-4} M, 1.2×10^{-4} M, and 1.3×10^{-4} M PDS decomposed during the first, second and third cycle), confirming the unchanged activity of the BiOI photocatalyst.

2.7. Effect of Inorganic Ions and Biologically Treated Domestic Wastewater as Matrix

The effect of matrix and matrix components is important for practical applicability; thus, pH, Cl^- (3.4×10^{-3} M), HCO_3^- (8.6×10^{-3} M), and biologically treated domestic wastewater on the efficiency was investigated (Figure 10). The concentration of Cl^- and HCO_3^- was adjusted to their concentration in the biologically treated wastewater. Table S1 shows the parameters of the matrix.

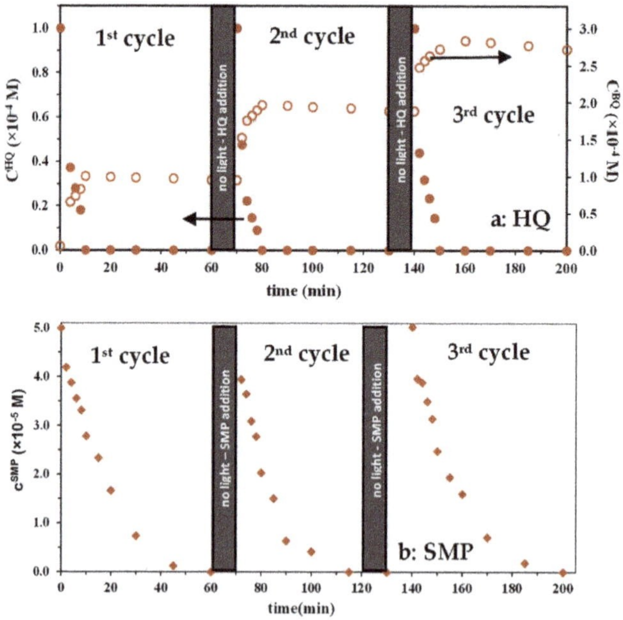

Figure 9. The transformation of HQ and formation of BQ (**a**) and the transformation of SMP (**b**) during three consecutive cycles in aerated BiOI suspensions, irradiated with Vis LED (c_0(HQ) = 1.0×10^{-4} M; c_0(SMP) = 5.0×10^{-5} M; c_0(PDS) = 5.0×10^{-4} M).

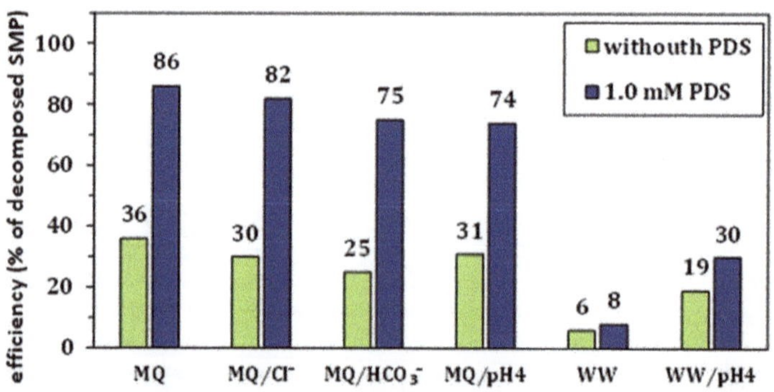

Figure 10. The effect of biologically treated domestic wastewater (ww), pH, Cl^- (3.4×10^{-3} M), and HCO_3^- (8.6×10^{-3} M) on the transformation of SMP (c_0(SMP) = 1.0×10^{-4} M; c_0(PDS) = 1.0×10^{-3} M) in BiOI suspension irradiated with Vis LED.

Biologically treated domestic wastewater almost wholly inhibited the transformation of both HQ and SMP, even in the presence of PDS. The matrix components may affect the surface properties of the photocatalyst, such as the surface charge and adsorption properties, or act as radical scavengers [66]. The effect of Cl^- (3.4×10^{-3} M) was negligible (Figure 10), even in PDS containing suspension; opposite that, Cl^- can scavenge $SO_4^{\bullet-}$ (7) [67]. Nevertheless, the formed Cl-containing reactive species (8–10) [68] can contribute to the transformation of organic substances, and their further reactions can even result in $^{\bullet}OH$ [69]. Consequently, the radical scavenging effect is not manifested in the decrease in the transformation rate [27].

$$Cl^- + SO_4^{\bullet-} \rightarrow Cl^{\bullet} + SO_4^{2-} \qquad k = 3.6 \times 10^8\ M^{-1}\ s^{-1} \tag{7}$$

$$Cl^{\bullet} + Cl^- \rightarrow Cl_2^{\bullet-} \qquad k = 7.8 \times 10^9\ M^{-1}\ s^{-1} \tag{8}$$

$$Cl^{\bullet}/Cl_2^{\bullet-} + H_2O \rightarrow ClOH^{\bullet-} + H^+/+ Cl^- \tag{9}$$

$$ClOH^{\bullet-} \rightarrow {}^{\bullet}OH + Cl^- \tag{10}$$

The HCO_3^- (8.6×10^{-3} M) can also react with $SO_4^{\bullet-}$ [70] and results in $CO_3^{\bullet-}$ (11), a less reactive and selective species [71]:

$$HCO_3^- + SO_4^{\bullet-} \rightarrow CO_3^{\bullet-} + SO_4^{2-} + H^+ \qquad k= 2.8 \times 10^6\ M^{-1}\ s^{-1} \tag{11}$$

HCO_3^- only slightly reduced the transformation rate (Figure 10), but the intensive orange color of the catalysts became white. XRD and DRS results proved that BiOI transforms into bismuth subcarbonate ($(BiO)_2CO_3$) (Figures S2–S4). The band gap of the formed, white material was 3.12 eV, close to the value reported for $(BiO)_2CO_3$ [72]. Surprisingly, however, this was not reflected in the cessation of activity since $(BiO)_2CO_3$ is also photoactive [73,74]. The Bi^{3+} entering the solution during the recrystallization could also activate the transformation of PDS. The harmful effect of HCO_3^- can be eliminated by changing the pH to 4. The change of the pH to 4 with sulfuric acid caused a slight decrease in the transformation rate compared to that determined in Milli-Q water. In biologically treated domestic wastewater, the transformation rate increased due to the elimination of HCO_3^- at acidic pH, but it was still significantly below the value measured in Milli-Q water (Figure 10). Probably the reason for this is not the effect of inorganic ions but organic compounds that are well adsorbed on the surface of BiOI. The TOC content of the wastewater decreased by 35% due to the addition of BiOI in the dark because of the adsorption of organic components. If they cannot transform, they block the transformation of the poorly adsorbed HQ, SMP, and PDS, and even the excitation of the BiOI. This assumption was confirmed by the observation; that the bleaching of BiOI in the matrix takes place much more slowly (Figure S4). This can also be interpreted by the adsorption of the organic components of the matrix on the surface, as they exert a protective effect against inorganic ions, i.e., the HCO_3^-.

3. Materials and Methods

3.1. Materials

Bismuth nitrate pentahydrate ($Bi(NO_3)_3 \times 5\ H_2O$), NaI, NaCl and NaBr, ethylene glycol, and ethanol were used for synthesizing photocatalysts. For photocatalytic test reactions, 1,4-hydroquinone (HQ), sulfamethoxypyridazine (SMP), and trimethoprim (TRIM) were used as target substances. Air or N_2 gas was applied to control the dissolved O_2 concentration of the treated solution or suspension. For investigation of the effect of peroxydisulfate ion, $Na_2S_2O_8$ (PDS) solution was added to the suspension. The pH was adjusted with H_2SO_4 or NaOH solutions. $Fe_2(SO_4)_3$, potassium oxalate, ammonium-reineckate salt, and 1,10-phenanthroline were used for the actinometric measurements. Table S2 shows the data of the materials used in this work. Each material was used without further purification. Table S1 shows the date of the biologically treated domestic wastewater (from the water treatment plant, Szeged, Hungary) used as a matrix.

3.2. Preparation of BiOX Photocatalysts

The BiOX photocatalysts were prepared via a solvothermal method, as described in the literature [13]. The $Bi(NO_3)_3 \times 5\ H_2O$ (2.07 g) and KI (0.7067 g), KBr (0.5066 g), or KCl (0.3174 g) were dissolved in 50 mL ethylene glycol with continuous stirring and heating (up to 45 °C). The suspension was heat-treated for 3 h at 120 °C in a PTFE-coated steel autoclave. The solid material was washed with distilled water and ethanol, then vacuum-filtered with a 0.1 μm pore size filter (Durapore®, hydrophilic PVDF) and dried for 24 h at 40 °C.

3.3. Photocatalytic Test Reactions

Two commercial LED tapes were used as light sources: the UV-LED emitting at 398 nm (LEDmaster, $\lambda_{emission}$= 398 ± 10 nm, 288 lumens, 4.6 W) and Vis-LED emitting warm white light (LEDmaster, $\lambda_{emission}$= 400–700 nm, 600 lumens, 4.6 W). 1.0 m of LED tape (60 LED/meter) was fixed on the inside of the aluminum, double-walled reactor having a 66 mm inner diameter (Figure 11a,b). The reactors were equipped with a water-cooling system to ensure the LEDs' constant light output. The electrical power required to operate the LEDs was the same (4.6 W) in all cases; thus, the efficiency of the photocatalysts was determined at the same electrical energy input.

Figure 11. The schematic figure (**a**) and photo (**b**) of the reactors and the emission spectra of the LED light sources (**c**).

In each case, 200 cm^3 suspension was irradiated in a cylindrical glass reactor (inner diameter: 40 mm) placed in the center of the double-walled reactor. The suspensions were stirred in the dark for 30 min to determine the amount of adsorbed target substance. O_2 or N_2 gas was continuously bubbled through the solution or suspension to keep constants the dissolved O_2 concentration; the bubbling was started 15 min before starting the reaction. The experiments were started by turning on the light source and adding PDS solution to the suspension containing photocatalysts and organic target substance. The initial concentration of HQ, BQ, SMP, and TRIM was 5.0×10^{-5} M or 1.0×10^{-4} M, while the BiOI photocatalyst dosage was 1.0 g dm^{-3}. The concentration of PDS was changed within the range of 0.2–2.0 $\times$ 10^{-3} M.

After sampling, 25 µL 0.3 M $Na_2S_2O_3$ solution was added to the sample to decompose the remaining PDS. The photocatalyst samples were centrifuged immediately (Dragonlab, 15,000 RPM) and filtered with syringe filters (0.22 µm, FilterBiO, PVDF-L) before further analysis.

3.4. Characterization of the Light Sources

The emission spectra of the LEDs (Figure 11c) were recorded using a two-channel fiber-optic CCD spectrometer (AvaSpec-FT2048) in the 180–880 nm wavelength range. The photon flux of the light sources was determined using Reinecke's salt [75,76] and the widely applied ferrioxalate [77,78] actinometry. Reinecke's salt actinometry can be applied in the visible range, in the 400–600 nm region, while ferrioxalate actinometry can be used in the UV and near-UV regions (254–500 nm) [78].

The UV LED's photon flux determined by Reinecke's salt actinometry was $(5.81 \pm 0.03) \times 10^{-6}$ mol_{photon} s^{-1}, and a slightly lower value, $(5.12 \pm 0.02) \times 10^{-6}$ mol_{photon} s^{-1} was obtained by ferrioxalate actinometry [77]. The photon flux was $(3.25 \pm 0.25) \times 10^{-6}$ mol_{photon} s^{-1} for the Vis-LEDs, obtained by Reinecke's salt actinometry. For calculating the apparent quantum yield of the transformation, the photon flux determined by Reinecke's salt actinometry was applied.

3.5. Characterization of Photocatalysts

The synthesized catalysts were characterized using powder X-ray diffractometry (XRD) (Rigaku Miniflex II, Cu Kα radiation source, 5.0–90.0 2Theta° range, with 4.0 2Theta° min^{-1} resolution). The surface area was determined via N_2 adsorption/desorption isotherms using a Quantachrome NOVA 2200 analyzer. Diffuse reflectance spectroscopy (DRS) was performed using an Ocean Optics DH-2000 light source and Ocean Optics USB4000 detector. The band gap energy values were evaluated by the Kubelka-Munk approach and the Tauc plot [79] and reinforced by the first derivative approach method [80].

3.6. Analytical Methods

HPLC measurements were performed with an Agilent 1100 HPLC equipped with a diode array UV detector (DAD) to separate the intermediates and determine 1,4-HQ, 1,4-BQ, SMP, and TRIM concentration in the treated suspension. The stationary phase was a Kinetex 2.6u XB-C18 100A (Phenomenex) reverse phase column, and the mobile phase consisted of 10 $v/v\%$ acetonitrile and 90% formic acid solution (0.1 $v/v\%$) for the analysis of SMP containing samples. The flow rate of eluent was 0.70 mL min^{-1}. The same column was used for BQ and HQ containing samples; the mobile phase consisted of 50% methanol and 50% water. The flow rate of eluent was 1.0 mL min^{-1}. To analyze TRIM-containing solutions, Gemini 3u C6-phenyl 110A reverse phase column was used with 10% methanol and 90% ammonium-formate buffer (20 mM, pH = 3.7) as eluent; the flow rate was 0.4 mL min^{-1}. The detection wavelength was 250, 290, 261, and 275 nm for BQ, HQ, SMP, and TRIM, respectively. The transformation efficiency of target substances was characterized by the initial degradation rate and percent of decomposed material after 10 and 60 min treatment. The initial transformation rate was obtained from linear regression fit to the concentration-time plot, generally up to 20% transformation. Some experiments were repeated three times to check the reproducibility of the experimental results.

The transformation of PDS was followed by the determination of SO_4^{2-}. The SO_4^{2-} concentration was measured using ion chromatography (Shimadzu Prominence LC-20AD). Shodex 5U-YS-50 column for cation detection (eluent contained 4.0 mM methanesulfonic acid and 2.5 mM phthalic acid), and Shodex NI-424 5U for anion detection (eluent: 2.3 mM aminomethane solution). The flow rate of the mobile phase was 1.0 mL min^{-1}.

The total organic carbon (TOC) content of the samples was determined by a Multi N/C 3100 analyzer (Analytik Jena, Jena, Germany) equipped with an NDIR detector. For spectrophotometric measurements, an Agilent 8453 UV-Vis spectrophotometer was used.

4. Conclusions

In this work, the effect of PDS on the transformation of various organic substances was investigated, in BiOCl, BiOBr, and BiOI suspensions, under UV (398 nm) and Vis radiation. The method described in the literature was used for the synthesis of the BiOX catalysts; their modification was not aimed. The parameters of the catalysts corresponded to the literature data; accordingly, 398 nm UV light is theoretically not suitable for the excitation of BiOCl; BiOBr can be excited mainly by UV light, while BiOI can also be efficiently excited by visible light. Due to PDS being an effective e_{cb}^- scavenger and the source of sulfate radical ion ($SO_4^{\bullet -}$), an enhanced transformation rate of organic compounds was expected.

Without organic matter, BiOI showed outstanding activity in PDS transformation; 15–20% of PDS was transformed even under Vis radiation, while for BiOBr and BiOCl only 3–4% of PDS transformed under 398 nm irradiation. The increased efficiency of HQ transformation in the presence of PDS can be partially attributed to PDS activation by $HBQ^{\bullet}$-this process contributed dominantly to the activation of PDS and the increase of HQ conversion for BiOCl and BiOBr, and additively for BiOI. BiOI showed the highest activity, especially under visible light radiation, most probably due to its outstanding activity in the activation of PDS among the tested BiOXs. For BiOBr and BiOCl, 75% conversion was reached during 60 min even under UV radiation in the presence of PDS. The efficiency was far exceeded by BiOI, even under Vis radiation; the HQ wholly transformed within 20 min. Similarly, for BiOI the SMP conversion increased to more than 90% from 40%. However, the conversion efficiency of SMP was still much lower than that of HQ when the $HBQ^{\bullet}$-initiated process also contributed to the enhanced efficiency.

Our results reflect that the activity of BiOX catalysts, through various factors, significantly depends on the model compound used-both in the presence and without PDS. For

BiOCl and BiOBr, most likely, photogenerated oxygen-vacancies and Bi^0 play an important role and influence the activity. In O_2-free, PDS-containing suspensions, the efficient transformation of HQ and SMP confirmed that the improved activity is primarily related to the enhanced efficiency of direct charge transfer and $SO_4^{\bullet-}$ formation.

The structure and activity of BiOI did not change during three consecutive cycles in the presence of PDS; however, in the biologically treated wastewater, the organic compounds of the matrix adsorbed on the BiOI surface blocked the processes taking place and significantly reduced the efficiency. Furthermore, BiOI was sensitive to HCO_3^-.

Our results proved that combining the PDS process with BiOI photocatalyst is a promising process that can be used for the selective conversion and degradation of organic pollutants since its effectiveness depends significantly on the properties of the compound to be removed. Some critical limiting factors in terms of efficiency and applicability are highlighted. In addition, many questions arose relating to the processes taking place on the surface of BiOX catalysts, as well as the formation and role of different reactive species, which motivate further research to obtain a more comprehensive picture of the activation of PDS with BiOX photocatalyst.

Supplementary Materials: The following supporting information can be downloaded at: https://www.mdpi.com/article/10.3390/catal13030513/s1, Table S1. The parameters of the biologically treated domestic wastewater; Table S2. The name, manufacturer, and purity of the used materials; Figure S1. The transformation of TRIM in aerated BiOI suspensions, irradiated with Vis LED (c0(TRIM) = 1.0 × 10−4 M; c0(SMP) = 1.0 × 10−4 M; NaF was added for desorption); Figure S2. The Effect of PDS, inorganic ions, and SMP on the XRD diffractogram of BiOI photocatalysts under UV and Vis radiation in aerated suspensions; Figure S3. The Effect of PDS and inorganic ions on the DRS spectra diffractogram of BiOI photocatalysts under UV and Vis radiation in aerated suspensions; Figure S4. The effect of pH and inorganic ions on the color of BiOI suspension (a: Milli-Q; b: biologically treated domestic wastewater, pH = 7.8; c: biologically treated domestic wastewater, pH = 4; d: HCO3− solution; e: the pH of the HCO3−solution was adjusted to 4 and stirred for 10 min. before suspending the BiOI).

Author Contributions: Conceptualization, writing—original draft preparation, review and editing, funding acquisition: T.A.; investigation: B.V., M.N. and L.F.; XRD and DRS measurements and data evaluation: Z.P.; visualization and TOC measurement, A.C. All authors have read and agreed to the published version of the manuscript.

Funding: This work was financed by the National Research, Development, and Innovation Office—NKFI Fund OTKA, project number FK132742. Tünde Alapi, thanks for the support of the János Bolyai Research Scholarship of the Hungarian Academy of Sciences. Luca Farkas, thanks for the financial support of the ÚNKP-22-3-SZTE-398 New National Excellence Program of the Ministry for Culture and Innovation from the source of the National Research, Development, and Innovation Fund.

Data Availability Statement: The data is included in the article or Supplementary Material.

Acknowledgments: The authors thank the Szegedi Vízmű Zrt. for providing the biologically treated domestic wastewater as a matrix.

Conflicts of Interest: The funders had no role in the design of the study; in the collection, analyses, or interpretation of data; in the writing of the manuscript; or in the decision to publish the results.

References

1. Morales-Paredes, C.A.; Rodríguez-Díaz, J.M.; Boluda-Botella, N. Pharmaceutical compounds used in the COVID-19 pandemic: A review of their presence in water and treatment techniques for their elimination. *Sci. Total Environ.* **2022**, *814*, 152691. [CrossRef] [PubMed]
2. Rizvi, S.G.; Ahammad, S.Z. COVID-19 and antimicrobial resistance: A cross-study. *Sci. Total Environ.* **2022**, *807*, 150873. [CrossRef] [PubMed]
3. Cipriani-Avila, I.; Molinero, J.; Cabrera, M.; Medina-Villamizar, E.J.; Capparelli, M.V.; Jara-Negrete, E.; Pinos-Velez, V.; Acosta, S.; Andrade, D.L.; Barrado, M.; et al. Occurrence of emerging contaminants in surface water bodies of a coastal province in Ecuador and possible influence of tourism decline caused by COVID-19 lockdown. *Sci. Total Environ.* **2023**, *866*, 161340. [CrossRef] [PubMed]

4. Thangadurai, P.; Beura, R.; Kumar, J.S. *Nanomaterials with Different Morphologies for Photocatalysis*; Springer: Berlin/Heidelberg, Germany, 2020; ISBN 9783540228608.
5. Friedmann, D. A General Overview of Heterogeneous Photocatalysis as a Remediation Technology for Wastewaters Containing Pharmaceutical Compounds. *Water* **2022**, *14*, 3588. [CrossRef]
6. Shurbaji, S.; Huong, P.T.; Altahtamouni, T.M. Review on the Visible Light Photocatalysis for the Decomposition of Ciprofloxacin, Norfloxacin, Tetracyclines, and Sulfonamides Antibiotics in Wastewater. *Catalysts* **2021**, *11*, 437. [CrossRef]
7. You, J.; Guo, Y.; Guo, R.; Liu, X. A review of visible light-active photocatalysts for water disinfection: Features and prospects. *Chem. Eng. J.* **2019**, *373*, 624–641. [CrossRef]
8. Djurišić, A.B.; He, Y.; Ng, A.M.C. Visible-light photocatalysts: Prospects and challenges. *APL Mater.* **2020**, *8*, 030903. [CrossRef]
9. Chen, R.; Fan, F.; Li, C. Unraveling Charge-Separation Mechanisms in Photocatalyst Particles by Spatially Resolved Surface Photovoltage Techniques. *Angew. Chem. Int. Ed.* **2022**, *61*, e202117567. [CrossRef]
10. Lv, X.; Lam, F.L.Y.; Hu, X. A Review on Bismuth Oxyhalide (BiOX, X=Cl, Br, I) Based Photocatalysts for Wastewater Remediation. *Front. Catal.* **2022**, *2*, 839072. [CrossRef]
11. Cheng, H.; Huang, B.; Dai, Y. Engineering BiOX (X = Cl, Br, I) nanostructures for highly efficient photocatalytic applications. *Nanoscale* **2014**, *6*, 2009. [CrossRef]
12. Ye, L.; Su, Y.; Jin, X.; Xie, H.; Zhang, C. Recent advances in BiOX (X = Cl, Br and I) photocatalysts: Synthesis, modification, facet effects and mechanisms. *Environ. Sci. Nano* **2014**, *1*, 90. [CrossRef]
13. Bárdos, E.; Király, A.K.; Pap, Z.; Baia, L.; Garg, S.; Hernádi, K. The effect of the synthesis temperature and duration on the morphology and photocatalytic activity of BiOX (X = Cl, Br, I) materials. *Appl. Surf. Sci.* **2019**, *479*, 745–756. [CrossRef]
14. Bárdos, E.; Márta, V.; Baia, L.; Todea, M.; Kovács, G.; Baán, K.; Garg, S.; Pap, Z.; Hernadi, K. Hydrothermal crystallization of bismuth oxybromide (BiOBr) in the presence of different shape controlling agents. *Appl. Surf. Sci.* **2020**, *518*, 146184. [CrossRef]
15. Chen, X.; Chen, P.; Yang, S.; Gao, H. Recent advances in bismuth oxyhalides photocatalysts and their applications. *Nanotechnology* **2023**, *34*, 052001. [CrossRef]
16. Singh, S.; Sharma, R.; Khanuja, M. A review and recent developments on strategies to improve the photocatalytic elimination of organic dye pollutants by BiOX (X=Cl, Br, I, F) nanostructures. *Korean J. Chem. Eng.* **2018**, *35*, 1955–1968. [CrossRef]
17. Ren, X.; Gao, M.; Zhang, Y.; Zhang, Z.; Cao, X.; Wang, B.; Wang, X. Photocatalytic reduction of CO2 on BiOX: Effect of halogen element type and surface oxygen vacancy mediated mechanism. *Appl. Catal. B Environ.* **2020**, *274*, 119063. [CrossRef]
18. Meng, X.; Zhang, Z. New insight into BiOX (X = Cl, Br, and I) hierarchical microspheres in photocatalysis. *Mater. Lett.* **2018**, *225*, 152–156. [CrossRef]
19. Lv, J.; Hu, Q.; Cao, C.; Zhao, Y. Modulation of valence band maximum edge and photocatalytic activity of BiOX by incorporation of halides. *Chemosphere* **2018**, *191*, 427–437. [CrossRef]
20. Zhang, L.; Jaroniec, M. Fundamentals of adsorption for photocatalysis. *Interface Sci. Technol.* **2020**, *31*, 39–62. [CrossRef]
21. Wu, L.; Zhang, Q.; Li, Z.; Liu, X. Mechanochemical syntheses of a series of bismuth oxyhalide composites to progressively enhance the visible-light responsive activities for the degradation of bisphenol-A. *Mater. Sci. Semicond. Process.* **2020**, *105*, 104733. [CrossRef]
22. Xiao, X.; Liu, C.; Zuo, X.; Liu, J.; Nan, J. Microwave Synthesis of Hierarchical BiOCl Microspheres as a Green Adsorbent for the pH-Dependent Adsorption of Methylene Blue. *J. Nanosci. Nanotechnol.* **2016**, *16*, 12517–12525. [CrossRef]
23. Zhang, Y.; Zhang, J.; Tang, W.; Wang, N.; Liu, Y.; Guo, H. Preparation and characterization of hierarchical BiO0.5Cl0.5 with excellent adsorption and photocatalytic abilities for removal of aquatic dyes. *Desalin. WATER Treat.* **2020**, *201*, 356–368. [CrossRef]
24. Liu, Z.; Wang, Q.; Tan, X.; Wang, Y.; Jin, R.; Gao, S. Enhanced photocatalytic performance of TiO2 NTs decorated with chrysanthemum-like BiOI nanoflowers. *Sep. Purif. Technol.* **2019**, *215*, 565–572. [CrossRef]
25. Lashuk, B.; Yargeau, V. A review of ecotoxicity reduction in contaminated waters by heterogeneous photocatalytic ozonation. *Sci. Total Environ.* **2021**, *787*, 147645. [CrossRef] [PubMed]
26. Wolfrum, E.J.; Ollis, D.F. Hydrogen Peroxide in Heterogeneous Photocatalysis. In *Aquatic and Surface Photochemistry*; CRC Press: Boca Raton, FL, USA, 2018; pp. 451–466. ISBN 9781351069847.
27. Karpel Vel Leitner, N. Sulfate radical ion—Based AOPs. In *Advanced Oxidation Processes for Water Treatment*; Stefan, M.I., Ed.; IWA Publishing: London, UK, 2017; pp. 429–460.
28. Lee, J.; von Gunten, U.; Kim, J.-H.H. Persulfate-Based Advanced Oxidation: Critical Assessment of Opportunities and Roadblocks. *Environ. Sci. Technol.* **2020**, *54*, 3064–3081. [CrossRef] [PubMed]
29. Wacławek, S.; Lutze, H.V.; Grübel, K.; Padil, V.V.T.; Černík, M.; Dionysiou, D.D. Chemistry of persulfates in water and wastewater treatment: A review. *Chem. Eng. J.* **2017**, *330*, 44–62. [CrossRef]
30. Ike, I.A.; Linden, K.G.; Orbell, J.D.; Duke, M. Critical review of the science and sustainability of persulphate advanced oxidation processes. *Chem. Eng. J.* **2018**, *338*, 651–669. [CrossRef]
31. Wang, J.; Wang, S. Activation of persulfate (PS) and peroxymonosulfate (PMS) and application for the degradation of emerging contaminants. *Chem. Eng. J.* **2018**, *334*, 1502–1517. [CrossRef]
32. Honarmandrad, Z.; Sun, X.; Wang, Z.; Naushad, M.; Boczkaj, G. Activated persulfate and peroxymonosulfate based advanced oxidation processes (AOPs) for antibiotics degradation—A review. *Water Resour. Ind.* **2023**, *29*, 100194. [CrossRef]
33. Ge, M.; Hu, Z.; Wei, J.; He, Q.; He, Z. Recent advances in persulfate-assisted TiO2-based photocatalysis for wastewater treatment: Performances, mechanism and perspectives. *J. Alloys Compd.* **2021**, *888*, 161625. [CrossRef]

34. He, S.; Chen, Y.; Li, X.; Zeng, L.; Zhu, M. Heterogeneous Photocatalytic Activation of Persulfate for the Removal of Organic Contaminants in Water: A Critical Review. *ACS EST Eng.* **2022**, *2*, 527–546. [CrossRef]
35. Wang, H.; Li, X.; Zhao, X.; Li, C.; Song, X.; Zhang, P.; Huo, P.; Li, X. A review on heterogeneous photocatalysis for environmental remediation: From semiconductors to modification strategies. *Chinese J. Catal.* **2022**, *43*, 178–214. [CrossRef]
36. Tian, D.; Zhou, H.; Zhang, H.; Zhou, P.; You, J.; Yao, G.; Pan, Z.; Liu, Y.; Lai, B. Heterogeneous photocatalyst-driven persulfate activation process under visible light irradiation: From basic catalyst design principles to novel enhancement strategies. *Chem. Eng. J.* **2022**, *428*, 131166. [CrossRef]
37. Hasija, V.; Nguyen, V.-H.H.; Kumar, A.; Raizada, P.; Krishnan, V.; Khan, A.A.P.; Singh, P.; Lichtfouse, E.; Wang, C.; Thi Huong, P. Advanced activation of persulfate by polymeric g-C3N4 based photocatalysts for environmental remediation: A review. *J. Hazard. Mater.* **2021**, *413*, 125324. [CrossRef] [PubMed]
38. Chen, H.; Zhang, X.; Jiang, L.; Yuan, X.; Liang, J.; Zhang, J.; Yu, H.; Chu, W.; Wu, Z.; Li, H.; et al. Strategic combination of nitrogen-doped carbon quantum dots and g-C3N4: Efficient photocatalytic peroxydisulfate for the degradation of tetracycline hydrochloride and mechanism insight. *Sep. Purif. Technol.* **2021**, *272*, 118947. [CrossRef]
39. John, D.; Jose, J.; Bhat, S.G.; Achari, V.S. Integration of heterogeneous photocatalysis and persulfate based oxidation using TiO2-reduced graphene oxide for water decontamination and disinfection. *Heliyon* **2021**, *7*, e07451. [CrossRef] [PubMed]
40. Peng, J.; Wang, Z.; Wang, S.; Liu, J.; Zhang, Y.; Wang, B.; Gong, Z.; Wang, M.; Dong, H.; Shi, J.; et al. Enhanced removal of methylparaben mediated by cobalt/carbon nanotubes (Co/CNTs) activated peroxymonosulfate in chloride-containing water: Reaction kinetics, mechanisms and pathways. *Chem. Eng. J.* **2021**, *409*, 128176. [CrossRef]
41. Qi, C.; Liu, X.; Ma, J.; Lin, C.; Li, X.; Zhang, H. Activation of peroxymonosulfate by base: Implications for the degradation of organic pollutants. *Chemosphere* **2016**, *151*, 280–288. [CrossRef]
42. Qi, C.; Wen, Y.; Zhao, Y.; Dai, Y.; Li, Y.; Xu, C.; Yang, S.; He, H. Enhanced degradation of organic contaminants by Fe(III)/peroxymonosulfate process with l-cysteine. *Chinese Chem. Lett.* **2022**, *33*, 2125–2128. [CrossRef]
43. Bu, Y.; Li, H.; Yu, W.; Pan, Y.; Li, L.; Wang, Y.; Pu, L.; Ding, J.; Gao, G.; Pan, B. Peroxydisulfate Activation and Singlet Oxygen Generation by Oxygen Vacancy for Degradation of Contaminants. *Environ. Sci. Technol.* **2021**, *55*, 2110–2120. [CrossRef]
44. Kohantorabi, M.; Moussavi, G.; Oulego, P.; Giannakis, S. Radical-based degradation of sulfamethoxazole via UVA/PMS-assisted photocatalysis, driven by magnetically separable Fe3O4@CeO2@BiOI nanospheres. *Sep. Purif. Technol.* **2021**, *267*, 118665. [CrossRef]
45. Tong, Y.; Zhou, P.; Liu, Y.; Wang, N.; Li, W.; Cheng, F.; Yang, B.; Liang, J.; Zhang, Y.; Lai, B. Strongly enhanced Fenton-like oxidation (Fe/peroxydisulfate) by BiOI under visible light irradiation: A novel and green strategy for Fe(III) reduction. *J. Hazard. Mater.* **2022**, *428*, 128202. [CrossRef] [PubMed]
46. Lv, Y.; Liu, Y.; Wei, J.; Li, M.; Xu, D.; Lai, B. Bisphenol S degradation by visible light assisted peroxymonosulfate process based on BiOI/B4C photocatalysts with Z-scheme heterojunction. *Chem. Eng. J.* **2021**, *417*, 129188. [CrossRef]
47. Ganose, A.M.; Cuff, M.; Butler, K.T.; Walsh, A.; Scanlon, D.O. Interplay of Orbital and Relativistic Effects in Bismuth Oxyhalides: BiOF, BiOCl, BiOBr, and BiOI. *Chem. Mater.* **2016**, *28*, 1980–1984. [CrossRef] [PubMed]
48. Náfrádi, M.; Hernadi, K.; Kónya, Z.; Alapi, T. Investigation of the efficiency of BiOI/BiOCl composite photocatalysts using UV, cool and warm white LED light sources—Photon efficiency, toxicity, reusability, matrix effect, and energy consumption. *Chemosphere* **2021**, *280*, 130636. [CrossRef] [PubMed]
49. Fónagy, O.; Szabó-Bárdos, E.; Horváth, O. 1,4-Benzoquinone and 1,4-hydroquinone based determination of electron and superoxide radical formed in heterogeneous photocatalytic systems. *J. Photochem. Photobiol. A Chem.* **2021**, *407*, 113057. [CrossRef]
50. Zamadar, M.; Cook, A.R.; Lewandowska-Andralojc, A.; Holroyd, R.; Jiang, Y.; Bikalis, J.; Miller, J.R. Electron Transfer by Excited Benzoquinone Anions: Slow Rates for Two-Electron Transitions. *J. Phys. Chem. A* **2013**, *117*, 8360–8367. [CrossRef] [PubMed]
51. Rahimi, S.; Ayati, B.; Rezaee, A. Kinetic modeling and determination role of sono/photo nanocatalyst-generated radical species on degradation of hydroquinone in aqueous solution. *Environ. Sci. Pollut. Res.* **2016**, *23*, 12185–12198. [CrossRef]
52. Rao, P.S.; Hayon, E. Experimental determination of the redox potential of the superoxide radical $\bullet O_2^-$. *Biochem. Biophys. Res. Commun.* **1973**, *51*, 468–473. [CrossRef]
53. Boreen, A.L.; Arnold, W.A.; McNeill, K. Triplet-Sensitized Photodegradation of Sulfa Drugs Containing Six-Membered Heterocyclic Groups: Identification of an SO_2 Extrusion Photoproduct. *Environ. Sci. Technol.* **2005**, *39*, 3630–3638. [CrossRef]
54. Mezyk, S.P.; Neubauer, T.J.; Cooper, W.J.; Peller, J.R. Free-Radical-Induced Oxidative and Reductive Degradation of Sulfa Drugs in Water: Absolute Kinetics and Efficiencies of Hydroxyl Radical and Hydrated Electron Reactions. *J. Phys. Chem. A* **2007**, *111*, 9019–9024. [CrossRef] [PubMed]
55. Náfrádi, M.; Alapi, T.; Farkas, L.; Bencsik, G.; Kozma, G.; Hernádi, K. Wavelength Dependence of the Transformation Mechanism of Sulfonamides Using Different LED Light Sources and TiO_2 and ZnO Photocatalysts. *Materials* **2021**, *15*, 49. [CrossRef] [PubMed]
56. Fang, Z.; Qi, J.; Xu, Y.; Liu, Y.; Qi, T.; Xing, L.; Dai, Q.; Wang, L. Promoted generation of singlet oxygen by hollow-shell CoS/g-C3N4 catalyst for sulfonamides degradation. *Chem. Eng. J.* **2022**, *441*, 136051. [CrossRef]
57. Zhang, G.; Cai, L.; Zhang, Y.; Wei, Y. Bi^{5+}, $Bi^{(3-x)+}$, and Oxygen Vacancy Induced $BiOCl_x I_{1-x}$ Solid Solution toward Promoting Visible-Light Driven Photocatalytic Activity. *Chem. A Eur. J.* **2018**, *24*, 7434–7444. [CrossRef]
58. He, Q.; Ge, M.; Yu, Q. Synergistic degradation of tetracycline by BiOBr microspheres combined with peroxydisulfate under visible LED light irradiation. *J. Chem. Sci.* **2021**, *133*, 98. [CrossRef]

59. Liu, C.; Mao, S.; Wang, H.; Wu, Y.; Wang, F.; Xia, M.; Chen, Q. Peroxymonosulfate-assisted for facilitating photocatalytic degradation performance of 2D/2D WO3/BiOBr S-scheme heterojunction. *Chem. Eng. J.* **2022**, *430*, 132806. [CrossRef]
60. Zhao, Q.; Xing, Y.; Liu, Z.; Ouyang, J.; Du, C. Synthesis and Characterization of Modified BiOCl and Their Application in Adsorption of Low-Concentration Dyes from Aqueous Solution. *Nanoscale Res. Lett.* **2018**, *13*, 69. [CrossRef] [PubMed]
61. Zhou, Y.; Jiang, J.; Gao, Y.; Ma, J.; Pang, S.-Y.; Li, J.; Lu, X.-T.; Yuan, L.-P. Activation of Peroxymonosulfate by Benzoquinone: A Novel Nonradical Oxidation Process. *Environ. Sci. Technol.* **2015**, *49*, 12941–12950. [CrossRef]
62. Fang, G.; Gao, J.; Dionysiou, D.D.; Liu, C.; Zhou, D. Activation of Persulfate by Quinones: Free Radical Reactions and Implication for the Degradation of PCBs. *Environ. Sci. Technol.* **2013**, *47*, 4605–4611. [CrossRef]
63. Ahmad, M.; Teel, A.L.; Watts, R.J. Mechanism of Persulfate Activation by Phenols. *Environ. Sci. Technol.* **2013**, *47*, 5864–5871. [CrossRef] [PubMed]
64. Herrera-Ordonez, J. The role of sulfate radicals and pH in the decomposition of persulfate in aqueous medium: A step towards prediction. *Chem. Eng. J. Adv.* **2022**, *11*, 100331. [CrossRef]
65. Xia, X.; Zhu, F.; Li, J.; Yang, H.; Wei, L.; Li, Q.; Jiang, J.; Zhang, G.; Zhao, Q. A Review Study on Sulfate-Radical-Based Advanced Oxidation Processes for Domestic/Industrial Wastewater Treatment: Degradation, Efficiency, and Mechanism. *Front. Chem.* **2020**, *8*, 592056. [CrossRef]
66. Náfrádi, M.; Alapi, T.; Bencsik, G.; Janáky, C. Impact of Reaction Parameters and Water Matrices on the Removal of Organic Pollutants by TiO2/LED and ZnO/LED Heterogeneous Photocatalysis Using 365 and 398 nm Radiation. *Nanomaterials* **2021**, *12*, 5. [CrossRef] [PubMed]
67. Zuo, Z.; Katsumura, Y.; Ueda, K.; Ishigure, K. Reactions between some inorganic radicals and oxychlorides studied by pulse radiolysis and laser photolysis. *J. Chem. Soc. Faraday Trans.* **1997**, *93*, 1885–1891. [CrossRef]
68. Yu, X.-Y.; Barker, J.R. Hydrogen Peroxide Photolysis in Acidic Aqueous Solutions Containing Chloride Ions. I. Chemical Mechanism. *J. Phys. Chem. A* **2003**, *107*, 1313–1324. [CrossRef]
69. De Laat, J.; Stefan, M.I. UV/Chlorine process in Advanced Oxidation Processes for Water Treatment: Fundamentals and Applications. In *Vanced Oxidation Processes for Water Treatment*; Stefan, M.I., Ed.; IWA Publishing: London, UK, 2017; pp. 383–424.
70. Huie, R.E.; Clifton, C.L. Temperature dependence of the rate constants for reactions of the sulfate radical, SO4-, with anions. *J. Phys. Chem.* **1990**, *94*, 8561–8567. [CrossRef]
71. Wojnárovits, L.; Tóth, T.; Takács, E. Rate constants of carbonate radical anion reactions with molecules of environmental interest in aqueous solution: A review. *Sci. Total Environ.* **2020**, *717*, 137219. [CrossRef]
72. Ni, Z.; Sun, Y.; Zhang, Y.; Dong, F. Fabrication, modification and application of $(BiO)_2CO_3$-based photocatalysts: A review. *Appl. Surf. Sci.* **2016**, *365*, 314–335. [CrossRef]
73. Xu, C.; Qiu, P.; Li, L.; Chen, H.; Jiang, F.; Wang, X. Bismuth Subcarbonate with Designer Defects for Broad-Spectrum Photocatalytic Nitrogen Fixation. *ACS Appl. Mater. Interfaces* **2018**, *10*, 25321–25328. [CrossRef]
74. Tang, J.; Cheng, G.; Zhou, H.; Yang, H.; Lu, Z.; Chen, R. Shape-Dependent Photocatalytic Activities of Bismuth Subcarbonate Nanostructures. *J. Nanosci. Nanotechnol.* **2012**, *12*, 4028–4034. [CrossRef]
75. Wegner, E.E.; Adamson, A.W. Photochemistry of Complex Ions. III. Absolute Quantum Yields for the Photolysis of Some Aqueous Chromium(III) Complexes. Chemical Actinometry in the Long Wavelength Visible Region. *J. Am. Chem. Soc.* **1966**, *88*, 394–404. [CrossRef]
76. Cornet, J.-F.J.F.; Marty, A.; Gros, J.B.J.-B. Revised Technique for the Determination of Mean Incident Light Fluxes on Photobioreactors. *Biotechnol. Prog.* **1997**, *13*, 408–415. [CrossRef]
77. Hatchard, C.G.; Parker, C.A. A new sensitive chemical actinometer—II. Potassium ferrioxalate as a standard chemical actinometer. *Proc. R. Soc. Lond. Ser. A Math. Phys. Sci.* **1956**, *235*, 518–536. [CrossRef]
78. Kuhn, H.J.; Braslavsky, S.E.; Schmidt, R. International UNION OF pure and applied chemistry—Chemical actinometry. *IUPAC Tech. Rep.* **2004**, *76*, 2105–2146.
79. Tauc, J. Optical properties and electronic structure of amorphous Ge and Si. *Mater. Res. Bull.* **1968**, *3*, 37–46. [CrossRef]
80. Flak, D.; Braun, A.; Mun, B.S.; Park, J.B.; Parlinska-Wojtan, M.; Graule, T.; Rekas, M. Spectroscopic assessment of the role of hydrogen in surface defects, in the electronic structure and transport properties of TiO_2, ZnO and SnO_2 nanoparticles. *Phys. Chem. Chem. Phys.* **2013**, *15*, 1417–1430. [CrossRef]

catalysts

MDPI

Article

Undoped and Fe-Doped Anatase/Brookite TiO_2 Mixed Phases, Obtained by a Simple Template-Free Synthesis Method: Physico-Chemical Characterization and Photocatalytic Activity towards Simazine Degradation

Stefano Gervasi [1,†], Nicola Blangetti [1,†], Francesca S. Freyria [1,*], Salvatore Guastella [1] and Barbara Bonelli [1,2,*]

1 Department of Applied Science and Technology and INSTM Unit of Torino-Politecnico, Corso Duca degli Abruzzi 24, I-10129 Torino, Italy
2 Interdepartmental Centre Polito[BIO]Med Lab, Corso Duca degli Abruzzi 24, I-10129 Torino, Italy
* Correspondence: francesca.freyria@polito.it (F.S.F.); barbara.bonelli@polito.it (B.B.); Tel.: +39-011-0904719 (B.B.)
† These authors contributed equally to this work.

Abstract: For the first time, Fe-doping (0.05, 1.0, and 2.5 wt.% Fe) was performed on a high-surface-area anatase/brookite TiO_2 by adopting a simple template-free sol-gel synthesis followed by calcination at a mild temperature. The powders' textural and surface properties were characterized by following a multi-technique approach. XRD analysis showed that the anatase/brookite ratio slightly varied in the Fe-doped TiO_2 (from 76.9/23.1 to 79.3/22.7); Fe doping noticeably affected the cell volume of the brookite phase, which decreased, likely due to Fe^{3+} ions occupying interstitial positions, and retarded the crystallite growth. N_2 sorption at −196 °C showed the occurrence of samples with disordered interparticle mesopores, with an increase in the specific surface area from 236 $m^2\ g^{-1}$ (undoped TiO_2) to 263 $m^2\ g^{-1}$ (2.5 wt.% Fe). Diffuse Reflectance UV-Vis spectroscopy showed a progressive decrease in the bandgap energy from 3.10 eV (undoped TiO_2) to 2.85 eV (2.5 wt.% Fe). XPS analysis showed the presence of some surface Fe species only at 2.5 wt.% Fe, and accordingly, the ζ-potential measurements showed small changes in the pH at the isoelectric point. The photocatalytic degradation of simazine (a persistent water contaminant) both under UV and simulated solar light was performed as a probe reaction. Under UV light, Fe-doping improved simazine degradation in the sample at 0.05 wt.% Fe, capable of degrading ca. 77% simazine. Interestingly, the undoped TiO_2 was also active both under UV and 1 SUN. This is likely due to the occurrence of anatase/brookite heterojunctions, which help stabilize the photogenerated electrons/holes.

Keywords: TiO_2 polymorphs; anatase/brookite heterojunctions; Fe-doped TiO_2; simazine degradation; detoxification

1. Introduction

Along with ZnO, TiO_2 is one of the most studied semiconductors, with actual and potential applications in various fields. These applications include photocatalytic processes, during which the following events take place upon TiO_2 interaction with the light of proper energy [1,2]:

$$TiO_2 + h\nu \rightarrow e^-_{CB} + h^+_{VB} \quad (1)$$

$$e^-_{CB} + O_2 + H^+ \rightarrow HO_2\bullet \quad (2)$$

$$e^-_{CB} + HO_2\bullet + H^+ \rightarrow H_2O_2 \quad (3)$$

$$2e^-_{CB} + O_2 + 2H^+ \rightarrow H_2O_2 \quad (4)$$

$$h^+_{VB} + H_2O \rightarrow HO\bullet + H^+ \quad (5)$$

$$2HO\bullet \rightarrow H_2O_2 \quad (6)$$

$$H_2O_2 + h\nu \rightarrow 2HO\bullet \quad (7)$$

$$H_2O_2 + e^-_{CB} + H^+ \rightarrow HO\bullet + H_2O \quad (8)$$

According to Reactions (1)–(8), valence band (VB) electrons are promoted to the conduction band (CB, e^-_{CB}), and simultaneously holes are produced in the VB (h^+_{VB}): the so photogenerated charge carriers (e^-_{CB} and h^+_{VB}) undergo several reactions producing (highly oxidizing) radical species (e.g., HO•).

From a practical point of view, photocatalytic applications of TiO_2 are usually hampered by two major drawbacks: (i) the fast recombination of photogenerated charge carriers [3], which decreases the overall yield [4,5] and (ii) the poor absorption of the solar spectrum, which contains only a minor fraction of UV light (ca. 4%), whereas the three most common polymorphs of TiO_2, namely anatase, brookite, and rutile, have average bandgap energies (E_g) in the UV range (of ca. 3.2, 3.4 and 3.0 eV, respectively [6,7]). By tailoring the semiconductor properties, different strategies can be adopted to cope with such issues: for instance, (i) the production of composites containing heterojunctions between TiO_2 and other substances [8,9] or between TiO_2 polymorphs [10,11] can improve the lifetime of photogenerated charge carriers, and (ii) the doping with heteroatoms can red shift the TiO_2 absorption edge and improve its absorption of solar light [12,13].

Concerning the presence of inter-polymorph heterojunctions to stabilize the charge carriers in undoped TiO_2, anatase/rutile heterojunctions are generally considered responsible for the superior photocatalytic activity of Degussa P25, the commercial product used as a benchmark in most studies [14,15]. Currently, there is a growing interest in anatase/brookite heterojunctions [16–18]. Indeed, we have observed that such heterojunctions can be responsible for the fair photocatalytic activity of an anatase/brookite TiO_2 mixed phase towards the degradation of the herbicide N-phenylurea, both under UV and simulated solar light [11].

Concerning doping with heteroatoms, Fe is an earth-abundant, non-toxic element, and can be successfully introduced to different types of materials, such as alumino-silicates, due to the similar charge to radius ratio of Fe^{3+} and Al^{3+} [19–21]. Concerning semiconductors, the effects of Fe doping are exploited, for instance, in ZnO and TiO_2. With ZnO, several effects are observed, such as the formation of Fe^{2+} species, the segregation of Zn/Fe mixed spinels, and the occurrence of interesting optical and magnetic properties [22–24]. Concerning the effect of Fe doping in TiO_2, which is the subject of this study, it is known to shift the absorption edge towards the visible range [25–28] by forming new levels within the TiO_2 bandgap (Scheme 1).

In the bulk, the Fe species may also act as traps for the photogenerated e^-_{CB} and h^+_{VB} (Scheme 1), avoiding their recombination and giving rise to the following reactions [29]:

$$Fe^{3+} + e^-_{CB} \rightarrow Fe^{2+} \quad (9)$$

$$Fe^{2+} + Ti^{4+} \rightarrow Fe^{3+} + Ti^{3+} \quad (10)$$

$$Fe^{3+} + h^+_{VB} \rightarrow Fe^{4+} \quad (11)$$

$$Fe^{4+} + OH^- \rightarrow Fe^{3+} + HO\bullet \quad (12)$$

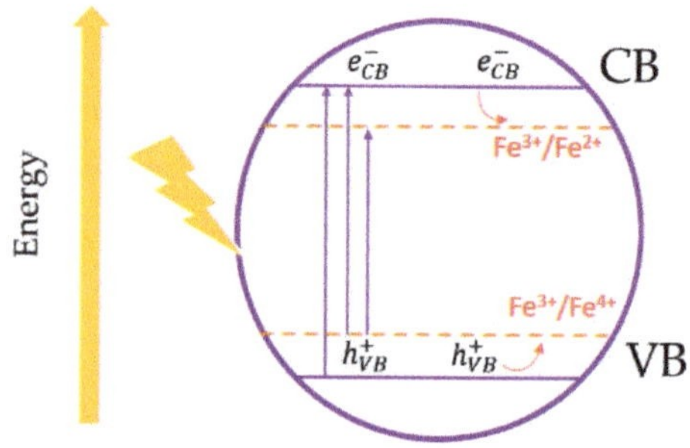

Scheme 1. Processes occurring in Fe-doped TiO_2 upon interaction with light (adapted from Ref. [29]).

In the presence of TiO_2, the photogenerated h^+_{VB} can react with Fe^{3+} species to produce (unstable) Fe^{4+} species, which can provide additional HO• species, according to Equations (11) and (12).

Moreover, Fe-species on the TiO_2 surface in the presence of H_2O_2 can undergo Fenton- and Fenton-like processes, well known in the homogeneous phase but possible in the heterogeneous phase as well. In such processes, both Fe^{3+} (Equation (13)) and Fe^{2+} (Equation (14)) ions may react with H_2O_2 to form radicals [30,31], which further react with H_2O_2 according to the following reactions that can be exploited to degrade, for instance, water pollutants:

$$Fe^{3+} + H_2O_2 \rightarrow Fe^{2+} + H^+ + HO_2\bullet \quad (13)$$

$$Fe^{2+} + H_2O_2 \rightarrow Fe^{3+} + OH^- + HO\bullet \quad (14)$$

$$HO\bullet + H_2O_2 \rightarrow H_2O + HO_2\bullet \quad (15)$$

$$HO_2\bullet + H_2O_2 \rightarrow HO\bullet + H_2O + O_2 \quad (16)$$

$$\text{Pollutant} + HO\bullet \rightarrow \text{degradationproducts} \quad (17)$$

Typically, Fe^{3+} species reacts more slowly (Equation (13)) than Fe^{2+} species (Equation (14)) [32], but in the presence of light with $\lambda < 600$ nm, the photo-Fenton process takes place, during which Fe^{3+} species is reduced to (more active) Fe^{2+} species, yielding HO• radicals [20,33]:

$$Fe(OH)^{2+} + h\nu \rightarrow Fe^{2+} + HO\bullet \quad (18)$$

$$Fe^{3+} + H_2O + h\nu \rightarrow Fe^{2+} + HO\bullet + H^+ \quad (19)$$

$$H_2O_2 + h\nu \rightarrow 2HO\bullet \quad (20)$$

Photocatalysis and Fenton-like reactions belong to the class of the so-called Advanced Oxidations Processes (AOPs), i.e., those methods that, through the production of very oxidizing species, allow degradation of several organic (and inorganic) substances, yielding by-products or mineralization [1,2]. By these methods, it would be possible to also treat persistent pollutants and/or micropollutants (present at very low concentrations), which can escape traditional water treatment plants [31,34–36], such as simazine (Scheme 2). Simazine is a herbicide of the triazine family. Although it has been banned since 1991 in Italy and Germany [37] and since 2004 in most of the other European countries [38], Simazine is still detected in some surface and ground waters, as traditional water treatment plants are unable to remove it. Recently, along with atrazine, a pesticide of the same family (Scheme 2), simazine has been banned in Hawaii and other U.S. territories [39], as several reports pointed out that it is a potential endocrine disruptor, which can lead to reproductive disorders and be toxic to humans and wildlife [38].

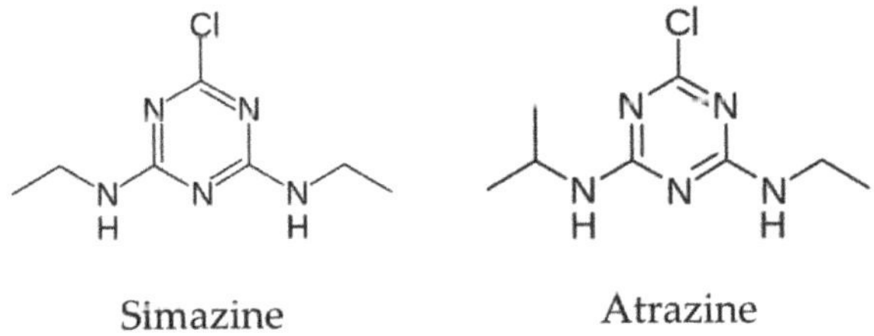

Scheme 2. The chemical structures of simazine (IUPAC name: 6-chloro-N2,N4-diethyl-1,3,5-triazine-2,4-diamine) and atrazine (IUPAC name: 6-Chloro-N2-ethyl-N4-(propan-2-yl)-1,3,5-triazine-2,4-diamine).

Due to its chemical stability and persistence [40], simazine is the second most detected herbicide/pesticide in waters of the Western world. Concerning the distribution of the pollutant in Europe, in 2017, the analysis of over 7000 sites including both surface and groundwater all over Europe showed that simazine was the third among herbicides/pesticides detected in surface waters and the second among those detected in groundwater [41]. The persistence of simazine in soils or a natural habit can make it a dangerous pollutant. Indeed, it is considered a pollutant of environmental concern [42], due to its persistence, stability, environmental distribution, very high toxicity for aquatic life and suspected carcinogenicity.

Concerning non-reactive methods for simazine removal, adsorption on widely available, low-cost adsorbents [43,44] suffers from the drawback of their regeneration. Concerning reactive methods, different solutions have been exploited. For instance, Sun et al. [45] studied the chemical degradation of simazine by diatomite-supported NZVI (Nanoscale Zero Valent Iron). For the photocatalytic removal of simazine through AOPs, only a few studies imply the use of visible or solar light and engineered composites of graphene-like materials [46,47], whereas most studies implying the use of TiO_2 are carried out under UV light [48–50].

In this work, for the first time, we synthesized and characterized samples of Fe-doped anatase/brookite high-surface-area mesoporous TiO_2 through a simple procedure based on pH control followed by calcination at a mild temperature (namely 200 °C); an undoped anatase/brookite (AB sample) TiO_2 was also obtained for comparison by slightly modifying a literature procedure [51]. On the one side, the adopted synthesis is more sustainable than other methods, implying the use of organic templates and the need for high calcination temperature to obtain high-surface-area mesoporous TiO_2 [10,52–55]; on the other side, it allows obtaining an anatase/brookite mixed phase that we already proved active under both UV and solar illumination [11]. To the best of our knowledge, it is the first time such a method [11,51] has been applied to obtain Fe-doped TiO_2: by adding proper amounts of $FeCl_3 \times 6H_2O$, three nominal Fe-contents have been studied, corresponding to 0.05, 1.0, and 2.5 wt.% Fe in the AB_0.05_Fe, AB_1.0_Fe, and AB_2.5_Fe samples, respectively. After the physico-chemical characterization of their textural and surface properties, the photocatalytic degradation of simazine under UV and solar light was used as a probe reaction in different conditions, namely the absence/presence of H_2O_2 in stoichiometric amount, as too high H_2O_2 concentration may lead to adverse effects, by consuming HO• radicals and producing less reactive HO_2• radicals (Equation (15)) [55].

2. Results and Discussion

2.1. Physico-Chemical Characterization

Figure 1a reports the X-ray Diffraction (XRD) patterns of the studied samples, and the corresponding results of the Quantitative Phase Analysis (QPA) is reported in Table 1. The undoped AB sample shows the peaks of both anatase (A, main peaks at ca. 25.2 and ca. 48.1 2θ values, ICDD card no. 01-073-1764) and brookite (B, main peaks at ca. 25.2, 25.5 and 30.6 2θ values, COD card No 96-900-4141), with a relative abundance of 77.7 wt.% and 22.3 wt.%, respectively. Doping with Fe does not lead to the appearance of other phases besides anatase and brookite nor to a consistent change in their abundance (Table 1), likely due to the low Fe content [52–55].

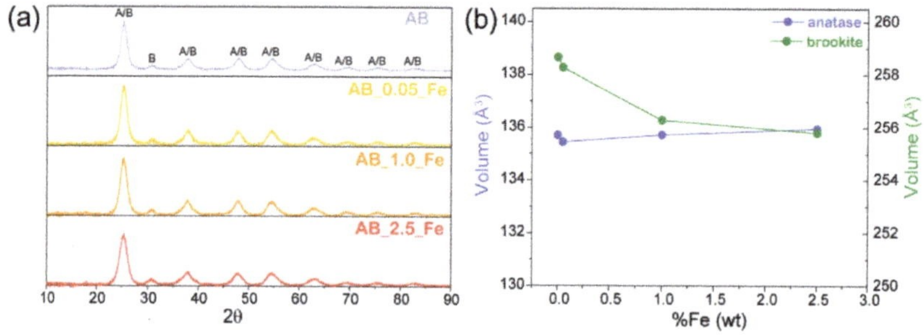

Figure 1. XRD patterns of the studied powder samples. The main peaks of anatase and brookite are labelled as A and B, respectively (**a**); values of the cell volumes of anatase and brookite versus the nominal Fe content (wt.%) (**b**).

Table 1. Some relevant physico-chemical properties of the studied samples, as obtained using XRD followed by Rietveld refinement [a] and Williamson-Hall method, [b] N_2 isotherms at −196 °C, [c] and DR UV-Vis spectroscopy. [d] A = anatase; B = brookite.

Sample	QPA Results: (Phase wt.%) [a]	Crystallite Size (±nm) [b]	BET SSA ($m^2\ g^{-1}$) [c]	V_{tot} ($cm^3\ g^{-1}$) [c]	Extrapolated/ Tauc's Plot Determined E_g Values (eV) [d]
AB	77.7 A 22.3 B	5.2 (0.2) A 7.1(0.1) B	236	0.293	3.30/ 3.10
AB_0.05_Fe	76.9 A 23.1 B	5.0 (0.1) A 5.6 (0.2) B	233	0.286	3.22/ 3.07
AB_1.0_Fe	79.3 A 20.7 B	4.7 (0.1) A 6.0 (0.2) B	243	0.253	3.10/ 2.92
AB_2.5_Fe	78.4 A 21.6 B	4.4 (0.1) A 5.7 (0.2) B	263	0.245	3.01/ 2.85

The broad peaks show that the AB powder has a low crystallinity degree, likely due to the low calcination temperature (namely 200 °C), and accordingly, the crystallite sizes are small (ca. 5.2 and 7.1 nm for anatase and brookite, respectively). As shown in Table 1, the crystallite size of both phases slightly decreases in the Fe-doped samples, in agreement with previous work on Fe-doped anatase/brookite TiO_2, where the presence of Fe^{3+} ions was found to retard the crystallite growth [56].

Figure 1b, instead, shows that the cell volume of brookite decreases with the Fe content, whereas the anatase cell volume is less affected by doping (Figure 1b) indicating that the effects of Fe-doping are more noticeable on the brookite phase. The crystallite size of both phases slightly decreases (Table 1), especially at the minimum amount of doping. Concerning the type of process when Fe^{3+} ions are inserted into the TiO_2 lattice, in a previous work on Fe-doped anatase/brookite/rutile TiO_2, using a reverse-micelle sol-gel method to incorporate Fe in both anatase and brookite [53], an initial expansion of both the anatase and brookite cell volumes occurred at 1.0 wt.% Fe, and then the cell volumes decreased at a Fe content above 1.0 wt.%. The former phenomenon was ascribed to the isomorphic substitution of Ti^{4+} by Fe^{3+} ions, while the latter was caused by the occupation of interstitial sites by (extra) Fe^{3+} ions. In this case, the peculiar synthesis procedure seems to help the latter process, which, in turn, mainly affects the brookite phase, characterized by a more disordered and open structure than anatase [7].

Figure 2 reports some selected FESEM micrographs of the studied powders: small nanoparticles are observed with four samples, which are all characterized by a high aggregation/agglomeration degree. The nanoparticles size and shape are mostly unaffected by the presence of Fe, and their size agrees with a previous microscopic characterization of an undoped anatase/brookite TiO_2, obtained by the same synthesis method of the AB sample, showing that the nanoparticles size is of the order of the crystallite size [11]. The corresponding EDX-determined elemental maps are reported in Figure S1 for the Fe-doped samples, showing a homogeneous distribution of Fe in the synthesized powders.

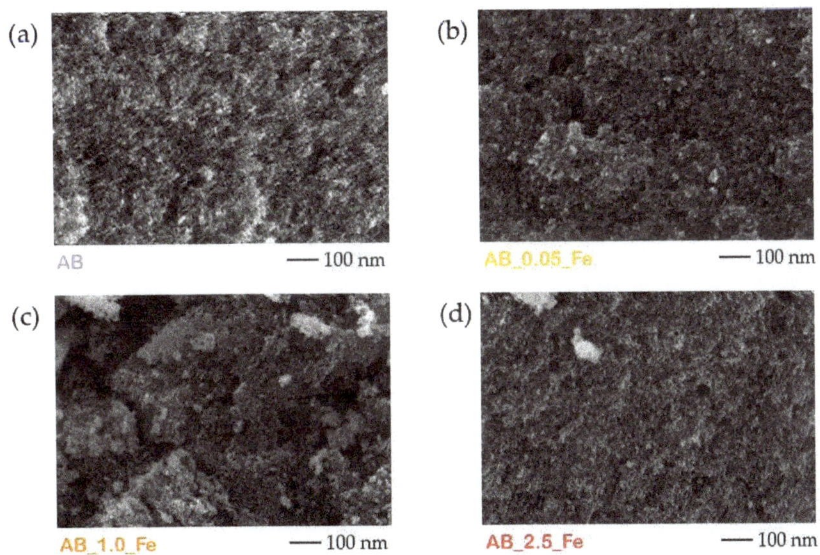

Figure 2. Selected FESEM micrographs of the AB (**a**), AB_0.05_Fe (**b**), AB_1.0_Fe (**c**) and AB_2.5_Fe (**d**) powder samples, showing nanoparticles with similar size and morphology, as well as a high degree of agglomeration/aggregation.

Figure 3 reports the N_2 adsorption/desorption isotherms measured at −196 °C on the studied samples: all the studied powders showed type IV isotherms with H2 hysteresis loops, indicating the occurrence of interconnected mesopores of different shapes and sizes. The adopted synthesis procedure likely leads to disordered inter-particle mesopores, as also shown by the Pore Size Distribution (PSD) curves reported in the insets. The AB, AB_1.0_Fe, and AB_2.5_Fe samples have a sharp PSD with a maximum at ca. 4.0 nm, whereas the AB_0.05_Fe sample has a broader PSD, likely due to some aggregation/agglomeration phenomena (Figure 2). The PSD curves show that the samples' pore diameters largely exceed simazine's molecular dimensions (ca. 1.1×0.75 nm [57]), thus diffusion problems during the photocatalytic tests should be avoided. Interestingly, the four powders are characterized by high specific surface area (SSA) values (Table 1). Except sample AB_0.05_Fe, which has an SSA very close to that of the (undoped) AB sample likely due to the low Fe content, the SSA is seen to increase with Fe doping.

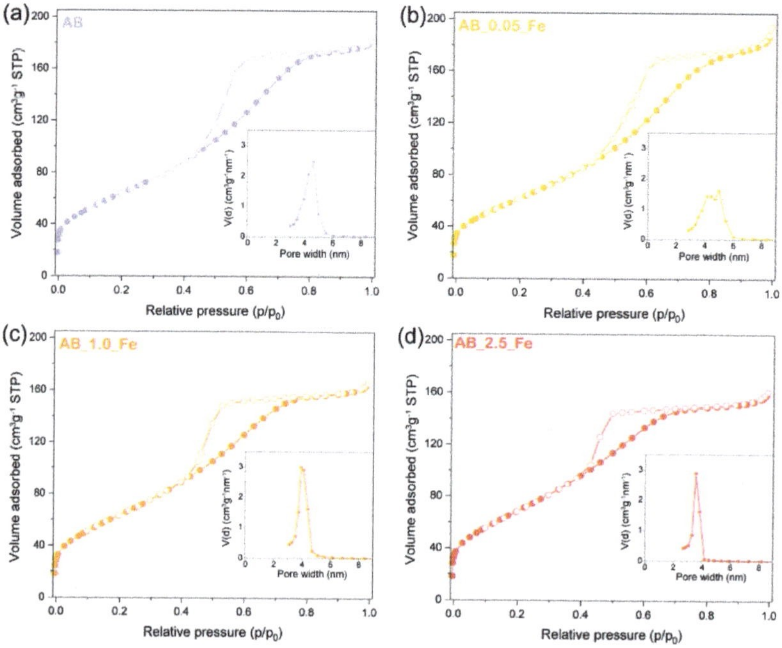

Figure 3. N_2 adsorption/desorption isotherms at − 196 °C on the AB (**a**), AB_0.05_Fe (**b**), AB_1.0_Fe (**c**) and AB_2.5_Fe (**d**) samples; full and empty symbols refer to adsorption and desorption run, respectively. Insets: PSD curves, as obtained by applying the BJH method to the isotherm's desorption branch.

Figure 4a reports the DR UV-Vis spectra of the studied samples: with the AB sample, the typical electronic transitions due to the charge transfer (CT) from O^{2-} to Ti^{4+} in undoped TiO_2 are observed, along with an Urbach tail (inset), which will be discussed later. With the AB_0.05_Fe sample, a band at ca. 235 nm can be assigned to CT from O^{2-} to Ti^{4} ions in an environment perturbed by the presence of Fe^{3+} ions [58]. Such an effect is not so noticeable with the other two powders and could be due to a stronger effect of Fe substitution on the TiO_2 matrix at low Fe content. Doping with Fe brings a red shift of the absorption onset, surely related to bulk Fe doping, in agreement with the reported XRD patterns, which also brings about a decrease of the energy gap, vide infra. The formation of interband states due to Fe doping can be appreciated in the UV-Vis spectra of the doped samples (inset to Figure 4a).

With Fe-doped TiO_2 obtained by different (template-assisted) sol-gel techniques, the formation of surface Fe-oxohydroxo cluster was observed, especially at 2.5 wt.% Fe. Correspondingly, XPS was able to detect the presence of some surface Fe species [53]. With the present samples, instead, XPS analysis detected the presence of surface Fe species only with the AB_2.5_Fe sample (Figure S2). As an Fe/Ti atomic ratio equal to 0.071 was measured by XPS, higher than the nominal value of 0.037, this indicates that at higher Fe content, some Fe oxyhydroxide species may occur; indeed, a signal was observed at ca. 480 nm (asterisk, Figure 4a), corresponding to a signal at 2.55 eV in the related Tauc's plot (asterisk, Figure 4b). XPS also detected some Cl traces (0.6 at.%; Cl/Fe atomic ratio of 0.46).

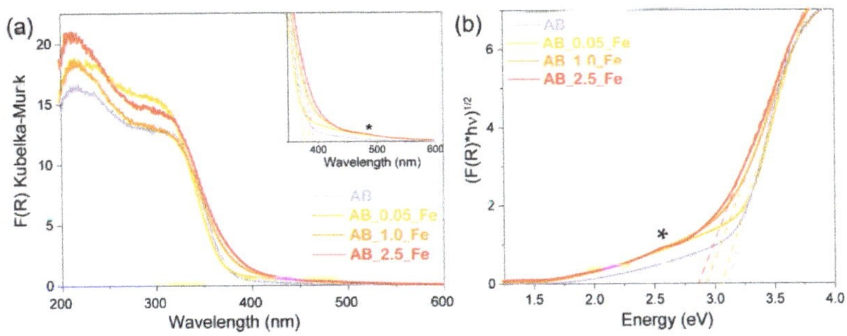

Figure 4. DR UV-Vis spectra of the studied powders (**a**) and corresponding Tauc's plot, as obtained by assuming a behaviour of indirect semiconductors (**b**). Asterisks at 480 nm (inset to Figure 4a) and at 2.55 eV (Figure 4b) refer to evidence of the formation of Fe oxyhydroxide species.

Concerning the bandgap energy, since the samples contain not only anatase (an indirect semiconductor) but also brookite (a direct semiconductor), the bandgap values (E_g) of the powders have been calculated both by applying the Tauc's plot method (Figure 4b) for an indirect semiconductor $((F(R) \times h\nu)^{1/2}$ and by extrapolation of the spectra absorption onset (dashed lines in Figure 4a). The obtained E_g values are reported in Table 1: slightly smaller values are obtained by Tauc's plot method, likely due to the indirect semiconductor assumption, but generally Fe brings about a decrease in the E_g values. Analysis of the Tauc's plot curves shows that the AB (undoped) sample is characterized by an Urbach tail, namely an exponential part in the energy spectrum of the absorption coefficient that may appear near the optical band edge in amorphous, disordered, and crystalline materials, including undoped TiO_2 mixed phases [59]. Figure S3 reports a comparison of the AB spectrum with a spectrum of Degussa P25, showing that the former sample extends its absorption towards longer wavelengths. Such phenomenon allows absorption of a broader range of light energy; therefore, it is useful in photocatalytic applications, especially under solar illumination (vide infra).

Figure 5 reports the ζ-potential curves of the studied powders in water: they are positively and negatively charged, respectively, above and below a pH of ca. 5.00–6.00. According to the literature, the pH at the isoelectric point (pH_{IEP}) of brookite and anatase is around 5.10 and 6.00, respectively [60], and Fe_2O_3 has pH_{IEP} values in the 5.80–6.20 range [61]. The pH_{IEP} of the undoped AB sample is very close to that of brookite, and the pH_{IEP} of the AB_1.0_Fe and AB_2.5_Fe samples is not much affected by Fe-doping, in agreement with the low Fe content. Concerning the AB_0.05_Fe sample, the slightly higher pH_{IEP} value may be due to the presence of a higher abundance of exposed anatase nanoparticles, due to a kind of heterogeneity induced by the agglomeration/aggregation phenomena, already evidenced by FESEM (Figure 2). Such agglomeration/aggregation phenomena may also occur in aqueous phase: indeed, according to the literature, stable nanoparticles suspension are characterized by values of surface charges above +30 mV and below −30 mV (dotted lines) [62,63], whereas the studied powders have lower (in absolute value) surface charges in a broad pH range.

The physico-chemical characterization reported so far showed that the adopted synthesis procedure led to the production of mesoporous anatase/brookite TiO_2 samples, occurring as highly agglomerated/aggregated nanoparticles and that the Fe-doping mainly occurred in the TiO_2 bulk, especially at low Fe content, with a more noticeable effect on the crystallographic parameters of the brookite phase.

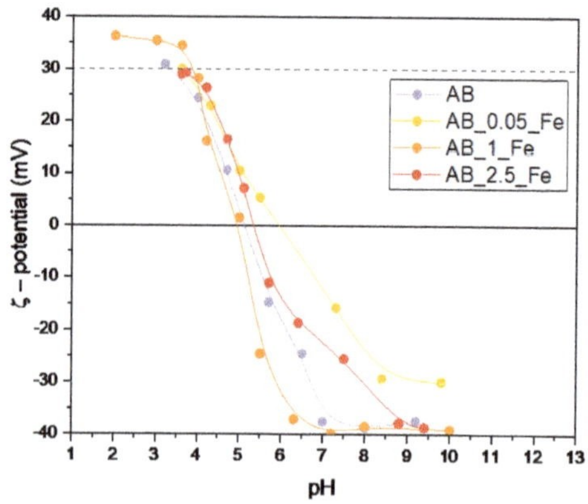

Figure 5. ζ-potential curves of the studied powders. The dashed lines at ζ = + 30 mV and −30 mV indicate the upper and lower values of surface charge above and below which stable suspensions of nanoparticles occur according to the literature [55,56].

2.2. Simazine Degradation under UV Light

Figure 6 reports the results of the photocatalytic tests obtained after 3 h illumination under UV light. The parent simazine solution (spectra taken at t = 0 min) has two bands at 222 and 265 nm, assigned to $\pi \rightarrow \pi^*$ and $n \rightarrow \pi^*$ electronic transitions, respectively [64]; under UV illumination, several new bands also form in the presence of the (undoped) AB sample, showing that the studied samples are photocatalytically active towards simazine degradation. Here, it is considered as a probe reaction, and the mechanisms of simazine degradation have already been addressed in the literature and will be used as a reference in this paper.

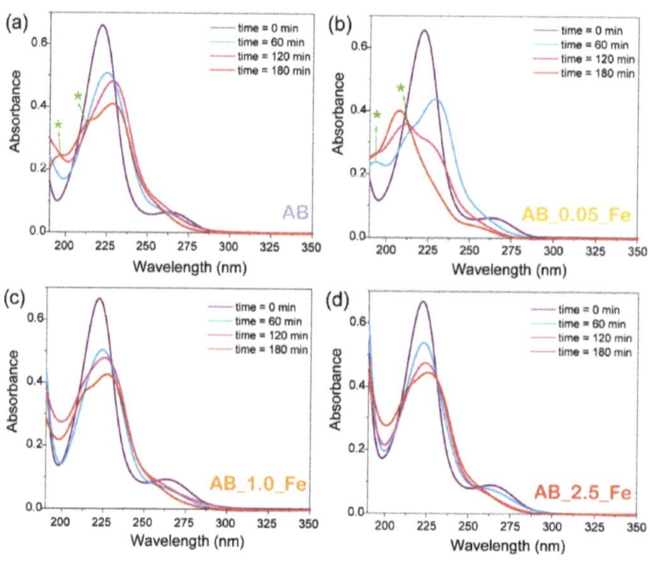

Figure 6. UV-Vis spectra of the supernatant solutions obtained after 60, 120 and 180 min under UV illumination in the presence of 1 g L^{-1} AB (**a**), AB_0.05_Fe (**b**), AB_1.0_Fe (**c**) and AB_2.5_Fe (**d**) photocatalyst (t = 0 min refers to the spectrum of the fresh 1.73 10^{-5} M simazine solution). Green asterisks indicate possible bands of cyanuric acid.

It has to be remarked that simazine is an extremely stable molecule, due to the energy gap between its frontiers orbitals [65] and, thus, its mineralization:

$$C_7H_{12}ClN_5 + 16O_2 \rightarrow 5HNO_3 + HCl + 7CO_2 + 3H_2O \quad (21)$$

is hardly achieved. Indeed, AOPs mostly lead to the formation of cyanuric acid (Scheme 3), which is not as toxic as the parent molecule [66] and has a characteristic UV spectrum in water, with a more intense band at 213 nm and a shoulder at 193 nm [67] (green asterisks in Figure 6). Typically, before the formation of cyanuric acid, simazine molecules (and triazines, in general) undergo dealkylation (i.e., loss of the lateral chains) and dechlorination (i.e., substitution of the Cl atom by an OH group) [50]. In the aqueous phase, dealkylation brings about a blue shift of the simazine 222 nm band (and the disappearance of the 265 nm band), whereas dechlorination brings about a red shift of the 222 nm band. The substitution of Cl, although without leading to mineralization, is considered a strategy of detoxification, i.e., the production of less toxic products [68,69], which can be a potential solution when mineralization is hard to achieve.

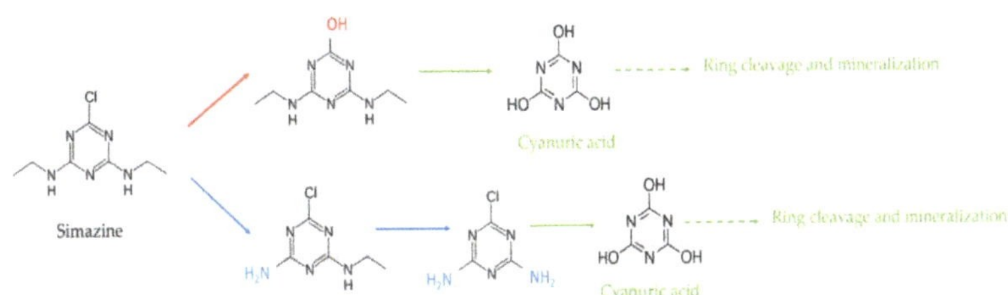

Scheme 3. Oxidative degradation pathways of simazine adapted from refs. [59,61].

In the following discussion, due to the complicated interpretation of UV-Vis spectra and since we are considering simazine degradation simply as a probe reaction to further characterize the studied samples, we will consider the intensity of the spectra at 222 nm: obviously, simazine by-products could absorb in that spectral range and, thus, the overall percentage of degraded simazine could be underestimated. It has to be remarked that if the intensity of the (broad) 266 nm band was considered, higher simazine degradation percentages were calculated, and we will consider an error of ±10% on the percentage of simazine degradation.

Figure S4a reports the results of the tests of UV photolysis, showing that in the adopted experimental conditions ca. 41% simazine was degraded, but the bands of cyanuric acid were not observed. Figure S4b reports the photocatalytic results in the presence of 1 g/L mesoporous anatase with a SSA of 150 $m^2\ g^{-1}$ and a bandgap of 3.20 eV [11,60], which was taken as a benchmark as our samples had anatase contents in the 77.7–79.3% range (Table 1). With the mesoporous anatase, dechlorination took place, likely due to the formation of 4,6-diamino-2-hydroxy-1,3,5-triazine (band at 229 nm [70]), which then transforms into a more hydroxylated product (isosbestic point), but the bands of cyanuric acid are not clearly distinguishable. With the AB powder (Figure 6a), firstly a red shift of the 222 nm band is observed, indicating the occurrence of a dechlorination process. Then, the bands of cyanuric acid (green asterisks) appear after 180 min illumination, indicating that with the undoped sample, the formation of cyanuric acid is accelerated, as compared to mere photolysis. With the AB_0.05_Fe sample, cyanuric acid forms. However, after 60 min, it is further degraded probably by ring opening, indicating not only that Fe-doping accelerates the conversion to cyanuric acid but also that the AB_0.05_Fe sample can further promote cyanuric acid degradation. The other two Fe-doped samples did not lead to any significant improvement in simazine degradation, indicating that a very small amount of Fe can improve the photocatalytic performance of the studied anatase/brookite TiO_2. Such a result agrees with previous literature studies on Fe-doped TiO_2, reporting that when Fe-doping mainly occurs in the TiO_2 bulk, rather than near the surface, as with our samples, Fe sites may act as recombination centres, lowering the yield of the photocatalytic process [29].

To test the possible photo–Fenton activity of the samples under UV light, the degradation of simazine was studied in the presence of a stoichiometric amount of H_2O_2, i.e., with a simazine/H_2O_2 molar ratio of 1/20, according to Equation (22):

$$C_7H_{12}ClN_5 + 20\ H_2O_2 \rightarrow 7CO_2 + 26H_2O + 5/2N_2 + \frac{1}{2}Cl_2 \quad (22)$$

The UV-Vis spectra (not reported for the sake of brevity) showed, again, a red shift of the 222 nm band, i.e., the occurrence of the detoxication process by substitution of Cl. However, as a whole, the samples did not improve their activity in the UV/H_2O_2 system, likely due to the small amount of surface Fe. Figure 7 compares the results as a percentage of simazine degradation after 180 min in the presence of UV and UV/H_2O_2, showing that the most active sample was, again, the AB_0.05_Fe sample.

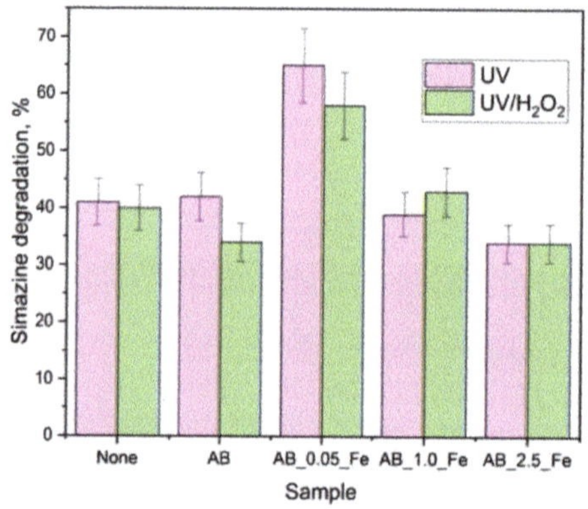

Figure 7. The percentage of simazine degradation after 180 min under UV light (purple histograms) and in the presence of UV and a stoichiometric amount of H_2O_2 (green histograms), as obtained by considering the intensity at 222 nm. An error of ±10% on the percentage of simazine degradation has been considered.

Some recent studies concern the degradation of simazine by Fenton and photo–Fenton processes in homogeneous phase: interestingly, the kinetic constant of simazine degradation by a UV/H_2O_2 system is higher than in the presence of UV/H_2O_2/Fe, likely due to a synergic effect of UV and H_2O_2, allowing a fast production of hydroxyls radicals [71]. Catalkaya et al. studied the degradation of simazine by Fenton reagent and found that the complete disappearance of simazine was achieved within 6 min reaction, but only 32% of simazine was mineralized after 15 min. This result indicates the formation of some intermediate products, which were not completely degraded to CO_2 and H_2O [34], in agreement with the extreme stability of the molecule. As a whole, our results in the heterogenous phase are promising, especially as far as the AB and the AB_0.05_Fe samples are concerned, in that we have observed that dechlorination and cyanuric formation occur, which can be useful in the perspective of detoxification processes, with the advantages of the heterogenous phase as compared to the homogeneous one [1,2].

2.3. Simazine Degradation under Simulated Solar Light

Figure S5a reports the UV-Vis spectra of the simazine solution after 5 h under 1 SUN, showing that the structure of simazine was practically unaffected by solar light and almost nihil degradation occurred. Figure S5b shows the results obtained with the mesoporous anatase used as a benchmark, with which ca. 12% simazine degradation was reached.

Figure 8 reports the spectra obtained after 5 h under 1 SUN in the presence of the 4 studied powders: with the AB sample (Figure 8a), simulated solar light brings about the decrease in the 222 nm band, from which an overall degradation of ca. 20% simazine can be reasonably estimated in this case (since no other bands formed). This result indicates that the sample can effectively exploit solar light, likely due to its absorption spectrum, (Figure 4a). This is in agreement with previous work on the degradation of N-phenylurea [11], where the occurrence of anatase/brookite heterojunctions was considered responsible of the fair activity of the anatase/brookite undoped TiO_2 under simulated solar light.

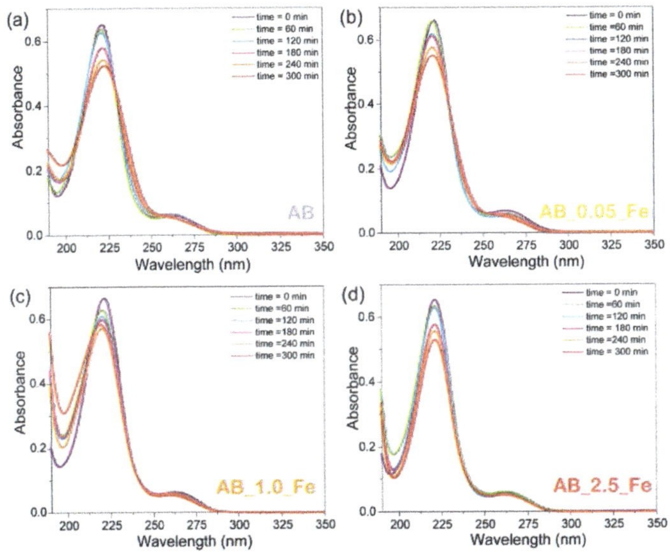

Figure 8. UV-Vis spectra of the supernatant solutions obtained after 60, 120, 180, 240 and 300 min under 1 SUN in the presence of 1 g L^{-1} AB (**a**), AB_0.05_Fe (**b**), AB_1.0_Fe (**c**) and AB_2.5_Fe (**d**) photocatalyst (t = 0 min refers to the spectrum of the fresh 1.73×10^{-5} M simazine solution)

It should be remarked that other materials, such as composites with relevant electrocatalytic [72] and photocatalytic applications [73,74], also present such property, i.e., the capability of accelerating interfacial charge separation, thus enhancing their electro- or photocatalytic efficiency. In the present work, this is simply obtained by a favourable alignment of the anatase and the brookite bands, as already shown in the literature with anatase and rutile polymorphs [12,75].

Unfortunately, doping with Fe does not boost the photocatalytic degradation, in that doped samples were able to degrade ca. 15–20% simazine (Figure 8b–d). To test a possible effect of surface Fe-doping in the presence of H_2O_2, ancillary experiments (not reported for the sake of brevity) were carried out under simulated solar light in the presence of a stoichiometric amount of H_2O_2, but no significant improvement was observed as compared to the tests in the absence of H_2O_2. It has to be remarked, however, that due to the high stability of the simazine molecule, most literature studies deal with its degradation under UV light in the presence of strong oxidants, such as ozone [66], and only a few papers deal with its degradation under solar light, thus the obtained results obtained under solar illumination are promising.

Interestingly, some recent literature studies concern the degradation of atrazine (a molecule similar to simazine, Scheme 2) by non-metal-doped TiO_2. With B-doped anatase/rutile mixed phases, where anatase/rutile heterojunctions occur [75], the non-metal helps stabilize the photogenerated charges; with ZnO/N-doped TiO_2 composites, where heterojunctions can stabilize the photogenerated charges [76], the non-metal helps extend the semiconductor absorption towards the visible range. On the ground of such recent works, our results are particularly interesting in terms of the detoxification of simazine by a TiO_2 mixed phase with anatase/brookite heterojunctions and a low amount of Fe doping having a positive effect on simazine degradation. This is especially true considering the importance of detoxification, i.e., the production of less toxic molecules, in agreement with the findings in Ref. [77] where a careful toxicity evaluation was also carried out.

3. Materials and Methods

3.1. Materials Synthesis

For the synthesis of both undoped and Fe-doped TiO_2, ACS-grade chemicals purchased by Merck-Sigma-Aldrich, Schnelldorf Distribution, Schnelldorf, Germany) were used.

The undoped anatase/brookite TiO_2 mixed phase (sample AB) has been synthesized according to a template-free sol-gel method under pH control [44]. In a 150 mL beaker, 10.0 mL $Ti(OPr)_4$ (titanium(IV) isopropoxide, 97%) was mixed with 10.0 mL isopropyl alcohol (≥99.8%): the solution was then stirred for 20 min at 500 rpm, by adding 100 mL bi-distilled water during stirring. The $Ti(OPr)_4$/isopropyl alcohol/water mixture was poured into a Teflon autoclave, heated inside a stove for 5 h at 80 °C, and cooled down to room temperature. The solution pH was then changed by dropwise adding 1.0 M HNO_3 solution until a pH equal to 2.0 was obtained. At variance with Mutuma et al. [51], the acidic solution was stirred at room temperature for 20 h (instead of 24 h) to obtain a gel, which was repeatedly washed with bi-distilled water and isopropanol and centrifuged 3 times for 12 min at 12,000 rpm. Then, it was dried at 100 °C for 12 h in a stove and calcined at 200 °C for 2 h (temperature ramp = 5 °C min^{-1}) before cooling down to r.t. (temperature ramp = 5 °C min^{-1}), obtaining the AB sample. Finally, the powder was washed four times by using an ethanol/water (1/3) mixture, centrifuged for 10 min at 8000 rpm, and dried for 24 h at 60°C.

Fe-doping (samples AB_0.05, AB_1.0_Fe, and AB_2.5_Fe)was obtained by direct synthesis, with modifying the aforementioned procedure by adding proper amounts of $FeCl_3 \times 6H_2O$ (Iron(III) chloride hexahydrate, 97%) to the bi-distilled water before adding the solution to the $Ti(OPr)_4$/isopropyl alcohol mixture. The solution underwent the same thermal treatment as for the undoped sample, and finally, the pH was adjusted with 1.0 M HNO_3 solution, when necessary, depending on the acidity of the solution of $FeCl_3 \times 6H_2O$ in bi-distilled water.

A 100% anatase mesoporous TiO_2, fully characterized in previous works [11,55], was used as benchmark for the photocatalytic tests.

3.2. Materials Characterization

Powders X-ray diffraction (XRD) patterns were measured on an X'Pert Philips PW3040 diffractometer (Panalytical, Almelo, The Netherlands) employing Cu Kα radiation (2θ range = 10–100°; step = 0.026° 2θ; time per step = 0.8 s). The obtained XRD patterns were indexed by referring to the Powder Data File database (PDF 2000, International Centre of Diffraction Data, Newtown Square, PA, USA). The QPA (Quantitative Phase Analysis) was obtained according to the full-profile Rietveld method (X'Pert High Score Plus 3.0e software). The crystallite average size (D) was determined according to the Williamson-Hall plot (X'Pert High Score Plus 3.0e software).

N_2 adsorption/desorption isotherms were measured at −196 °C on powders pre-outgassed at 150 °C for 4 h to remove water and other atmospheric contaminants (Micromeritics ASAP 2020Plus, Micromeritics, Norcross, GA, USA). The Brunauer–Emmett–Teller (BET) method and the Barrett–Joyner–Halenda (BJH) method were adopted to determine the samples' Specific Surface Area (SSA) and Pore Size Distribution (PSD), respectively.

FE-SEM (Field Emission Scanning Electron Microscopy) micrographs were collected on a Merlin FESEM instrument (Carl-Zeiss AG, Oberkochen, Germany), equipped with an

EDX (Energy Dispersive X-ray Analysis) probe (Oxford instruments, Abingdon, UK) for semi-quantitative elemental analysis.

The Diffuse Reflectance (DR) UV-Vis spectra of the powders were collected on a Cary 5000 UV-Vis-NIR spectrophotometer (Varian instruments, Palo Alto, CA, USA), equipped with a DR sphere.

X-Ray Photoelectron Spectroscopy (XPS) analysis was performed on a PHI 5000 VersaProbe equipment (Physical Electronics, Feldkirchen, Germany) adopting monochromatic Al Kα radiation (1486.6 eV) as an X-ray source and a pass-energy of 187.75 eV. The spectral line shift of the C 1 s binding energy (BE) value at 284.8 eV was used to eliminate possible sample charging effects.

The powders' electrophoretic mobility was measured as a function of pH by dynamic light scattering (DLS, a Zetasizer Nano-ZS Malvern Instruments, Worcestershire, UK). In a typical experiment to determine the ζ-potential, the powders were suspended in ultrapure water and sonicated for 2 min (10 W/mL, 20 kHz, Sonoplus, Bandelin, Berlin, Germany) or magnetically stirred for 5 min. The pH was adjusted by adding either 0.1 M NaOH or 0.1 M HCl.

3.3. Simazine Degradation Tests

All the photocatalytic tests were carried out by adding a proper amount of photocatalyst (corresponding to 1.0 g L^{-1} concentration) to 40 mL of 1.73 10^{-5} M aqueous solution of simazine (natural pH = 5.5).

For the tests under UV light, a medium-pressure Hg lamp (light intensity of 55 $mW \cdot cm^{-2}$, LC3, Hamamatsu Photonic, Hamamatsu, Japan) was used. A second set of experiments was made by adding a stoichiometric amount of H_2O_2.

For the photocatalytic tests under 1 SUN, a plasma lamp (LIFI STA-40, LUXIM, Santa Clara, CA, USA) was used to simulate solar light (AM 1.5 G, 100 mW cm^{-2}) under the following illumination conditions: 1 SUN, i.e., ~1000 W m^{-2} in the visible range and ~22 W m^{-2} in the UV range.

During the tests, the liquid/solid suspension inside the testing tube has been continuously stirred using a magnetic stirrer at ca. 300 rpm, as reported elsewhere [55]. The reaction mixture was not de-aerated although the reacting tube was sealed, with the consequence that the atmospheric O_2 was always present. At constant time intervals, aliquots of the suspension were withdrawn and immediately centrifuged twice at 12,000 rpm for 12 min (Thermo Fisher Scientific SL 16R centrifuge, Thermo Electron LED GmbH, Osterode am Harz, Germany) to obtain the supernatant solution, which was analysed on a UV-Vis-NIR spectrophotometer (Cary 5000, Varian Instruments, Mulgrave, Australia).

4. Conclusions

A simple, template-free sol-gel synthesis method based on pH control and mild calcination temperature was developed to dope an anatase/brookite TiO_2 mixed phase with Fe, an earth-abundant non-toxic metal. The physico-chemical characterization showed that the adopted synthesis procedure led to the production of high-surface-area mesoporous TiO_2 nanoparticles. Fe doping mostly occurred in the bulk of the TiO_2 matrix, in that XRD showed a noticeable change in the cell volume of the brookite phase, and XPS analysis was able to detect Fe at the TiO_2 surface only with the 2.5 wt.% Fe sample.

The photocatalytic degradation of simazine, a herbicide known for being a persistent pollutant with suspected toxicity to humans and the environment, has been studied as a probe reaction. Under UV light, different simazine by-products were observed as compared to mere photolysis: the best results in terms of simazine removal were obtained with the sample at 0.05% Fe, in agreement with the literature showing that higher Fe contents in the bulk of doped TiO_2 may lead to undesired recombination phenomena. In the presence of UV/H_2O_2, the samples photo–Fenton activity was very limited, likely due to the scarcity of surface Fe species.

Very interesting results were observed under simulated solar light, in that 4 samples were able to provide degradation of 15–20% simazine, after 5 h illumination, whereas solar light was unable to degrade simazine by mere photolysis. Moreover, the undoped anatase/brookite phase was the most active one under simulated solar light, likely due to the presence of an Urbach tail in its UV-Vis spectrum and to the presence of anatase/brookite heterojunctions, allowing effective exploitation of the photogenerated charge carriers, as already observed by some of us in previous work on the photocatalytic degradation of N-phenylurea.

Concerning the material optimization, future work will focus on the study of Fe contents in the 0.05–1.0 wt.% range, to test the possible effect on the samples' photocatalytic properties, as different Fe contents may also lead to noticeable effects on the anatase cell volume in terms, for instance, of isomorphic versus interstitial substitution. Regarding simazine degradation, future work will concern the study of the effect of varying pH, as it could affect the production of hydroxyl radicals and/or the simazine/surface interactions and include the study of the process in actual water samples, with emphasis on the search for optimal conditions to achieve detoxification.

Supplementary Materials: The following supporting information can be downloaded at: https://www.mdpi.com/article/10.3390/catal13040667/s1, Figure S1: Overlapped and single-element EDX maps of Ti, O and Fe with the following samples: (a) AB_0.05_Fe; (b) AB_1.0_Fe; (c) AB_2.5_Fe. The maps of C and Cl are not reported, as the elements occurred in traces. Figure S2: XPS survey spectra of the studied samples, showing that surface Fe species were detected only with the AB_2.5_Fe sample. Figure S3: DR UV-Vis spectra of the AB powder and of a Degussa P25 commercial powder, showing that the former has an Urbach tail, absent with the latter. Figure S4: UV-Vis spectra of the supernatant solutions obtained after 60, 120, and 180 min under UV illumination in the absence of any photocatalyst (a) and in the presence of 1 g/L mesoporous anatase (b). In both graphs, time = 0 min refers to the spectrum of the fresh 1.73×10^{-5} M simazine solution. Figure S5: UV-Vis spectra of the supernatant solutions obtained after 60, 120, 180, 240, and 300 min under 1 SUN in the absence of any photocatalyst (a) and in the presence of 1 g/L mesoporous anatase (b). In both graphs, time = 0 min refers to the spectrum of the fresh $1.73\ 10^{-5}$ M simazine solution.

Author Contributions: Conceptualization, B.B. and F.S.F.; methodology, B.B.; formal analysis, S.G. (Stefano Gervasi) and S.G. (Salvatore Guastella); investigation, S.G. (Stefano Gervasi), N.B., S.G. (Salvatore Guastella) and F.S.F.; data curation, S.G. (Stefano Gervasi), F.S.F. and N.B.; writing—original draft preparation, B.B.; writing—review and editing N.B. and F.S.F.; supervision, N.B., F.S.F. and B.B.; funding acquisition, B.B. All authors have read and agreed to the published version of the manuscript.

Funding: This research was funded by the program PNRR Mac-2 NODES Nord Ovest Digitale e Sostenibile and Agritech National Research Center and received funding from the European Union Next-Generation EU (PIANO NAZIONALE DI RIPRESA E RESILIENZA (PNRR)—MISSIONE 4 COMPONENTE 2, INVESTIMENTO 1.4—D.D. 1032 17/06/2022, CN00000022). This manuscript reflects only the authors' views and opinions, neither the European Union nor the European Commission can be considered responsible for them

Data Availability Statement: Not applicable.

Acknowledgments: The authors thank M. Sangermano (Department of Applied Science and Technology, Politecnico di Torino) for lending the UV lamp, S. Hernandez (Department of Applied Science and Technology, Politecnico di Torino) for lending the solar lamp, and T. Tosco (Department of Environment, Land and Infrastructure Engineering, Politecnico di Torino) for providing access to electrophoretic measurements.

Conflicts of Interest: The authors declare no conflict of interest.

References

1. Freyria, F.S.; Armandi, M.; Compagnoni, M.; Ramis, G.; Rossetti, I.; Bonelli, B. Catalytic and Photocatalytic Processes for the Abatement of N-Containing Pollutants from Wastewater. Part 2: Organic Pollutants. *J. Nanosci. Nanotechnol.* **2017**, *17*, 3654–3672. [CrossRef]
2. Compagnoni, M.; Ramis, G.; Freyria, F.S.; Armandi, M.; Bonelli, B.; Rossetti, I. Photocatalytic Processes for the Abatement of N-Containing Pollutants from Waste Water. Part 1: Inorganic Pollutants. *J. Nanosci. Nanotechnol.* **2017**, *17*, 3632–3653. [CrossRef]
3. Ozawa, K.; Emori, M.; Yamamoto, S.; Yukawa, R.; Yamamoto, S.; Hobara, R.; Fujikawa, K.; Sakama, H.; Matsuda, I. Electron-hole recombination time at TiO_2 single-crystal surfaces: Influence of surface band bending. *J. Phys. Chem. Lett.* **2014**, *5*, 1953–1957. [CrossRef] [PubMed]
4. Guayaquil-Sosa, J.F.; Serrano-Rosales, B.; Valadés-Pelayo, P.J.; de Lasa, H. Photocatalytic hydrogen production using mesoporous TiO_2 doped with Pt. *Appl. Catal. B Environ.* **2017**, *211*, 337–348. [CrossRef]
5. Lavorato, C.; Argurio, P.; Mastropietro, T.F.; Pirri, G.; Poerio, T.; Molinari, R. Pd/TiO_2 doped faujasite photocatalysts for acetophenone transfer hydrogenation in a photocatalytic membrane reactor. *J. Catal.* **2017**, *353*, 152–161. [CrossRef]
6. Di Paola, A.; Bellardita, M.; Palmisano, L. Brookite, the Least Known TiO_2 Photocatalyst. *Catalysts* **2013**, *3*, 36–73. [CrossRef]
7. Manzoli, M.; Freyria, F.S.; Blangetti, N.; Bonelli, B. Brookite, a sometimes under evaluated TiO_2 polymorph. *RSC Adv.* **2022**, *12*, 3322–3334. [CrossRef]
8. Liu, Y.; Wang, Z.; Huang, W. Influences of TiO_2 phase structures on the structures and photocatalytic hydrogen production of CuO_x/TiO_2 photocatalysts. *Appl. Surf. Sci.* **2016**, *389*, 760–767. [CrossRef]
9. Liu, Y.; Ye, Z.; Li, D.; Wang, M.; Zhang, Y.; Huang, W. Tuning CuO_x-TiO_2 interaction and photocatalytic hydrogen production of CuO_x/TiO_2 photocatalysts via TiO_2 morphology engineering. *Appl. Surf. Sci.* **2019**, *473*, 500–510. [CrossRef]
10. Bonelli, B.; Esposito, S.; Freyria, F.S. Mesoporous Titania: Synthesis, properties and comparison with non-porous titania. In *Titanium Dioxide*; Janus, M., Ed.; Intech: London, UK, 2017; ISBN 978-953-51-5493-8.
11. Freyria, F.S.; Blangetti, N.; Esposito, S.; Nasi, R.; Armandi, M.; Annelio, V.; Bonelli, B. Effects of the Brookite Phase on the Properties of Different Nanostructured TiO_2 Phases Photocatalytically Active Towards the Degradation of N-Phenylurea. *ChemistryOpen* **2020**, *9*, 903–912. [CrossRef]
12. Fang, J.; Wang, F.; Qian, K.; Bao, H.; Jiang, Z.; Huang, W. Bifunctional N-Doped Mesoporous TiO_2 Photocatalysts. *J. Phys. Chem. C* **2008**, *112*, 18150–18156. [CrossRef]
13. Hossain, M.A.; Elias, M.; Sarker, D.R.; Diba, Z.R.; Mithun, J.M.; Azad, M.A.K.; Siddiquey, I.A.; Rahman, M.M.; Uddin, J.; Uddin, M.N. Synthesis of Fe- or Ag-doped TiO_2–MWCNT nanocomposite thin films and their visible-light-induced catalysis of dye degradation and antibacterial activity. *Res. Chem. Intermed.* **2018**, *44*, 2667–2683. [CrossRef]
14. Ohno, T.; Tokieda, K.; Higashida, S.; Matsumura, M. Synergism between rutile and anatase TiO_2 particles in photocatalytic oxidation of naphthalene. *Appl. Catal. A Gen.* **2003**, *244*, 383–391. [CrossRef]
15. Jiang, X.; Manawan, M.; Feng, T.; Qian, R.; Zhao, T.; Zhou, G.; Kong, F.; Wang, Q.; Dai, S.; Pan, J.H. Anatase and rutile in evonik aeroxide P25: Heterojunctioned or individual nanoparticles? *Catal. Today* **2018**, *300*, 12–17. [CrossRef]
16. Manga Raju, I.; Siva Rao, T.; Divya Lakshmi, K.V.; Ravi Chandra, M.; Swathi Padmaja, J.; Divya, G. Poly 3-Thenoic acid sensitized, Copper doped anatase/brookite TiO_2 nanohybrids for enhanced photocatalytic degradation of an organophosphorus pesticide. *J. Environ. Chem. Eng.* **2019**, *7*, 103211. [CrossRef]
17. Tay, Q.; Wang, X.; Zhao, X.; Hong, J.; Zhang, Q.; Xu, R.; Chen, Z. Enhanced visible light hydrogen production via a multiple heterojunction structure with defect-engineered g-C_3N_4 and two-phase anatase/brookite TiO_2. *J. Catal.* **2016**, *342*, 55–62. [CrossRef]
18. Zhao, B.; Chen, F.; Jiao, Y.; Yang, H.; Zhang, J. Ag^0-loaded brookite/anatase composite with enhanced photocatalytic performance towards the degradation of methyl orange. *J. Mol. Catal. A Chem.* **2011**, *348*, 114–119. [CrossRef]
19. Panov, G.I.; Sheveleva, G.A.; Kharitonov, A.S.; Romannikov, V.N.; Vostrikova, L.A. Oxidation of benzene to phenol by nitrous oxide over Fe-ZSM-5 zeolites. *Appl. Catal. A Gen.* **1992**, *82*, 31–36. [CrossRef]
20. Bahadori, E.; Vaiano, V.; Esposito, S.; Armandi, M.; Sannino, D.; Bonelli, B. Photo-activated degradation of tartrazine by H_2O_2 as catalyzed by both bare and Fe-doped methyl-imogolite nanotubes. *Catal. Today* **2018**, *304*, 199–207. [CrossRef]
21. Shafia, E.; Esposito, S.; Armandi, M.; Bahadori, E.; Garrone, E.; Bonelli, B. Reactivity of bare and Fe-doped alumino-silicate nanotubes (imogolite) with H_2O_2 and the azo-dye Acid Orange 7. *Catal. Today* **2016**, *277*, 89–96. [CrossRef]
22. Mardani, H.R.; Forouzani, M.; Ziari, M.; Biparva, P. Visible light photo-degradation of methylene blue over Fe or Cu promoted ZnO nanoparticles. *Spectrochim. Acta—Part A Mol. Biomol. Spectrosc.* **2015**, *141*, 27–33. [CrossRef] [PubMed]
23. Zhang, W.; Zhao, J.; Liu, Z.; Liu, Z. Structural, optical and magnetic properties of Zn_{1-x} Fe_x O powders by sol-gel method. *Appl. Surf. Sci.* **2013**, *284*, 49–52. [CrossRef]
24. Ciciliati, M.A.; Silva, M.F.; Fernandes, D.M.; De Melo, M.A.C.; Hechenleitner, A.A.W.; Pineda, E.A.G. Fe-doped ZnO nanoparticles: Synthesis by a modified sol-gel method and characterization. *Mater. Lett.* **2015**, *159*, 84–86. [CrossRef]
25. Yu, J.; Xiang, Q.; Zhou, M. Preparation, characterization and visible-light-driven photocatalytic activity of Fe-doped titania nanorods and first-principles study for electronic structures. *Appl. Catal. B Environ.* **2009**, *90*, 595–602. [CrossRef]
26. Tong, T.; Zhang, J.; Tian, B.; Chen, F.; He, D. Preparation of $Fe3^{3+}$-doped TiO_2 catalysts by controlled hydrolysis of titanium alkoxide and study on their photocatalytic activity for methyl orange degradation. *J. Hazard. Mater.* **2008**, *155*, 572–579. [CrossRef] [PubMed]

27. Khan, H.; Khan Swati, I. Fe^{3+}-doped Anatase TiO_2 with d−d Transition, Oxygen Vacancies and Ti^{3+} Centers: Synthesis, Characterization, UV−vis Photocatalytic and Mechanistic Studies. *Ind. Eng. Chem. Res.* **2016**, *55*, 6619–6633. [CrossRef]
28. Sood, S.; Umar, A.; Mehta, S.K.; Kansal, S.K. Highly effective Fe-doped TiO_2 nanoparticles photocatalysts for visible-light driven photocatalytic degradation of toxic organic compounds. *J. Colloid Interface Sci.* **2015**, *450*, 213–223. [CrossRef]
29. Komaraiah, D.; Radha, E.; Sivakumar, J.; Sayanna, R. Influence of Fe^{3+}ions on the optical properties and photocatalytic ability of spin coated Fe^{3+}doped brookite TiO_2 thin films. *Funct. Compos. Struct.* **2022**, *4*, 15010. [CrossRef]
30. Ramirez, J.H.; Maldonado-Hódar, F.J.; Pérez-Cadenas, A.F.; Moreno-Castilla, C.; Costa, C.A.; Madeira, L.M. Azo-dye Orange II degradation by heterogeneous Fenton-like reaction using carbon-Fe catalysts. *Appl. Catal. B Environ.* **2007**, *75*, 312–323. [CrossRef]
31. Neyens, E.; Baeyens, J. A review of classic Fenton's peroxidation as an advanced oxidation technique. *J. Hazard. Mater.* **2003**, *98*, 33–50. [CrossRef]
32. Pignatello, J.J.; Oliveros, E.; MacKay, A. Advanced oxidation processes for organic contaminant destruction based on the fenton reaction and related chemistry. *Crit. Rev. Environ. Sci. Technol.* **2006**, *36*, 1–84. [CrossRef]
33. Rahim Pouran, S.; Abdul Aziz, A.R.; Wan Daud, W.M.A. Review on the main advances in photo-Fenton oxidation system for recalcitrant wastewaters. *J. Ind. Eng. Chem.* **2015**, *21*, 53–69. [CrossRef]
34. Catalkaya, E.C.; Kargi, F. Advanced oxidation and mineralization of simazine using Fenton's reagent. *J. Hazard. Mater.* **2009**, *168*, 688–694. [CrossRef] [PubMed]
35. Klavarioti, M.; Mantzavinos, D.; Kassinos, D. Removal of residual pharmaceuticals from aqueous systems by advanced oxidation processes. *Environ. Int.* **2009**, *35*, 402–417. [CrossRef] [PubMed]
36. Catalkaya, E.C.; Kargi, F. Degradation and Mineralization of Simazine in Aqueous Solution by Ozone/Hydrogen Peroxide Advanced Oxidation. *J. Environ. Eng.* **2009**, *135*, 1357–1364. [CrossRef]
37. European Union. Council Directive 91/414/EEC of 15 July 1991 concerning the placing of plant protection products on the market. *Off. J. Eur. Union* **2010**, 1–32.
38. Grasselli, F.; Bussolati, S.; Ramoni, R.; Grolli, S.; Basini, G. Simazine, a triazine herbicide, disrupts swine granulosa cell functions. *Anim. Reprod.* **2018**, *15*, 3–11. [CrossRef]
39. Endocrine-Disrupting Pesticide Atrazine to Be Banned in Hawaii, Five U.S. Territories, Prohibited on Conifers, Roadsides. Available online: https://biologicaldiversity.org/w/news/press-releases/endocrine-disrupting-pesticide-atrazine-be-banned-hawaii-five-us-territories-prohibited-conifers-roadsides-2020-09-23/ (accessed on 23 September 2020).
40. Rao, Y.; Chu, W. Visible light-induced photodegradation of simazine in aqueous TiO_2 suspension. *Ind. Eng. Chem. Res.* **2013**, *52*, 13580–13586. [CrossRef]
41. Mohaupt, V.; Völker, J.; Altenburger, R.; Birk, S.; Kirst, I.; Kühnel, D.; Küster, E.; Semerádová, S.; Šubelj, G.; Whalley, C. *Pesticides in European Rivers, Lakes and Groundwaters—Data Assessment*; ETC/ICM: Magdeburg, Germany, 2020.
42. Stefano, P.H.P.; Roisenberg, A.; Santos, M.R.; Dias, M.A.; Montagner, C.C. Unraveling the occurrence of contaminants of emerging concern in groundwater from urban setting: A combined multidisciplinary approach and self-organizing maps. *Chemosphere* **2022**, *299*, 134395. [CrossRef]
43. Esquerdo, A.A.; Rico, D.P.; Galvañ, P.J.V.; Gadea, I.S. Activated carbon and ozone to reduce simazine in water. *Water* **2020**, *12*, 2900. [CrossRef]
44. Loiseau, T.; Serre, C.; Huguenard, C.; Fink, G.; Taulelle, F.; Henry, M.; Bataille, T.; Férey, G. A Rationale for the Large Breathing of the Porous Aluminum Terephthalate (MIL-53) Upon Hydration. *Chem.—A Eur. J.* **2004**, *10*, 1373–1382. [CrossRef] [PubMed]
45. Sun, Z.; Zheng, S.; Ayoko, G.A.; Frost, R.L.; Xi, Y. Degradation of simazine from aqueous solutions by diatomite-supported nanosized zero-valent iron composite materials. *J. Hazard. Mater.* **2013**, *263*, 768–777. [CrossRef] [PubMed]
46. Flores, K.; Valdes, C.; Ramirez, D.; Eubanks, T.M.; Lopez, J.; Hernandez, C.; Alcoutlabi, M.; Parsons, J.G. The effect of hybrid zinc oxide/graphene oxide (ZnO/GO) nano-catalysts on the photocatalytic degradation of simazine. *Chemosphere* **2020**, *259*, 127414. [CrossRef] [PubMed]
47. Boruah, P.K.; Darabdhara, G.; Das, M.R. Polydopamine functionalized graphene sheets decorated with magnetic metal oxide nanoparticles as efficient nanozyme for the detection and degradation of harmful triazine pesticides. *Chemosphere* **2021**, *268*, 129328. [CrossRef]
48. Suhaimy, S.H.M.; Lai, C.W.; Tajuddin, H.A.; Samsudin, E.M.; Johan, M.R. Impact of TiO_2 nanotubes' morphology on the photocatalytic degradation of simazine pollutant. *Materials* **2018**, *11*, 2066. [CrossRef]
49. Meriam Suhaimy, S.; Abd Hamid, S.; Lai, C.; Hasan, M.; Johan, M. TiO_2 Nanotubes Supported Cu Nanoparticles for Improving Photocatalytic Degradation of Simazine under UV Illumination. *Catalysts* **2016**, *6*, 167. [CrossRef]
50. Evgenidou, E.; Fytianos, K. Photodegradation of Triazine Herbicides in Aqueous Solutions and Natural Waters. *J. Agric. Food Chem.* **2002**, *50*, 6423–6427. [CrossRef]
51. Mutuma, B.K.; Shao, G.N.; Kim, W.D.; Kim, H.T. Sol–gel synthesis of mesoporous anatase–brookite and anatase–brookite–rutile TiO_2 nanoparticles and their photocatalytic properties. *J. Colloid Interface Sci.* **2015**, *442*, 1–7. [CrossRef]
52. Mancuso, A.; Sacco, O.; Vaiano, V.; Bonelli, B.; Esposito, S.; Freyria, F.S.; Blangetti, N.; Sannino, D. Visible Light-Driven Photocatalytic Activity and Kinetics of Fe-Doped TiO_2 Prepared by a Three-Block Copolymer Templating Approach. *Materials* **2021**, *14*, 3105. [CrossRef]

53. Mancuso, A.; Blangetti, N.; Sacco, O.; Freyria, F.S.; Bonelli, B.; Esposito, S.; Sannino, D.; Vaiano, V. Photocatalytic Degradation of Crystal Violet Dye under Visible Light by Fe-Doped TiO_2 Prepared by Reverse-Micelle Sol–Gel Method. *Nanomaterials* **2023**, *13*, 270. [CrossRef]
54. Piumetti, M.; Freyria, F.S.; Armandi, M.; Geobaldo, F.; Garrone, E.; Bonelli, B. Fe- and V-doped mesoporous titania prepared by direct synthesis: Characterization and role in the oxidation of AO7 by H_2O_2 in the dark. *Catal. Today* **2014**, *227*, 71–79. [CrossRef]
55. Freyria, F.; Compagnoni, M.; Ditaranto, N.; Rossetti, I.; Piumetti, M.; Ramis, G.; Bonelli, B. Pure and Fe-Doped Mesoporous Titania Catalyse the Oxidation of Acid Orange 7 by H_2O_2 under Different Illumination Conditions: Fe Doping Improves Photocatalytic Activity under Simulated Solar Light. *Catalysts* **2017**, *7*, 213. [CrossRef]
56. Soo, C.W.; Juan, J.C.; Lai, C.W.; Hamid, S.B.A.; Yusop, R.M. Fe-doped mesoporous anatase-brookite titania in the solar-light-induced photodegradation of Reactive Black 5 dye. *J. Taiwan Inst. Chem. Eng.* **2016**, *68*, 153–161. [CrossRef]
57. Sannino, F.; Pansini, M.; Marocco, A.; Bonelli, B.; Garrone, E.; Esposito, S. The role of outer surface/inner bulk Brønsted acidic sites in the adsorption of a large basic molecule (simazine) on H-Y zeolite. *Phys. Chem. Chem. Phys.* **2015**, *17*, 28950–28957. [CrossRef]
58. López, R.; Gómez, R. Photocatalytic Degradation of 4-Nitrophenol on Well Characterized Sol-Gel Molybdenum Doped Titania Semiconductors. *Top. Catal.* **2011**, *54*, 504–511. [CrossRef]
59. Yaghoubi, H.; Li, Z.; Chen, Y.; Ngo, H.T.; Bhethanabotla, V.R.; Joseph, B.; Ma, S.; Schlaf, R.; Takshi, A. Toward a Visible Light-Driven Photocatalyst: The Effect of Midgap-States-Induced Energy Gap of Undoped TiO_2 Nanoparticles. *ACS Catal.* **2015**, *5*, 327–335. [CrossRef]
60. Blangetti, N.; Freyria, F.S.; Calviello, M.C.; Ditaranto, N.; Guastella, S.; Bonelli, B. Photocatalytic Degradation of Paracetamol under Simulated Sunlight by Four TiO_2 Commercial Powders: An Insight into the Performance of Two Sub-Micrometric Anatase and Rutile Powders and a Nanometric Brookite Powder. *Catalysts* **2023**, *13*, 434. [CrossRef]
61. Mustafa, S.; Tasleem, S.; Naeem, A. Surface charge properties of Fe_2O_3 in aqueous and alcoholic mixed solvents. *J. Colloid Interface Sci.* **2004**, *275*, 523–529. [CrossRef] [PubMed]
62. Holmberg, J.P.; Ahlberg, E.; Bergenholtz, J.; Hassellöv, M.; Abbas, Z. Surface charge and interfacial potential of titanium dioxide nanoparticles: Experimental and theoretical investigations. *J. Colloid Interface Sci.* **2013**, *407*, 168–176. [CrossRef]
63. Nasi, R.; Sannino, F.; Picot, P.; Thill, A.; Oliviero, O.; Esposito, S.; Armandi, M.; Bonelli, B. Hybrid organic-inorganic nanotubes effectively adsorb some organic pollutants in aqueous phase. *Appl. Clay Sci.* **2020**, *186*, 105449. [CrossRef]
64. Liu, Y.; Zhu, K.; Su, M.; Zhu, H.; Lu, J.; Wang, Y.; Dong, J.; Qin, H.; Wang, Y.; Zhang, Y. Influence of solution pH on degradation of atrazine during UV and UV/H_2O_2 oxidation: Kinetics, mechanism, and degradation pathways. *RSC Adv.* **2019**, *9*, 35847–35861. [CrossRef] [PubMed]
65. Hu, J.-Y.; Morita, T.; Magara, Y.; Aizawa, T. Evaluation of the Reactivity of Pesticides with Ozone in Water using the Energies of Frontier Molecular Orbitals. *Water Res.* **2000**, *34*, 2215–2222. [CrossRef]
66. Rivas, F.J.; Navarrete, V.; Beltrán, F.J.; García-Araya, J.F. Simazine Fenton's oxidation in a continuous reactor. *Appl. Catal. B Environ.* **2004**, *48*, 249–258. [CrossRef]
67. Cantú, R.; Evans, O.; Kawahara, F.K.; Shoemaker, J.A.; Dufour, A.P. An HPLC Method with UV Detection, pH Control, and Reductive Ascorbic Acid for Cyanuric Acid Analysis in Water. *Anal. Chem.* **2000**, *72*, 5820–5828. [CrossRef]
68. Mu, Y.; Chen, Y.; Fu, Q.; He, P.Y.; Sun, Q.; Zou, J.P.; Zhang, L.; Wang, D.; Luo, S. Transformation of Atrazine to Hydroxyatrazine with Alkali-H_2O_2 Treatment: An Efficient Dechlorination Strategy under Alkaline Conditions. *ACS Water* **2021**, *1*, 1868–1877. [CrossRef]
69. Mu, Y.; Zhan, G.; Huang, C.; Wang, X.; Ai, Z.; Zou, J.; Luo, S.; Zhang, L. Dechlorination-Hydroxylation of Atrazine to Hydroxyatrazine with Thiosulfate: A Detoxification Strategy in Seconds. *Environ. Sci. Technol.* **2019**, *53*, 3208–3216. [CrossRef] [PubMed]
70. Moreira, A.J.; Borges, A.C.; Gouvea, L.F.C.; MacLeod, T.C.O.; Freschi, G.P.G. The process of atrazine degradation, its mechanism, and the formation of metabolites using UV and UV/MW photolysis. *J. Photochem. Photobiol. A Chem.* **2017**, *347*, 160–167. [CrossRef]
71. Lojo-López, M.; Andrades, J.A.; Egea-Corbacho, A.; Coello, M.D.; Quiroga, J.M. Degradation of simazine by photolysis of hydrogen peroxide Fenton and photo-Fenton under darkness, sunlight and UV light. *J. Water Process Eng.* **2021**, *42*, 102115. [CrossRef]
72. Cao, Q.; Hao, S.; Wu, Y.; Pei, K.; You, W.; Che, R. Interfacial charge redistribution in interconnected network of Ni_2P–Co_2P boosting electrocatalytic hydrogen evolution in both acidic and alkaline conditions. *Chem. Eng. J.* **2021**, *424*, 130444. [CrossRef]
73. Cao, Q.; Li, Q.; Pi, Z.; Zhang, J.; Sun, L.-W.; Xu, J.; Cao, Y.; Cheng, J.; Bian, Y. Metal–Organic-Framework-Derived Ball-Flower-like Porous Co_3O_4/Fe_2O_3 Heterostructure with Enhanced Visible-Light-Driven Photocatalytic Activity. *Nanomaterials* **2022**, *12*, 904. [CrossRef]
74. Zhang, Z.; Zhang, Y.; Han, X.; Guo, L.; Wang, D.; Lv, K. Assembly of $CaIn_2S_4$ on Defect-Rich BiOCl for Acceleration of Interfacial Charge Separation and Photocatalytic Phenol Degradation via S-Scheme Electron Transfer Mechanism. *Catalysts* **2021**, *11*, 1130. [CrossRef]
75. Hurum, D.C.; Agrios, A.G.; Gray, K.A.; Rajh, T.; Thurnauer, M.C. Explaining the Enhanced Photocatalytic Activity of Degussa P25 Mixed-Phase TiO_2 Using EPR. *J. Phys. Chem. B* **2003**, *107*, 4545–4549. [CrossRef]

76. Navarra, W.; Ritacco, I.; Sacco, O.; Caporaso, L.; Farnesi Camellone, M.; Venditto, V.; Vaiano, V. Density Functional Theory Study and Photocatalytic Activity of ZnO/N-Doped TiO_2 Heterojunctions. *J. Phys. Chem. C* **2022**, *126*, 7000–7011. [CrossRef]
77. Navarra, W.; Sacco, O.; Daniel, C.; Venditto, V.; Vaiano, V.; Vignette, D.A.L.; Bojic, C.; Libralato, G.; Lofrano, G.; Carotenuto, M. Photocatalytic degradation of atrazine by an N-doped TiO_2/polymer composite: Catalytic efficiency and toxicity evaluation. *J. Environ. Chem. Eng.* **2022**, *10*, 108167. [CrossRef]